International Conference on Applied Sciences 2014

(ICAS2014)

IOP Conference Series: Materials Science and Engineering
Volume 85

Hunedoara, Romania
2-4 October 2014

ISBN: 978-1-5108-0749-5
ISSN: 1757-8981

Printed from e-media with permission by:

Curran Associates, Inc.
57 Morehouse Lane
Red Hook, NY 12571

Some format issues inherent in the e-media version may also appear in this print version.

Printed by Curran Associates, Inc. (2015)

For permission requests, please contact the Institute of Physics
at the address below.

Institute of Physics
Dirac House, Temple Back
Bristol BS1 6BE UK

Phone: 44 1 17 929 7481
Fax: 44 1 17 920 0979

techtracking@iop.org

Additional copies of this publication are available from:

Curran Associates, Inc.
57 Morehouse Lane
Red Hook, NY 12571 USA
Phone: 845-758-0400
Fax: 845-758-2634
Email: curran@proceedings.com
Web: www.proceedings.com

Table of contents

Volume 85

International Conference on Applied Sciences 2014 (ICAS2014)
2–4 October 2014, Hunedoara, Romania

Accepted papers received: 22 May 2015
Published online: 18 June 2015

Preface

Papers

IOP Conf. Series: Materials Science and Engineering **85** (2015) 011001 doi:10.1088/1757-899X/85/1/011001

International Conference on Applied Sciences ICAS2014

Hunedoara, Romania, October 2 – 4, 2014

Organizers:

Military Economics Academy of Wuhan and ***Politehnica University of Timişoara***

in cooperation with

Wuhan University
Huazhong University of Sciences and Technology
Naval University of Engineering
Zhongnan University of Economics and Law
Air Force Early Warning Academy
"Henri Coanda" Air Force Academy of Brasov
University of Medicine and Pharmacy "Victor Babes" Timisoara

and

Scientific Interdisciplinary Values for Education and Research Society
General Association of Romanian Engineers, Hunedoara Branch
National Society of Environmental Science and Engineering

IOP Conf. Series: Materials Science and Engineering **85** (2015) 011001 doi:10.1088/1757-899X/85/1/011001

The **International Conference on Applied Sciences (ICAS2014)** took place in Hunedoara, Romania from 2–4 October 2014 at the Engineering Faculty of Hunedoara. The conference takes place alternately in Romania and in P.R. China and is organized by "Politehnica" University of Timişoara, Romania, and Military Economics Academy of Wuhan, P.R. China, with the aim to serve as a platform for exchange of information between various areas of applied sciences and to promote the communication between scientists of different nations, countries and continents.

The **topics** of the conference covered a comprehensive spectrum of issues:

1. Economical Sciences
2. Engineering Sciences
3. Fundamental Sciences
4. Medical Sciences

The conference gathered qualified researchers whose expertise can be used to develop new engineering knowledge that has the potential for application in economics, defense, medicine, etc. There were nearly 100 registered participants from six countries, and four invited and 56 oral talks were delivered during the two days of the conference.

Based on the work presented at the conference, selected papers are included in this volume of *IOP Conference Series: Materials Science and Engineering*. These papers present new research in the various fields of Materials Engineering, Mechanical Engineering, Computer Engineering, and Mathematical Engineering. It is our great pleasure to present this volume of *IOP Conference Series: Materials Science and Engineering* to the scientific community to promote further research in these areas. We sincerely hope that the papers published in this volume will contribute to the advancement of knowledge in their respective fields.

Special thanks should be directed to the **Organizing Committees** for their tremendous efforts in organizing the conference:

International Organizing Committee
Co-Presidents
Serban Viorel-Aurel, Politehnica University of Timisoara
Zhou Laixin, Military Economics Academy of Wuhan
Members
Dan Daniel, Politehnica University of Timisoara
Fang Zhengqi, Military Economics Academy of Wuhan
Gherman Laurian, "Henri Coanda" Air Force Academy of Brasov
Guo Zhonghou, Military Economics Academy of Wuhan
Lin Pinghua, Huazhong University of Sciences and Technology
Panoiu Caius, Politehnica University of Timisoara
Popa Mircea, Politehnica University of Timisoara
Zhou Yanming, Military Economics Academy of Wuhan

Local Organizing Committee
Coordinator
Lemle Ludovic Dan, Politehnica University of Timisoara
Members
Alexa Vasile, Politehnica University of Timisoara
Alic Carmen, Politehnica University of Timisoara
Benea Marius, Politehnica University of Timisoara

Bistrian Diana, Politehnica University of Timisoara
Bleoca Adrian, Politehnica University of Timisoara
Deaconu Sorin, Politehnica University of Timisoara
Grigore Constantin, Politehnica University of Timisoara
Harau Carmen, Politehnica University of Timisoara
Jitian Simion, Politehnica University of Timisoara
Ratiu Sorin, Politehnica University of Timisoara
Socalici Ana, Politehnica University of Timisoara
Stoica Romulus, Politehnica University of Timisoara
Tirian Ovidiu, Politehnica University of Timisoara
Secretariat
Berdie Adela, Politehnica University of Timisoara
Rob Raluca, Politehnica University of Timisoara

The members of the **Scientific Committee** who selected and reviewed the papers included in the Proceedings of ICAS2014 are:

Bordeasu Ilare, Politehnica University of Timisoara
Chen Hua, Wuhan University
Cioaca Catalin, "Henri Coanda" Air Force Academy of Brasov
Constantinescu George, "Henri Coanda" Air Force Academy of Brasov
Dehelean Cristina, University of Medicine and Pharmacy "Victor Babes" Timisoara
Fang Guojia, Wuhan University
Gao Liang, Huazhong University of Science and Technology
Guan Zailin, Huazhong University of Science and Technology
Guo Deyin, Wuhan University
Herban Sorin, Politehnica University of Timisoara
Huang Kun, Huazhong University of Science and Technology
Kiss Imre, Politehnica University of Timisoara
Lesenciuc Adrian, "Henri Coanda" Air Force Academy of Brasov
Li Baofeng, Huazhong University of Science and Technology
Li Xuewu, Military Economics Academy of Wuhan
Li Yan, Wuhan University
Luculescu Doru, "Henri Coanda" Air Force Academy of Brasov
Luo Min, Military Economics Academy of Wuhan
Ma Huijun, Military Economics Academy of Wuhan
Miron Liliana Ecaterina, "Henri Coanda" Air Force Academy of Brasov
Miu Xiaoping, Huazhong University of Science and Technology
Motoc Andrei, University of Medicine and Pharmacy "Victor Babes" Timisoara
Muscalagiu Ionel, Politehnica University of Timisoara
Osaci Mihaela, Politehnica University of Timisoara
Pan Chunxu, Wuhan University
Panoiu Manuela, Politehnica University of Timisoara
Pater Lucian Flavius, Politehnica University of Timisoara
Popoviciu Octavian Mircea, Politehnica University of Timisoara
Rosu Serban, University of Medicine and Pharmacy "Victor Babes" Timisoara
Todea Carmen, University of Medicine and Pharmacy "Victor Babes" Timisoara
Wan Min, Huazhong University of Science and Technology
Wang Gengsheng, Wuhan University
Wang Xianjia, Wuhan University
Wu Shaohua, Military Economics Academy of Wuhan

Wu Jun, Huazhong University of Science and Technology
Xu Yong, Military Economics Academy of Wuhan
Ye Yonggang, Wuhan University
Yu Jiajiang, Military Economics Academy of Wuhan
Zhang Chengjian, Huazhong University of Science and Technology
Zhu Qinglin, Military Economics Academy of Wuhan
Zou Yunsong, Military Economics Academy of Wuhan

We thank the authors for their contributions and we would also like to express our gratitude to everyone who contributed to this conference, especially for the generous support of the sponsors:

HUNEDOARA MicroMega *S.C. Micro-Mega HD S.A.*

ArcelorMittal ***ArcelorMittal S.A.***

DHS Viaţa în mişcare! ***S.C. Eurosport DHS S.A.***

BEE SPEED Automatizări ***BeeSpeed Automatizari S.R.L.***

opac Design System ***S.C. Opac Design System S.R.L.***

DUCODAN Since 1993 ***S.C. Ducodan S.R.L.***

Ludovic Dan Lemle and *Yiwen Jiang*
Coordinators of the Scientific Committee of ICAS2014

Peer review statement

All papers published in this volume of *IOP Conference Series: Materials Science and Engineering* have been peer reviewed through processes administered by the proceedings Editors. Reviews were conducted by expert referees to the professional and scientific standards expected of a proceedings journal published by IOP Publishing.

IOP Conf. Series: Materials Science and Engineering **85** (2015) 012001 doi:10.1088/1757-899X/85/1/012001

Experimental research on quality features of metallurgical coke

V Andrei[1] and N Constantin[1]

[1] University Politehnica of Bucharest, Materials Science and Engineering Faculty, 313 Splaiul Independentei, 060042 Bucharest, Romania

E-mail: victorandreil1@yahoo.com

Abstract. From all the solid fuels, the metallurgical coke is the most used in obtaining iron in the blast furnace. Together with the iron ore, manganese ore and fluxes, it constitutes the basis of raw materials and materials for elaborating pig iron. This paper presents the results of laboratory investigations by the authors to determine the most important quality characteristics of some types of coke used in the blast furnace charge.

1. Introduction

The metallurgical coke is a fuel obtained in the pyrogenic process (heating in the absence of air) at temperatures of 1000°-1100°C of agglutinated coal which are oily coals with 16-30% volatile (lately it's been used gas coals). At 950°C the volatiles are removed and dry residue is the only thing that remains (raw coke), in which is enclosed also the coal ashes. Out of 1 ton of coal, 0.6-0.8 tons of coke is obtained.

Metallurgical coke ensures the following functions in the blast furnace:

Fuel: its combustion has to provide the necessary heat for the processes in the blast furnace (ore reduction, carbonates decomposition, sulfides, water, forming of iron and slag, covering the heat losses etc.);

Reducing agent: takes part directly, through its Carbon content, but also indirectly through CO obtained from burned C in coke in the tuyeres, in reducing Fe, Mn, Si, P oxides in the ore;

Carburizing agent: in the crucible, gives away the C to the bain that is necessary to complete the irons composition (the carburizing of Fe begins in the tank at small temperatures of 1000°C to 1%C and continues at temperatures higher than 1100°C followed by the irons melting due to decreasing of t_{mel});

Refining agent: ensures the loads permeability with gas passing through (it occupies over 50% of the furnaces volume);

Sustaining the material column in the furnace, being the only one in a solid state near the tuyeres (coke does not melt, it burns and gasifies).

2. Coke properties

The most important properties of coke are porosity, electrical conductivity, grain size, mechanical resistance and reactivity.

Porosity is an important characteristic to determine the size of the active surfaces of coke, therefore the reaction power. It has values of 45-55% (total surface of 1 kg of coke = 40 m^2).

The apparent porosity (sum of open pores) is determined by introducing coke in water and vacuum degassing at a certain pressure.

The real porosity (sum of all pores) expresses the volume of goals in the coke piece and it is established using the formula:

$$P = \left(1 - \frac{\text{apparent specific weight}}{\text{real specific weight}}\right) \times 100, \% \quad (1)$$

Electrical conductivity, this value indicates if the coking process went as it was supposed to (a good coking leads to values up to 30-40 times higher of the electrical resistance). This property is considered especially in electrical furnaces.

Grain size determines the permeability of materials column in the blast furnace and the gas flow through the load, the coke occupies 50-55% of the furnace volume [1]. Optimal grain size is controversial, medium size recommended is 45-60 mm (3 times larger than the medium size of the ores). The following indices have been established:

D_{50} – Medium dimension (dimension of grading screen meshes at 50% pass and 50% denial);

D_{75}, D_{25} – Quarters dimension (75% pass and 25% denial);

$\frac{D75}{D25}$ – Homogeneity indices;

$$K = \frac{(40-80)mm}{(+80\ mm)+(25-40mm)}, \text{ Uniformity coefficient.} \quad (2)$$

Nowadays, in most enterprises, before loading in the furnace the coke is passed through a grading screen in order to separate the small pieces. In furnaces up to 1000 m^{3}Vu the grading screen has holes of 25 mm wide and in larger furnaces of 30-40 mm wide.

In general, the maximum dimension of the pieces of coke is not limited [1].

Basically, in furnaces that operate with well classed ore, the coke must have a dimension between 25-74 mm. In the case of ore insufficiently classed, especially if it contains pieces under 8-10 mm and pieces over 60-70 mm, the dimension of coke must be larger than 40-120 mm.

The operating results of the furnaces which work with classed ore loading and which shred the coke grain size over 60-70 mm simultaneously with the removal of small pieces (below 25-30 mm) have proven some advantages.

The coke that is under moderate mechanical stress followed by classing, improves its mechanical resistance, as well as its homogeneity, but the medium dimension becomes smaller.

Standardized method for determining the granulometric composition consists in passing a determined quantity of coke through a grading screen, each time weighting the remaining material on the screens, as well as passing through the last screen.

The results are expressed in mass percentage, in rest on the grading screen, as well as the passing through the last screen. The weight of the sample, depending on the cokes size, the correlation between them is shown in Table 1.

Table 1. The correlation between the weight of the sample and the cokes size:

	Coke with a large grain					Small coke	
Max dimension of coke pieces, (mm)	≥80	63	40	20	10	5-10	5
Min weight of the sample, (kg)	100	80	30	10	5	1.5	0.5

Mechanical resistance is the main characteristic on which the coke behaviour depend the most in the blast furnace where it undertakes falls and friction. These falls and frictions could lead to the shredding of the pieces, also leading to a worsening flow of gas through the load.

Coke shredding in the furnace leads to the worsening of the gas flow through the load and to accumulation of chopped coke, producing in some areas problems at the loads descent, tuyeres burning, increased specific coke consumption.

The cokes resistance represents the sum of the resistance at compression, shearing, friction, pressing, dynamic shock and it is determined using samples in which the coke is the subject of hitting and friction efforts through rolling in a tambour. The resistance is determined at the carbonization plant, as well as at the furnace (the resistance indices is greater, the coke being stabilized during transport and depositing). The resistance, in Micum drum, is an international standardized method, being the most used one, indices obtained in the test are:
M_{40}= mechanical resistance
M_{10} = abrasion
For a high quality coke these indices must have values between M_{40}= 80 ÷ 85% and M_{10} = 5 ÷ 7%.

In the recent years it has been used also a method of establishing the cokes mechanical resistance with chemical attack and heating. This way, the CSR (Coke Strength after Reaction) index is being determined. The CSR represents the resistance of coke after the reaction with CO_2. A high quality metallurgical coke must have a CSR of 55-60%.

The reactivity of coke is the ability to react with oxidizing gases in the furnace conditions, being of utmost importance the reaction of the coke with oxygen or injected air, the reaction of coke with CO_2, the coke reaction with H_2O. Reactions of carbon coke, before reaching the tuyeres, and O_2, CO_2, H_2O occur in the oxidizing area of the crucible.

Determination of reactivity is carried out when a stream of CO_2 passes through a granular mass of coke of 10 to 25 mm, at a temperature of approx. 900°C. The reactivity is expressed by the amount of CO_2 that had passed into CO after the reaction with carbon coke. The variation of cokes reactivity influences, in restricted ranges, the size of the combustion area from the crucible.

The reactivity of coke expresses, through a conventional index its affinity for oxygen, carbon dioxide and water vapour reactions [2].

Measurements are conventional, the results depending on porosity, grain size, the gas flow rate, the determination temperature and type of coal from which the coke was made.

Depending on the oxidizing agent the cokes property to react with CO_2 is determined $(C + CO_2 <,> 2CO)$ and rarely the cokes property to react with H_2O $(C + H_2O <,> CO + H_2)$ or the cokes property to react with O_2 $(C + O_2>, <CO_2)$. Coke reactivity determines in a more important manner the coke consumption in the furnace.

From the reactivity point of view, the cokes are classified as follows:
Cokes slightly reactive (R = 100-200%);
Cokes medium reactive (R = 50-100%);
Cokes heavily reactive (R < 50%).

Furnaces in Romania are currently working with several types of coke, cokes from their own coke plants in the case of Arcelor Mittal Galati, which represents approx. 50%, the rest consisting of various cokes import (China, Japan, Poland, Czech Republic and Russia) [3].

In general, the compositions of the coke used (technical analysis) are known, and followed systematically. Similarly is done also with the cokes grain, determinations are performed less frequently and with an informational purpose.

Taking into account the lack of data on the characteristics of reactivity of coke used and the structural characteristics that together determine their behavior at high temperatures in the furnace, it will be researched in the lab "Iron Metallurgy" in University Politehnica of Bucharest (UPB) selected representative samples from the coke used frequently in Arcelor Mittal Galati furnaces, in order to reach conclusions about how they might influence the reduction processes.

IOP Conf. Series: Materials Science and Engineering **85** (2015) 012001 doi:10.1088/1757-899X/85/1/012001

The best known method for determining the reactivity is the Koppers method. The determinations are done with equipment made of quartz tube with a diameter of 20 mm, electric furnace whose temperature can be adjusted and kept constant, CO_2 cylinder, pressure reducer and needle valve for fine tuning of the gas flow, tight valves binding the quartz tube to the gas line and gas analyzer.

Through the quartz tube filled with 34 cm^3 coke, grain size 0.5 - 1 mm, is passed for 15 minutes a stream of CO_2, which is introduced into the furnace.

The tube is connected to the circuit through metal reinforcements to the gas agent network whose flow rate is maintained constant at 10.8 L / h for 60 minutes. After 10 minutes of placing the tube in the oven the first sample of resulting gas is collected from the reaction monitored and analysed with an Orsat apparatus. The operation is repeated 5 times under the same conditions, collecting and analyzing the resulting gas each time.

The reactivity calculus reaction is:

$$R = \frac{CO}{CO_2 + 1/2CO} \times 100, [\%] \tag{3}$$

The reactivity is defined through a coefficient of the reaction velocity between the CO_2 and coke:

$$K_m = \frac{V_0}{m} \times 2\ln\left(\frac{2\frac{C1}{C0}}{1+\frac{C1}{C0}} + \frac{1-\frac{C1}{C0}}{1+\frac{C1}{C0}}\right) \tag{4}$$

C_1 = CO_2 concentration in the reacting gas [%]
C_0 = CO_2 concentration in the initial gas (practically 100%)
Vo = CO flow (cm^3/s)
m = cokes mass [g]

A quality coke for the furnace must have a low reactivity (~50%). The reactivity of coke is influenced by the catalytic effect of ash contained therein.

The deficiency of this method is that the determination is made for the laboratory samples of crushed coke, with a 0.5 - 1 mm or 1 - 3 mm grain and the pieces are introduced into the furnace 40 - 60 mm.

Such differences in behavior arise due to the pores and specific surface pieces. Since coke reactivity is a feature of its surface, the reaction taking place between C and CO_2 on grain surface and in the open pores, it should be reported to the sample surface. In this case, results from laboratory to industrial conditions should be very close.

To overcome this obstacle, researchers from the University POLITEHNICA of Bucharest (UPB), in collaboration with the researchers at ICEM Bucharest have developed a method for determining the reactivity of coke “to pieces".

Measurements are carried out on pieces of coke size 25 - 30 mm, using the operating principle and of the Koppers plant.

The adaptation to the new conditions is achieved by using a reaction tube having a diameter of 40 mm and a flow rate of 67.5 L / h to ensure the same reaction velocity in the same contact time „gas - solid". In order to maintain the sample at the same temperature zone of the oven, we were working with 50 grams coke sample that is placed in a reaction tube corresponding to the height occupied by the 34 cm^3 Koppers sample.

Values of the determinations made using this method retains the same meaning for different coke samples with values obtained by conventional methods.

3. Experimental research for the determination of the cokes mechanical resistance

In the United Kingdom and the USA is used for resistance research a method that consists of performing 4 drops on an iron board from a height of 1.83 m of a coke sample weighting 23 kg in pieces larger than 51 mm. The resistance is expressed depending on the cokes granulation after shredding (% coke larger than 51 mm and % coke over 38 mm). In Western Europe, IRSID is used, in the USA and Japan the ASTM method. In Romania, Micum drum test is being used, more rarely Sundgreen drum test. In the laboratory, a smaller drum is used, Nedelmann drum [2].

Experiments are being conducted with coke samples used nowadays in the loading of blast furnaces in Romania. This paper will pursue in comparison the mechanical resistances of different types of coke, also the correlations between the cokes technical analysis especially the ash (A) content, the volatile (V) and the mechanical resistance.

Endpoints of the installations for determining the mechanical strength of coke and the conditions imposed on the samples to be tested are presented in Table 2.

Table 2. Features of laboratory drums used to determine the mechanical strength:

Drum	Diameter [mm]	Length [mm]	Coke grain [mm]	Revolutions [rev/min]	Sample weight [g]
Nedelmann	250	750	15-25	930/1	100
Sinuk	300	1000	40-50	40/50	1000
MICUM	500	2000	25-40	50/10	1000

3.1. Experimental laboratory installation

The experimental installation used is a Nedelmann drum, with a reel of 180 mm in diameter, 70 mm wide and 3 bits on the shaft. The reel is driven in rotation via a motor – reducer group. The shaft with bits has a rotating movement in reverse of that of the reel. The rotation speed is of 18 revolutions / minute. This is the same installation as used in determining the ore resistance.

3.2. Research methodology

The sample used for analysing with a weight of 100 g and the grain size of 10-25 mm is introduced in the reel which is rotated for 2 minutes (36 rotations). Next the coke is passed through 10 mm and 1 mm square screens, and then it is determined by weighting the denial at the screens of 10 mm and the passing at the screens of 1 mm.

For the experimental research, the coke sample is loaded in a drum with the inner diameter of 1000 mm and the length of 1000 mm with 4 angle sections (100×100×10), displayed at equal distances throughout the length of the drum. Using a fork with 50 mm distance between the horns is taken the pile of 160 kg (consisting of 24 incremental samples each of 7 kg), large grained coke, making up to 3 samples of 50 kg for two parallel determinations and a spare.

The sample is then loaded in the drum by hand. After loading, the drum is closed and rotated at 25 revolutions / minute for 4 minutes.

Following cokes mechanical screening, the round mesh screens 40, 20 and 10 mm, arranged in cascade. The coke is characterized in terms of quality indices M_{40} and M_{10}.

Index M_{40} is the expression as a percentage of the fraction greater than 40 mm that is left in the drum.

M_{10} index is the percentage fraction less than 10 mm from the test resulting.

Research has shown the importance that the heat resistance to characterize the quality of coke in the blast furnace operating conditions [3].

For this determination different methods were developed, based on:

-Heating the coke drum where the determination is to be made;

-Heating the coke in a nearby furnace and moving it to a testing drum.

The researchers of UPB, IMOMM Department, have developed an installation consisting of an electric furnace with a tight cylindrical muffle, of which the lower flap can be opened from distance, together with a laboratory MICUM drum with 500 mm diameter, 2000 mm length with 3 angles on each generator of 35×35×4 mm, rotating at 50 revolutions / min.

For setting a resistance value closer to normal to the industrial MICUM drum, experiments have indicated the best results at 500 rpm. [4]

The coke sample used in the experiment is obtained from crushing and screening through a 40 mm and a 25 mm round mesh of the sample of air-dried industrial coke. An amount of 1 kg of the industrial coke sample is introduced into the drum and is subject to 500 rotations, for determining the resistance to cold.

1 kg sample used for determining the high temperature resistance is charged into the oven and heated for 3 to 4 hours by means of electrical resistances, up to the temperature of 1000°C, which is maintained for 1 hour. Proof seal of the sample in the retort prevents loss through combustion or partial gasification.

The hot sample is passed through the drum, the duration of the operation being 2 minutes, and the loss of temperature of 50-60°C. Once the lid is closed the drum starts rotating at 50 revolutions / minute for 10 minutes. The resistance is expressed as a percentage, after screening the sample through the mesh of 25 and 10 mm, the results obtained are presented in Table 3.

Table 3. The resistances of coke, at high and low temperatures for the MICUM laboratory sample:

Details		Coke Romania	Coke Russia	Coke Czech Republic	Coke briquettes ICEM
Resistance at low	> 25 mm	87,32	89,06	87,04	76,88
temperatures, %	10-25 mm	6,42	4,01	4,99	12,87
	< 10 mm	6,26	6,93	7,97	10,30
Resistance at high	> 25 mm	80,10	89,78	86,15	81,15
temperatures, %	10-25 mm	13,11	2,59	8,24	8,50
	<10 mm	6,79	7,63	5,61	10,35
Change, %	> 25 mm	- 8,28	0,83	- 1,02	5,63
	<10 mm	8,5	10,1	29,20	0,48
Loss	Weight, %	4,35	4,67	4,57	6,75

4. Experimental research for the determination of the cokes reactivity

4.1. Experimental laboratory installation

The Iron Metallurgy laboratory from UPB, allows the testing of samples of coke having the dimensions which they are normally in a furnace.

The experimental laboratory installation for the determination of the cokes reactivity is shown in Figure 1.

It should be noted that this method is used with good results in recent years in Italy at Italsider and in France by Chercer Morieau [3].

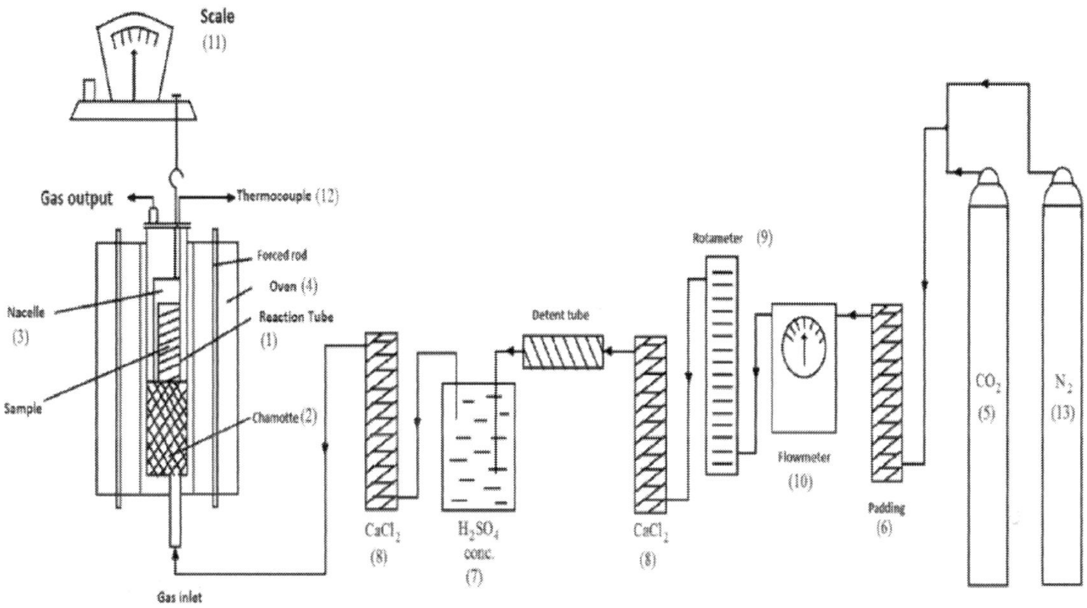

Figure 1. Schematic diagram of the experimental laboratory installation

4.2. Research methodology

It is used for coke samples weighing 400 grams and a grain size of 10 to 25 mm. The sample is loaded on the nacelle (3) of the stainless steel mesh with an inner diameter of 120 mm and a height of 300 mm. Platform coke sample is inserted into the stainless steel tube (reaction tube) (1) with a diameter of 140 mm, length 1000 mm, placed vertically inside the oven (4) heated with forced rod.

The determination is performed at a constant temperature of 900°C measured using thermocouple (12), Pt-Pt-Rh type introduced into the sample.

After ~ 5 minutes of nitrogen purging from the canister, CO_2 is blown from the bottle at a flow rate of 12 L / min for 30 minutes. Before reaching the incandescent coke sample, the gas passes through the flowmeter (9), the gas meter (10) and the cleaning vessel filled with cotton wool (6), sulphuric acid (7) and calcium chloride (8).

Sample weight variation during the blast of CO_2 balance is shown (11). After 30 minutes, the sample was again purged with nitrogen and then allowed to cool slowly to approx. 24 hours with a neutral atmosphere of nitrogen oven.

Next operation is extracting the sample, weighing it and screening it for determining the denial at 10 mm sieve.

Results for the cokes reactivity:

The results are determined by weighing the amount of C in coke which reacts with CO_2 as the difference between (M1) initial mass of the sample and (M2) final mass after reaction with CO_2. With laboratory work presented reactivity is determined by the relationship:

$$R = \frac{M1-M2}{M2} \times 100, \% \tag{5}$$

Hereby established after weighing the denial at meshes of 10 mm, the amount of coke (q2) retains the original dimensions. This percentage relates to the initial mass (q1).

The results of laboratory measurements are presented in Table 4.

Table 4. The results of laboratory measurements for cokes reactivity

Details	Coke Romania	Coke Russia	Coke Czech Republic	Coke briquettes ICEM
Reactivity	64	76	82	103

It is noted that all coke types, irrespective of their origin have low reactivity with respect to CO_2, (with the exception of coke briquettes), which means that it will retain the same structural strength to the complete combustion. A low reactivity of the combustion zone size increases, with effect on lowering the load.

5. Conclusions

Currently, the trend worldwide is to practice heating methods for determining the mechanical strength and coke reactivity.

Introduction in the industrial practice of these methods, has led to the use in blast furnace of coke with superior quality indices, thus achieving optimum operating indicators of the blast furnace.

The results of experimental research for the determination of the cokes mechanical resistance show that all types of coke tested can be used in blast furnaces, with the difference that the coke obtained in Romania has a lower resistance, a high temperature than the coke of the Czech Republic or Russia provenance.

The results of experimental research for the determination of the coke mechanical resistance show very large differences between types of coke analysed.

The best value for the coke reactivity index was from the coke with Czech Republic provenance, followed by coke from Russia and Romania.

Coke briquettes tested has a very high reactivity index, which causes a negative influence through the burning zone of the crucible, decreasing at the same time the rate of descent of the load in a blast furnace.

References

[1] Constantin N, Necşulescu C and Rucai V 1993 *Research on ways of increasing the intensity of the blast furnace operation by improving circulation gas*, Upgrading flows and processes the steel industry, Hunedoara, Romania, November 3-4, pp 42-46

[2] Burlacu J C and Constantin N 2012 Physical and chemical characteristics of raw materials used in blast furnaces in view of an improved efficiency, *Metalurgia International* **17**(5) 10-15

[3] Constantin N, Geantă V, Şaban R and Bunea D 1991 Determining through theoretical mathematical modeling the optimal coke consumption in blast furnaces, depending on the composition of the load, *Metalurgia* **6** 14-17

[4] Stanasila C, Constantin N, Stanasila O and Petrache R 2007 A new method to dry some granular materials, *U.P.B. Scientific Bulletin, Series B* **69**(4) 71

IOP Conf. Series: Materials Science and Engineering **85** (2015) 012002 doi:10.1088/1757-899X/85/1/012002

Technological aspects at continuous casting of semi-finished products with ϕ270mm

E Ardelean[1], M Ardelean[1], T Hepuț[1] and A Lăscuțoni[1]

[1] Polytechnic University Timisoara, Department of Engineering and Management, 5 Revolution Street, 331128 Hunedoara, Romania

E-mail: erika.ardelean@upt.ro

Abstract. Continuous casting installation especially appreciated because steel can be poured in a more varied assortment. The flexibility of the system is not sufficient if the casting parameters are not properly adopted and adapted to the specific brand of steel. This paper presents some technical aspects relative to continuous casting of semi-finished products with φ270mm section. Graphical dependencies obtained in Excel and analytical equations of this allows to specialists from industry to adopt values for the addicted parameters according to the independent ones, already known.

1. Introduction

Continuous casting of steel, originally developed for casting of non-ferrous metals and alloys, has been adapted for casting of steel, gradually replacing ingots casting. With the changes in installation design and technology, were replacement vertical installations with installations which have curved wire in different zones – Figure 1. [1]

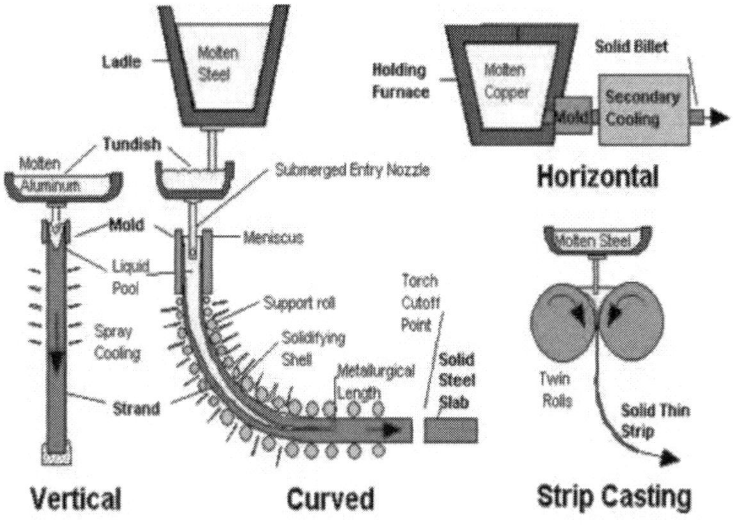

Figure 1. Types of continuous casting machines

IOP Conf. Series: Materials Science and Engineering **85** (2015) 012002 doi:10.1088/1757-899X/85/1/012002

Also, by the changes made to casting machines, range assortment has grown both in section and in shape – Figure 2. [2]

In extending of the continuous casting process in detriment of conventional casting has contributed improvements both the facility and the casting process: use immersion tubes [3], the coatings powders in tundish [4] and casting powders in crystallizer [5], [6], the electromagnetic stirring [7], properly designed tundish and crystallizer design [8], the possibility of casting simultaneously through the same system several sizes section or shape.

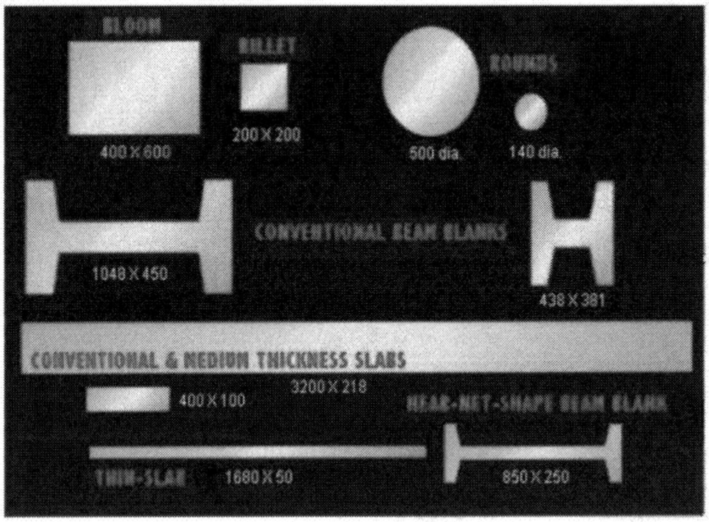

Figure 2. Typo dimensions of continuously cast semi-finished product

2. Experimental data's

To a specifically analyse of the steel continuous casting technology, were analysed a total of 20 charges of steel S235JR brand (according to EN 10025). It is noted that the charges have been produced in an electric arc furnace, treated in the LF secondary treatment installation (Ladle Furnace) and then continuously cast, on the 5-wire system as a billet section φ270mm (three-wire) and section φ310mm (two-wire). It is noted that on the continuous casting facility has been cast to the present billet section φ150mm, φ180mm, φ310mm bloom section and 240x270mm.

From the point of view of steel making process has been not found problems, at steel discharge from the LF that matching in terms of chemical composition [9].

At discharge of liquid steel from LF secondary treatment facility, steel temperature was within the limits 1586-1610^{0}C, noticing it an overheating compared to liquidus temperature with 64-94^{0}C – Figure 3. Overheating was necessary taking into account the time (and the route) to go ladle with liquid steel from steelmaking and secondary treatment hall to continuous casting plant hall.

During the continuous casting steel temperature is a technological parameter that is monitored continuously; as evidence, measurements made from 10 to 10 min relative to the temperature of the steel distributor, where continuously steel is in constant movement, being casted from the casting ladle and then in installation crystallizer. In most cases studied (Figure 4) there was a slight decrease of this parameter, which is normal because the tundish has no possibility of heating the steel. In one case there is a high temperature after 40-45min, but can be justified by entering into tundish of hot steel quantities of steel, brought by next charge of the casting sequence.

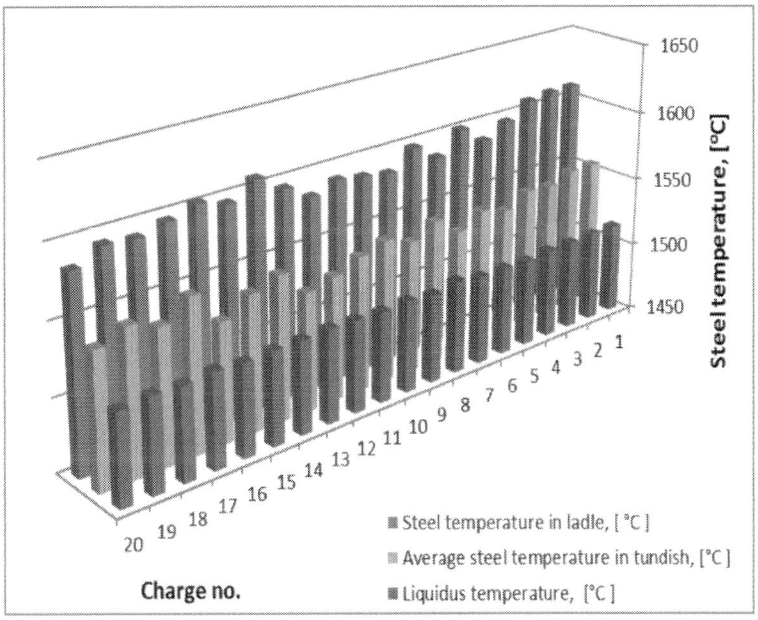

Figure 3. Comparative analysis of steel temperatures (liquidus, ladle and tundish)

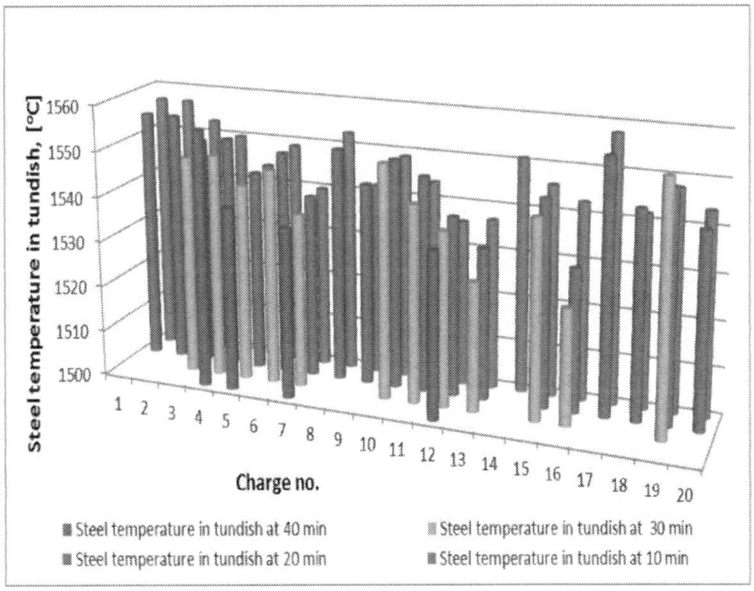

Figure 4. Changes in temperature of liquid steel in tundish during casting

That, the casting temperature caused the variation of another important parameter: the casting speed Figure 5. It is noted that with increasing of temperature are reduced the speed of continuous casting, so that the cooling of the billets be adequate and their quality is not negatively affected.

Figure 6 presents the variation in overheating degree of the steel in the tundish according to the casting speed. It is noted that casting speed adjustment was made through the entire process of continuous casting, as evidenced by the inverse dependences: an increased temperature gradient, the casting speed was slightly decreased in order to not induce additional thermal stress in the semi-finished product.

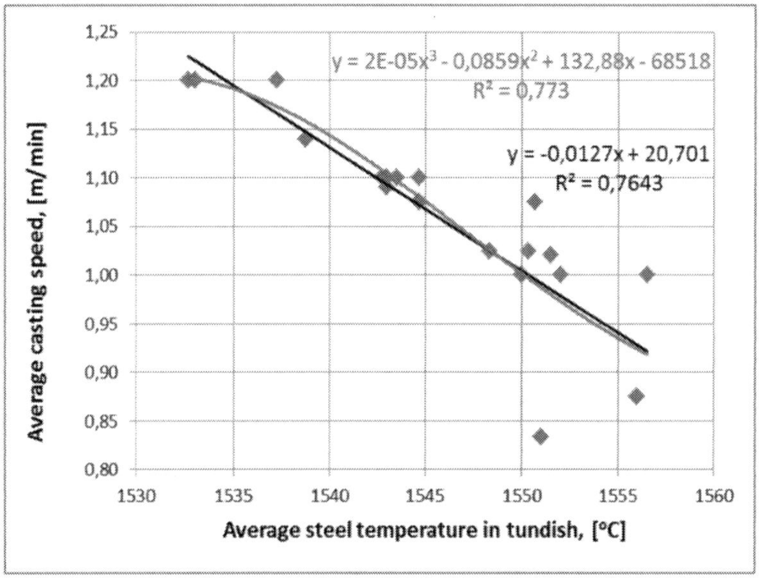

Figure 5. Dependence between casting speed and the average temperature of steel in tundish

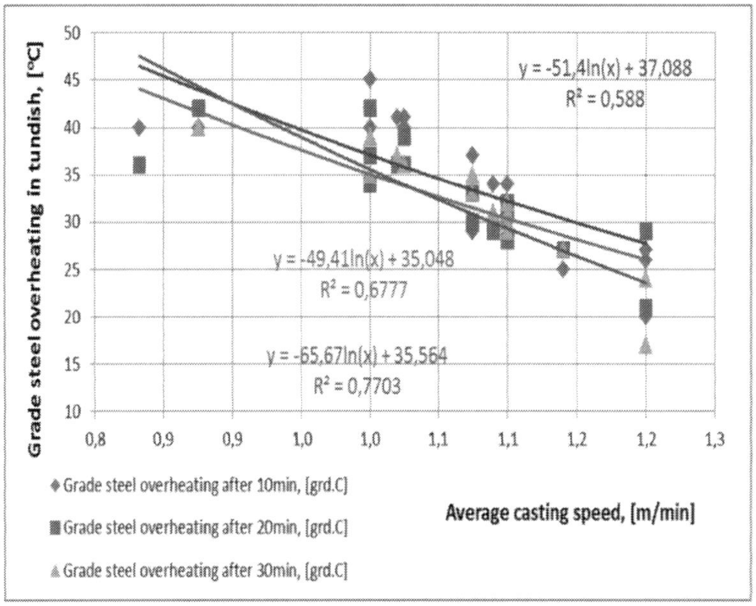

Figure 6. The dependence between overheating degree and average speed casting

Primary cooling takes place in the crystallizer; at studied charges water cooling values were as follows: pressure 8.9 to 9.3 bar, flow rate of 1800 l / min. In Figure 7 is presented the dependence between the average degree of overheating of steel in tundish and cooling water pressure in crystallizer. It can be seen directly proportional interrelation between the two parameters: with increases of steel temperature (the degree of overheating) increases also the water pressure in the crystallizer.

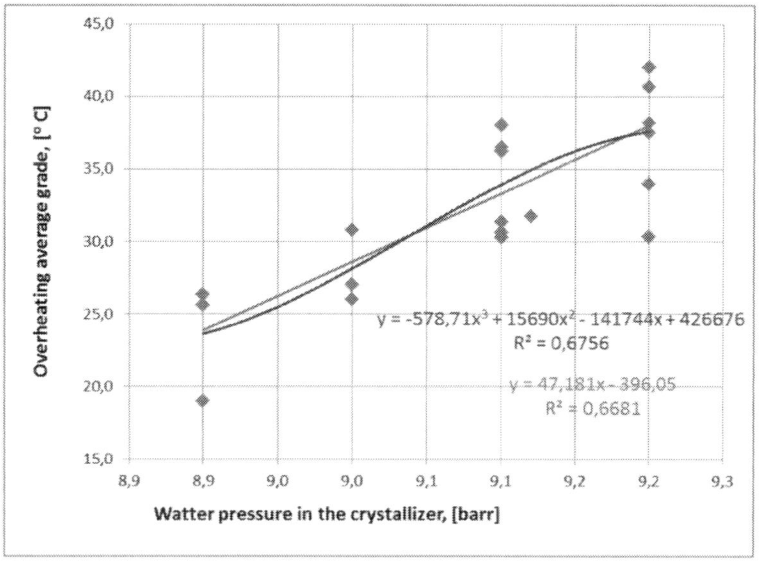

Figure 7. Changes in the overheating degree of steel according to the water pressure in the crystallizer

Cooling has continued later in the secondary cooling zones (divided into three zones according to the ratio 40:40:20), but even in this sector there is an adjustment of the cooling regime with liquid steel superheat degree – Figure 8. Secondary cooling water flow rates, presented in Figure 9, reveals identical values for Z_1 and Z_2 respectively slightly below the Z_3.

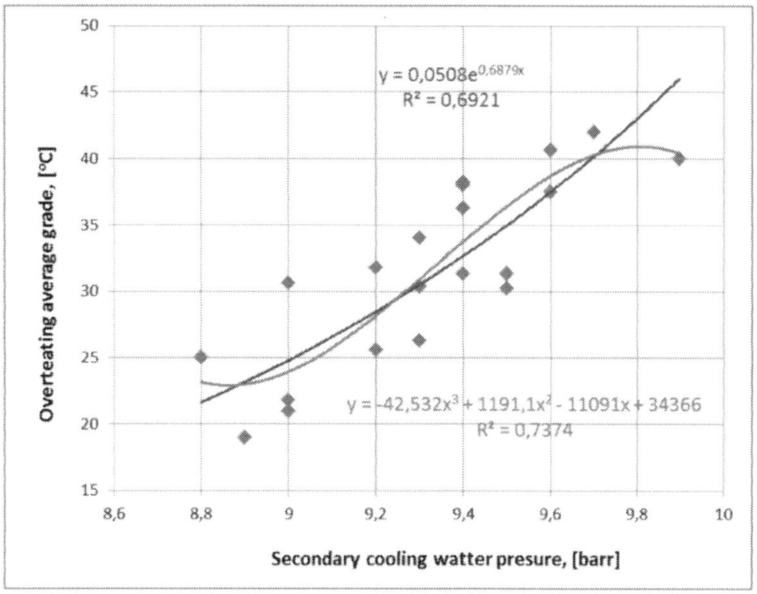

Figure 8. Changes in the degree of overheating depending on the secondary cooling water pressure

Finally it should be noted that the third parameter of influence to the continuous casting technology: casting time. At the studied charges it ranged between 39-69min, Figure 10. Both extremes are atypical: in the first case the cast was interrupted and steel was evacuated emergency ladle; in the second case, it not was casted on the all 5 wires, which artificially increased casting time.

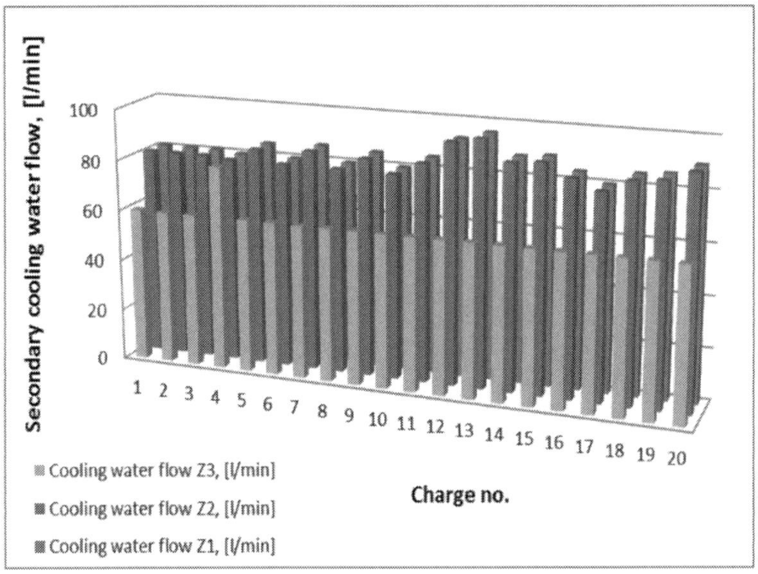

Figure 9. Changes in secondary cooling water flow

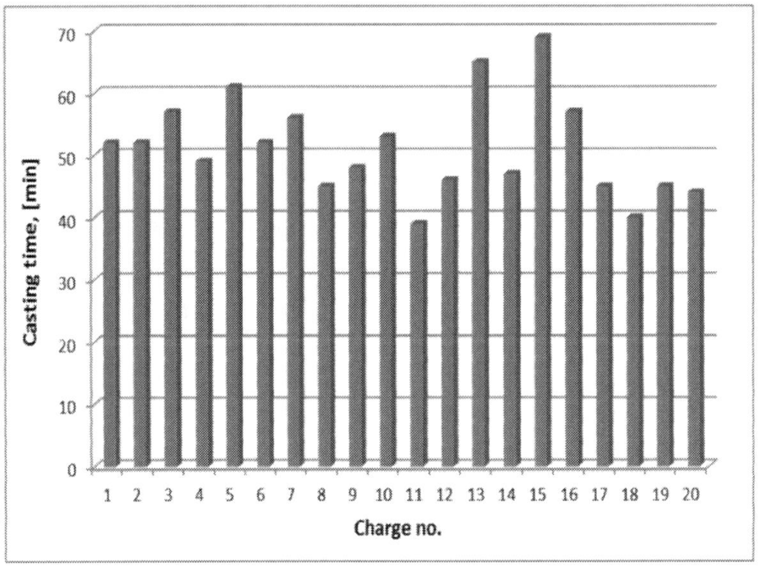

Figure 10. Changes in the casting time

3. Conclusions

From the above results the following conclusions:

- continuous casting of the profile with section φ270mm has no problems, any atypical values of some parameters are immediately remedied by changing the operating parameters of the facility;
- overheating degree of the steel in the casting ladle and then in tundish is slightly increased over the limits recommended by the literature (Mannesmann) - 45-60°C compared to liquidus temperature;

- casting speed allows an average value of 1,05m/min., casting speed adapted to the facility being studied. An increase in casting speed would have positive impacts on growth of the facility productivity, but must considered the effects to intensify of cooling on the quality of semifinished product;

- adaptation of the cooling parameters (flow and pressure in the crystallizer and in the secondary cooling zone) depending on the technological parameters is done by using cooling programs implemented at the studied facility.

Acknowledgments
This work ws partially supported by the strategic grant POSDRU/159/1.5/S/137070 (2014) of the Ministry of National Education, Romania, co-Financed by the European Social Fund - Investing in People, Sectorial Operational Programme within the Human Resources Development 2007-2013.

References

[1] ***http://ccc.illinois.edu
[2] ***http://www.steel.org
[3] Wondrak T, Eckert S, Gerbeth G, Klotsche K, Stefani F, Timmel K, Peyton A J, Terzija N and Yin W 2011 Combined Electromagnetic Tomography for Determining Two-phase Flow Characteristics in the Submerged Entry Nozzle and in the Mold of a Continuous Casting Model, *Metallurgical and Materials Transactions B* **42**(6) 1201-1210
[4] Ardelean E, Ardelean M, Socalici, A and Heput T 2007 Research concerning the obtaining from industrial wastes of a cover powder used in steel continuous casting, *Advanced Materials Research* **23** 329-332
[5] Kawamoto M, Tsukaguchi Y, Nishida N, Kanazawa T and Hiraki S 1997 Improvement of the Initial Stage of Solidification by Using Mild Cooling Mold Powder, *ISIJ International* **37**(2) 134-139
[6] Ardelean E 2010 Research on obtaining lubricating powder from various types of wastes, *Journal of Environmental Protection and Ecology* **2** 593-600
[7] Haiqi Y, Baofeng W, Huiqin L and Jianchao L 2008 Influence of electromagnetic brake on flow field of liquid steel in the slab continuous casting mold, *Journal of Materials Processing Technology* **202**(1-3) 179–187
[8] Solhed H, Jonsson L and Jönsson P 2002 A theoretical and experimental study of continuous-casting tundishes focusing on slag-steel interaction, *Metallurgical and Materials Transactions B* **33**(2) 173-185
[9] Ardelean E, Ardelean M, Josan A and Pinca-Bretotean C 2014 Contribution on the influence of steel ladle processing (LF) upon the nitrogen removal rate, *IOP Conference Series: Materials Science and Engineering* **57** 012001

IOP Conf. Series: Materials Science and Engineering **85** (2015) 012003 doi:10.1088/1757-899X/85/1/012003

Using HMI Weintek in command of an industrial robot arm

C Barz[1], T Latinovic[2], I B A Balan[1], A Pop-Vadean[3] and P P Pop[3]

[1] Technical University of Cluj-Napoca, North University Centre from Baia Mare, Department of Electrical, Electronic and Computer Engineering, 62A Victor Babes Street, 430083, Baia Mare, Romania
[2] University of Banja Luka, Faculty of Mechanical Engineering, 71 Stepanovica Street, Banja Luka, 78000, Republic of Srpska, Bosnia and Herzegovina
[3] Technical University of Cluj-Napoca, Department of Mechatronics and Machine Dynamics, 103-105 Muncii Street, Cluj-Napoca, 400641, Romania

E-mail: cristian.barz@cunbm.utcluj.ro

Abstract. The present paper intends to highlight the utility and importance of HMI in the control of the robotic arm, commanding a Siemens PLC. The touch screen HMI Weintek eMT3070a is the user interface in the process command of Siemens PLC, in which the distances and displacement speeds are introduced on the three axes. The interface includes monitoring robotic arm movement but also allows its command by incrementing step by step the motion over axis.

1. Introduction

The strong development of the information technology, which led to the appearance of more and more performing computer systems, to the realization of complex software applications and to the creation of new equipment and technology in general, allowed the introduction of modern automation solutions on a large scale.

The automation of the manufacturing processes was required by the general effort of the industrial manufacturers to obtain high productivity and to improve the quality and reliability of products. The decrease of manufacturing costs and the improvement of work conditions are aimed.

The industrial robots appeared as a response to the human's need to automate the manufacturing processes, especially the repetitive ones. The main activities that industrial robots can carry out are linked to the transporting and manipulating of objects and also to realizing some processes (painting, inspection, assembling, etc.). [1]

The utilization of PLCs presents a series of advantages, among which the reducing of manufacturing time and the decrease of costs are the most important. Mainly, any application that requires electric control needs a PLC. [2], [3]

The interface is necessary in the commanding and monitoring of the process realized by the PLCs, because they do not have a screen. EMT3070a is a touchscreen produced by Weintek Company, which facilitates the creation of a graphic interface for a high number of PLCs found on the market.

2. The structure of the industrial robot arm

Robots are one of the things indispensable to the functioning of society in the XXI century. Found in manufacturing processes (Figure 1), but also in other areas of interest, such as: entertainment, transportation, communications, medicine and agriculture, robots improve human performance from quantitative and qualitative point of view.

Figure 1. Robot arm.

At the moment, industrial robots represent the meeting point of the recent discoveries in several fields: mechanics, automation, electronics, computers and actuation systems. The complexity of this branch is reflected on the mechanical architecture, as well as on the management system. [4]

2.1. Siemens PLC

The Programmable Logic Controllers are command and adjustment programmable automates that are used for industrial machines and processes. Their programming is done using dedicated software, developed by each PLC manufacturer, but having as common point the use of the Ladder Diagram (command electric schemes). [8]

2.2. The stepper motor

The auctioning of the robotic arm is done with the help of Nema 42 stepper motor, which are placed on the 3 displacement axis.

The Nema 42 stepper motor (Figure 2) has been developed firstly for users who restricted the available installation space and who want an engine with increased torque. The Nema 42 uses advanced magnetic technologies to offer a significantly higher torque level than that of a standard engine device.

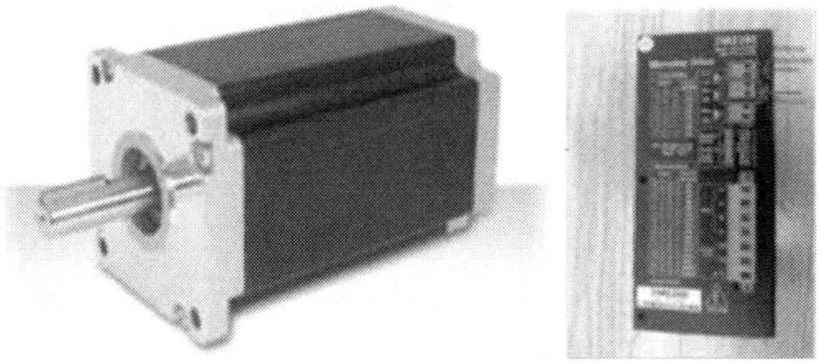

Figure 2. Step by step engine and Microstep Driver 2M2280.

IOP Conf. Series: Materials Science and Engineering **85** (2015) 012003 doi:10.1088/1757-899X/85/1/012003

2.3. Microstep Driver 2M2282.3

Stepper drivers normally work by chopping up a supply voltage using an embedded PWM chip. Microcontroller based steppers drivers can achieve very high rotation speeds in stepper motors.

Using a microcontroller, it is possible to have extreme control over exactly how each individual coil is energized inside the motor. [5]

The command of the step by step engines involves the application of impulses on their windings, impulses that are realized with the help of *Microstep Driver 2M2280* converter.

2.4. The Weintek interface

All PLCs do not dispose of a graphical interface control (Human Machine Interface - HMI), requiring the use of an HMI to enter input data and monitor progress made by the program.

Series eMT3070a is a new generation of HMI from Weintek Company and it has the possibility to program the PLC and also to transfer the date between similar dispositives. [6], [7]

Easy Builder Pro is software developed by Weintek for programming various types of graphical user interface to controlling and command the various types of PLCs. The software has a diverse library of images, buttons, graphics, and sound monitoring screen, but also allows the insertion of other elements that can be used to create interfaces, depending on the project conducted.

Interfaces allows creating different security levels that can be divided into categories of users. Each user access can be set to different interfaces of the program, you can set up to 12 users whom have individual levels of security:

- Administrator - has access to all buttons, windows and all functions available;
- User - has access to all the windows but no to all buttons and available functions;
- Guest - has access only to monitoring.

3. Robotic arm interface command

User interface that we have created for the simulation operation of robotic arm is made of five windows (main menu X axis, Y axis, Z axis and animation (monitoring)). At the time of simulate, navigation between windows is performed using specific buttons (Function Keys).

Main Menu panel (Figure 3) present major information to any user, while having windows (screens) for additional axes robotic arm that works. This panel consists of buttons with which we communicate with the other interface windows, buttons and switches automatic and manual start and stop the operation of the program.

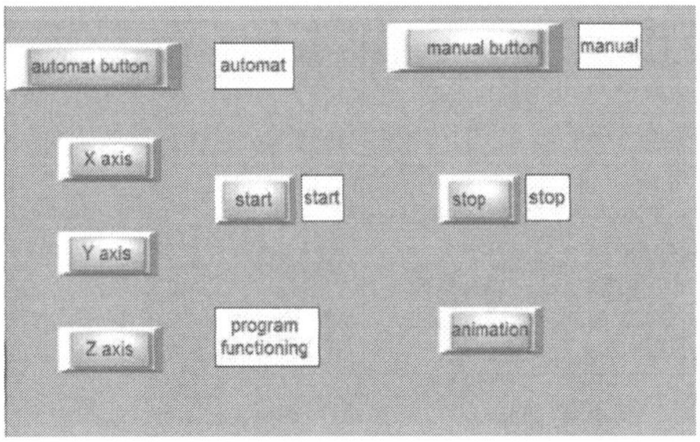

Figure 3. Main menu panel.

Start and stop buttons are buttons of starting and stopping set the programmed bit, we're still here and start and stop lights working distance when inputs I124.0 (start) and I124.1 (stop) operates the same from the PLC program.

3.1. X axis menu

Accessing window X axis is done by pressing the X axis in the Main Menu window. In this window you can find the main buttons for operating the robotic arm on direction of X (Figure 4).

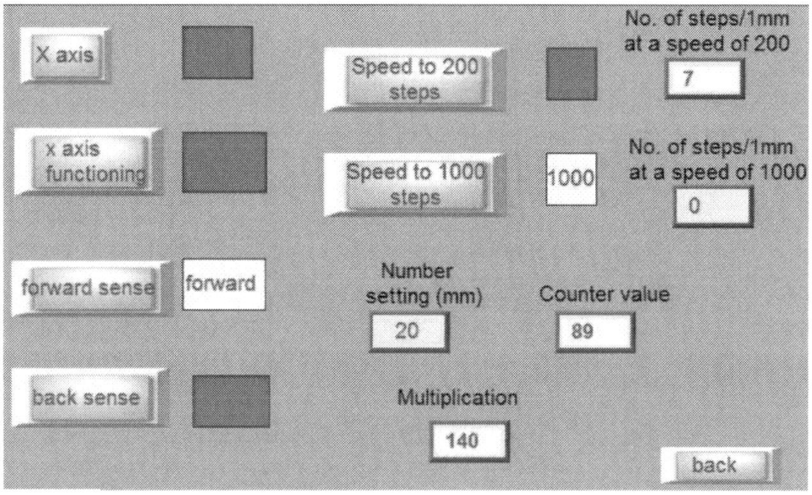

Figure 4. Main of X axis.

The numerical indicators, No. of steps/1mm at a speed of 200, Multiplication, Counter value, No. of steps/1mm at a speed of 1000 and numerical button Number setting in X axis windows, its will be finding also on the windows of Y and Z axis.

Setting the numeric button (MW16) with this button entering the distance (in millimeters) that we want to move the robotic arm (Figure 5)

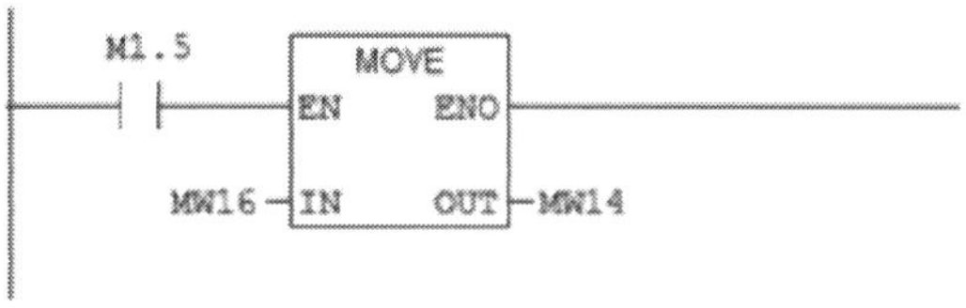

Figure 5. Entering the set of interface.

Numeric indicator *No. of steps/1mm at a speed of 200* (MW8) indicates how many steps must be at a speed of 200 steps in order to move a millimeter.

The button Axa X (M1.0) starts axis x along with relay output time (timer) and output counter.

LED Axa X (M1.0) indicates the function on X axis, the button works when the X axis button is pressed. The button *forward sense* (M0.3) controls robotic arm in the forward direction.

The interface consists of windows which make it possible to navigate between the X, Y and Z of the robotic arm. It performs the following commands:

- Establish starting and stopping application with Weintek panel,
- Set the distance you want to made the arm on the axes X, Y and Z,
- Setting impulses which they receive stepping motors, engines impulses have different configurations from 200 to 12800 pulses.
- Setting the direction of movement of the robotic arm in the X, Y and Z.

3.2. Animation (monitoring)

In this window is represented the functioning of the robot arm on the directions X and Y accessing is done from the Main Menu window. Window animation follows the movement of the robotic arm on the X and Y directions in both automatic and manual modes.

3.2.1. Automatic mode. Animation works automatically by default (Figure 6). This shows how the robotic arm moves in X and Y was set as the range of movement of the arm on axis.

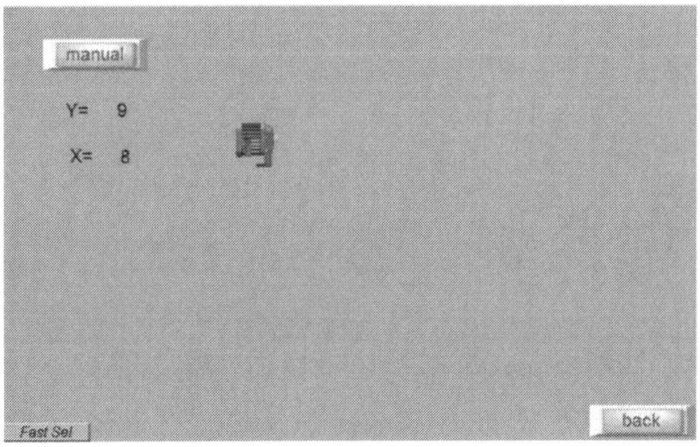

Figure 6. Automatic mode.

Manual button controls the animation in the desired mode, automatic mode is when the manual button is not activated.

3.2.2. Manual mode. The animation works in manual mode only after we manually switched OFF button in the ON state. 

Manual mode shows the arrow buttons with which move around the robotic arm on the X and Y.

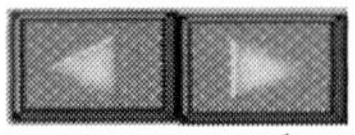

In manual mode (Figure 7) robotic arm moves in X and Y directions only when we action on the displacement arrows on the axis or enter the place where we want to move around the arm by entering a numeric value by pressing X or Y. Also in manual mode we can setup the value of how we want to operate in automatic mode.

Numeric buttons X and Y shows us the value of the arm on the X and Y, when it is in automatic or manual mode.

Figure 7. Manual mode.

4. Conclusions

The interface has been designed so that the displacement moves on X, Y and Z of the robot arm to be controlled by using the eMT3070a interface. It allows to start and stop the application on the panel Weintek, as well as monitor the movement on the axes.

PLC programming has the advantage of creating artificial intelligence by implementing control algorithms in the software and the ability to perform movements based on data provided by the sensors imposed by the developer.

References

[1] Kurfess T R 2005 *Robotics and automation handbook,* CRC Press, Washington D C, USA
[2] Mărgineanu I 2005 *Automate Programabile,* Editura Albastră, București, Romania
[3] Dunning G 2002 *Introduction to Programmable Logic Controllers,* TWI Press Inc., Terre Haute, Indiana, USA
[4] Barabás T and Vesselényi T 2004 *Robotică – Conducerea şi programarea roboţilor industriali,* Editura Universităţii din Oradea, Oradea, Romania
[5] Biro K A, Viorel I A, Szabo L and Henneberger G 2005 *Maşini electrice speciale,* Ed. Mediamira, Cluj-Napoca, Romania
[6] ***Weintek eMT3070a Manual
[7] ***www.weintek.com
[8] ***Simatic S7, Siemens documentation

IOP Conf. Series: Materials Science and Engineering **85** (2015) 012004 doi:10.1088/1757-899X/85/1/012004

Characterization of the WC coatings deposited by plasma spraying

M L Benea[1] and L P Benea[2]

[1] Politehnica University of Timisoara, Department of Engineering and Management, 5 Revolution Street, 331128 Hunedoara, Romania
[2] The National Institute of Applied Sciences Lyon

E-mail: laura.benea@fih.upt.ro

Abstract. Tungsten monocarbide (WC) is deposited using a plasma jet on the martensitic non-corrosive steel support (Z12CNDV12), in three different thicknesses.The characteristics of the coatings are determined by: its chemical composition, optical microscopy, RX analysis, tensile adhesion strength, Vickers hardness, the nature and the processing degree of the substrate and the deposition conditions.The method used for determining the behaviour in a corrosive environment of the WC coatings deposited by plasma spraying consists in measuring the electrochemical potential difference between the coating and the substrate, which are immersed in a solution containing NaCl as a corrosive agent. The experimental results are then mathematically processed in order to determine a law and the mechanisms involved.

1. Introduction

The WC – Co coatings are increasingly certain to be an alternative mean of effectively protecting metal structures and equipment exposed to corrosive and/or abrasive environments. [1], [2], [3].

The characteristics of the coatings are determined by: its chemical composition, optical microscopy, RX analysis, tensile adhesion strength, Vickers hardness, the nature and the processing degree of the substrate and the deposition conditions. The common techniques to attain these depositions are: Air Plasma Spray (APS), Vacuum Plasma Spray (VPS) and High Velocity Oxygen Fuel (HVOF) [4], [5], [6], [7].

The studied coatings in this paper have been deposited by Air Plasma Spray, on a martensitic stainless steel substrate, using a Metco 73 spraying powder.

In order to analyze the behavior of these coatings in a corrosive environment, in literature there are mentioned multiple examples of solutions that can be used in this purpose such as: the 0.1 M NaOH, 0.1 M H2SO4, but also NaCl [7]. We have used NaCl solutions of different concentrations.

Our objective is to characterize the WC coatings and to find a parameter to reflect in a sensitive and easy to quantify way the corrosive action of the corrosion agent upon the coating. This parameter represents the electrochemical potentials of the substrate and the coating immersed into the NaCl solution. Generally, in order to ensure the anti-corrosive action, the coating/substrate couples of choice are those in which the electrochemical potential of the coating is higher than that of the substrate [8], [9].

2. Experimental section

2.1. *Experimental conditions at the WC coatings spraying in plasma*

The metallic substrate on which the coatings were sprayed is the Z12CNDV12, a martensitic stainless steel whose chemical composition is presented in Table 1.

Table 1. Chemical composition of martensitic stainless steel Z12CNDV12.

Chemical composition (%wt)	C	Si	Mn	S	P	Cr	Ni	Mo	V	N2
	0,06 0,15	≤ 0,35	0,50 0,90	≤ 0,025	≤ 0,035	11,0 12,5	2,00 3,00	1,5 2,0	0,25 0,40	0,02 0,04

Before performing the ceramic coating, the surface of the metallic substrate was prepared in advance by mechanical procedures. During the experimentation, the preparation of the surfaces of the substrate was achieved by blasting aluminum oxide particles. The layer of coatings on the substrate prepared in this way was done almost immediately in order to avoid the contamination of surfaces, which can affect the adhesion between the coat and the substrate.

The coatings were made using a Metco 7MB equipment. The powder used is Metco 73, which contains 83% WC and 17% Co and has the size particle size ranging between 10 and 45 μm.

The plasma deposition was made using the following parameters:

- Primary plasma gas: argon, 0.62 MPa pressure;
- Secondary plasma gas: helium, 0.62 MPa pressure;
- Intensity of the current at the generator: 800A and 45÷55 V tension;
- Spraying distance: 80 mm;
- (deposited) powder flow: 2.7 kg/h.

2.2. Determination of coating adhesion

The tensile adhesion strength is a quantity which characterizes the force necessary to detach the thermally sprayed coating from the substrate.

The tensile test is a quasi-direct measure of the adhesion force. The device used is presented in Figure1.

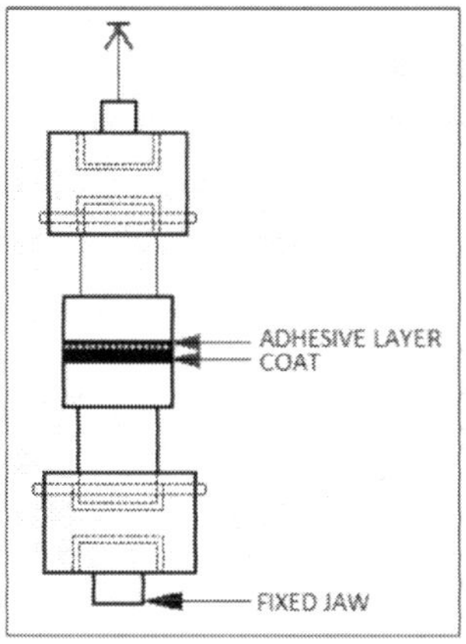

Figure 1. The static method of determining the adhesion of the sprayed coatings.

In the case of its application to plasma coatings, according to the region where rupture occurs, it is necessary to differentiate cohesive ruptures that occur inside the coating of adhesive fractures which appear at the interface.

The test consists in subjecting the coating on the substrate, by means of two glued test pieces on both sides of the coating and the substrate, to a force at a constant speed. The value of the tensile adhesion strength is the ratio between the maximum force achieved at breaking point and the surface of the coating, the stress needing to be evenly distributed over the surface. [10]

In practice, this test is advantageous because of its rapidity, but a poor alignment of the assembly can induce types of stresses other than traction, such as the shearing stress. In the case of very porous deposits, it is difficult to obtain precise results because of the adhesive which can touch the substrate, as the coating is thinner.

2.3. Corrosion resistance determined by the potentiostatic method

The couple made of the coating (WC-Co) and the substrate (stainless steel) has been immersed in NaCl solutions having concentrations between 0,172 and 2,85 M. In each case, the electrochemical potential between the coating and the substrate has been determined using as a reference the calomel electrode [11], [12].

The scheme of the installation used is shown in Figure 2.

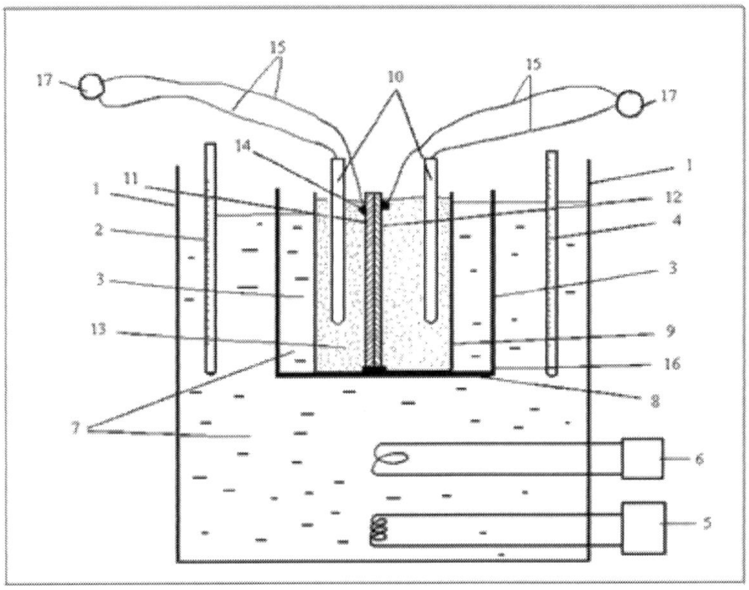

Figure 2. The assembly used for the determination of the corrosion resistance.
1-Thermostate; 2-Measurement thermometer; 3-Fixing device of the support; 4-Contact thermometer; 5-Heater; 6-Water recycling pump; 7-Water; 8-Support; 9-Berzelius glass; 10-Reference (calomel) electrodes; 1-Coating; 12-Metallic substrate; 13-Electrolite (NaCl solution); 14-Contact; 15-Conductors; 16-Silicon (adhesive for sealing); 17-Milivoltmeter

The measurements were made for three different values of the coat thickness (0,05 mm, 0,10 mm and 0,25 mm).

The difference between the electrochemical potential of the coating and that of the substrate represents the measure of the protection degree the coating can provide. The greater the difference, the better the protection of the metallic substrate.

3. Results and discussions

The powder used in plasma jet coatings was Metco 73, which has a grain size of 10-45 µm. For this powder, the granulometric distribution was also determined by means of microphotography obtained with an LMA 10 Carl Zeiss Jena optical microscope.

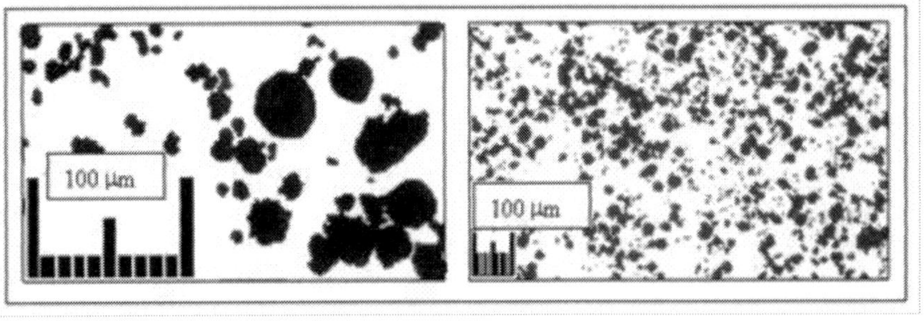

Figure 3. Microphotography of Metco 73 powder: a) x 256; b) x 64.

In order to determine the structure of the wolfram carbide covering layers, the photographies takes used 50x, 100x, 250x and 400x magnifying order, by means of a NEOPHOT-2 microscope; these photos are shown in Figure 4.

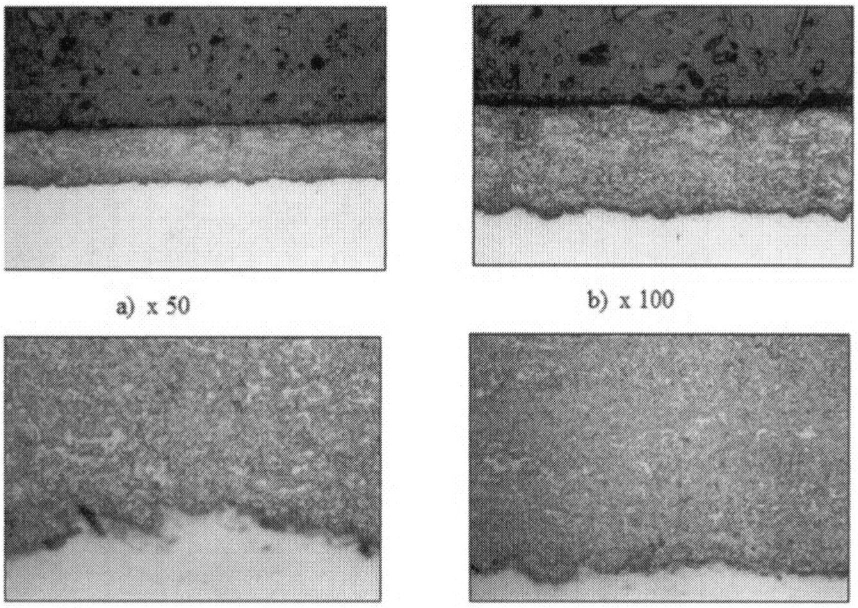

a) x 50 b) x 100

c) x 250 d) x 400

Figure 4. Microphotographies of the Metco 73 powder covering layer.

It is noticeable in all the images that the substrate (the martensitic stainless steel) has a very light color, while the polymer used in attaching the probe to the microscope, above the layer, is dark colored.

The covering layers have a blade structure.

The dark spots represent the pores in the substrate and one can notice that they have an irregular shape. The particles that are not melted can be easily noticed at 250x or 400x magnifying as they have a lighter color and preserve the rounded shape of the initial grains.

In the carbide strata the percentage of unmelted particles is low.

The determination of the crystalline structure of the covering layers was done by means of RX spectra obtained of a DRON-3 refractometer. Radiation Cu K_{α}, was used at a voltage of 30 kV and a current of 30 mA. The detector movement rate was 1°/min.

By comparing the RX spectra for the Metco 73 powder and for the covering layer obtained with this powder (Figure 5), one can notice that in the covering layer there are no more drops that characterize cobalt, therefore the covering layer contains nothing but wolfram carbide.

Vikersmicrohardness (HV 0.3) was determined by applying a 2.94 N force to the samples for a given period of time, using a square-based, straight pyramid diamond penetrator having the angle between the opposed sides of 136°. The arithmetic mean of the diagonals of the print was 0.025 mm and the values of the microhardness ranged between 800 and 950 N/mm2.

The adherence of the wolfram carbide layers was determined as described in paragraph 2.2. The resulting values of traction resistance ranged between 45-50 MPa. These values point to the fact that such coatings have a high adherence.

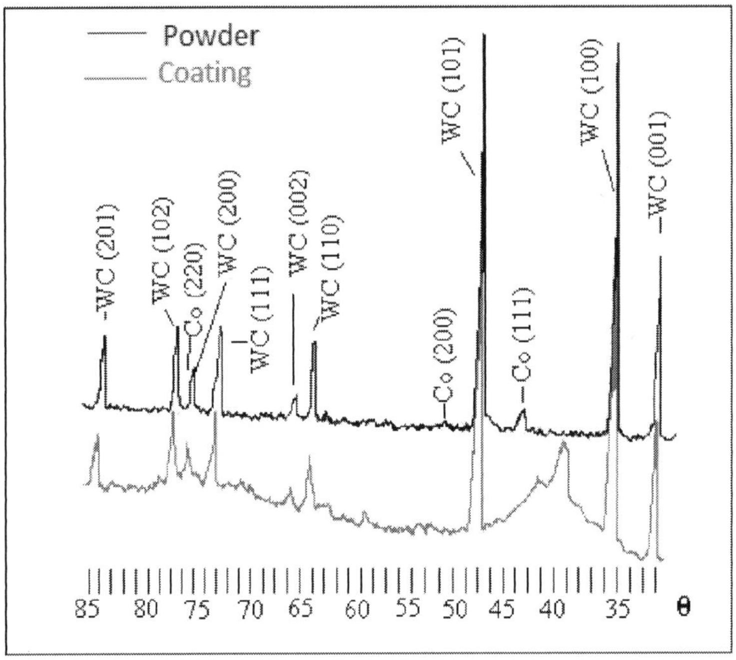

Figure 5. Comparison of the RX spectra obtained with the powder and covering layer achieved withMetco 73 (WCCo17).

The results of the experiments meant to determine the corrosion resistance by potentiostatic methods (in module) are shown in Table 2.

Table 2. The variation of the difference of the electrochemical potential [mV] between the coat and substrate, with the thickness of the coat and the concentration of the sodium chloride.

			0,05 mm			0,10 mm			0,25 mm		
c (%wt)	ρ ($g\cdot cm^{-3}$)	CM ($mol\cdot L^{-1}$)	ε_s^a (mV)	ε_d^b (mV)	Δε (mV)	ε_s^a (mV)	ε_d^b (mV)	Δε (mV)	ε_s^a (mV)	ε_d^b (mV)	Δε (mV)
1,00	1,0072	0,1722	253	281	28	73	326	253	3	361	358
2,00	1,0145	0,3468	278	330	52	106	370	264	105	406	301
3,00	1,0217	0,5239	335	346	12	146	393	247	83	416	333
4,00	1,0290	0,7036	348	360	12	228	415	187	88	416	328
5,00	1,0366	0,8860	360	376	16	246	396	150	115	420	305
6,00	1,0440	1,0707	376	386	10	221	406	185	128	423	295
7,00	1,0513	1,2580	385	391	6	246	426	180	141	431	290
8,00	1,0585	1,4475	380	398	18	231	415	184	188	433	245
9,00	1,0659	1,6398	393	401	8	225	416	191	241	433	192
10,00	1,0733	1,8347	391	393	2	230	413	183	216	425	209
11,00	1,0809	2,0325	396	403	7	230	420	190	221	415	194
12,00	1,0886	2,2330	401	410	9	233	423	190	243	428	185
13,00	1,0962	2,4360	405	411	6	236	420	184	198	426	228
14,00	1,1039	2,6418	406	411	5	240	425	185	215	437	222
15,00	1,1115	2,8500	406	411	5	251	428	177	235	440	205

[a] the electrochemical potential [mV] of the substrate
[b] the electrochemical potential [mV] of the coat

In order to obtain some mathematical correlations between the corrosion resistance, estimated by the electrochemical potential difference, and the concentration of the NaCl solution (for each thickness), the experimental data have been processed usingMatlab.The graphical representations are shown in Figures 6, 7 and 8. All measurements were made at a constant temperature of 20°C (293K).

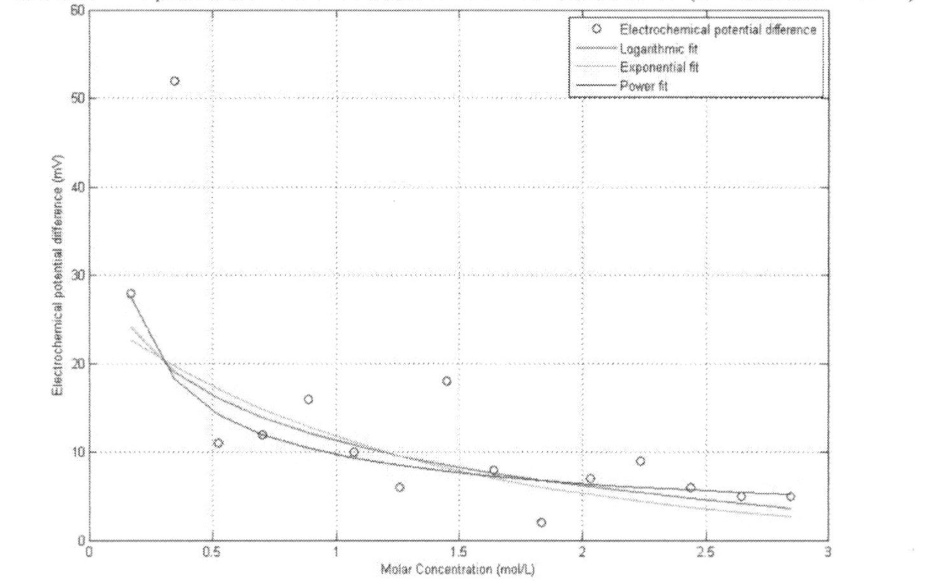

Figure 6. Electrochemical potential difference in function of the molar concentration(coating thickness 0,05 mm).

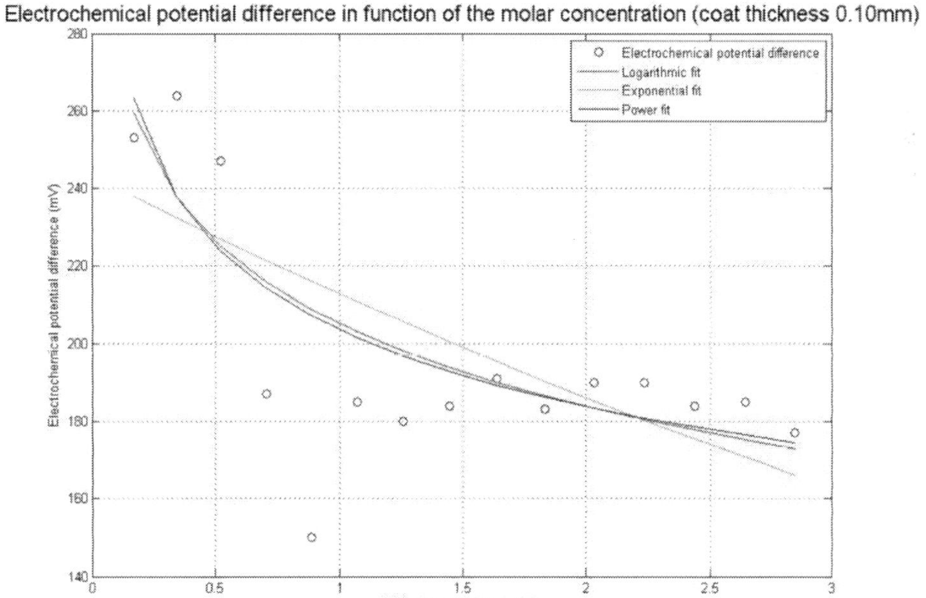

Figure 7. Electrochemical potential difference in function of the molar concentration (coating thickness 0,1 mm).

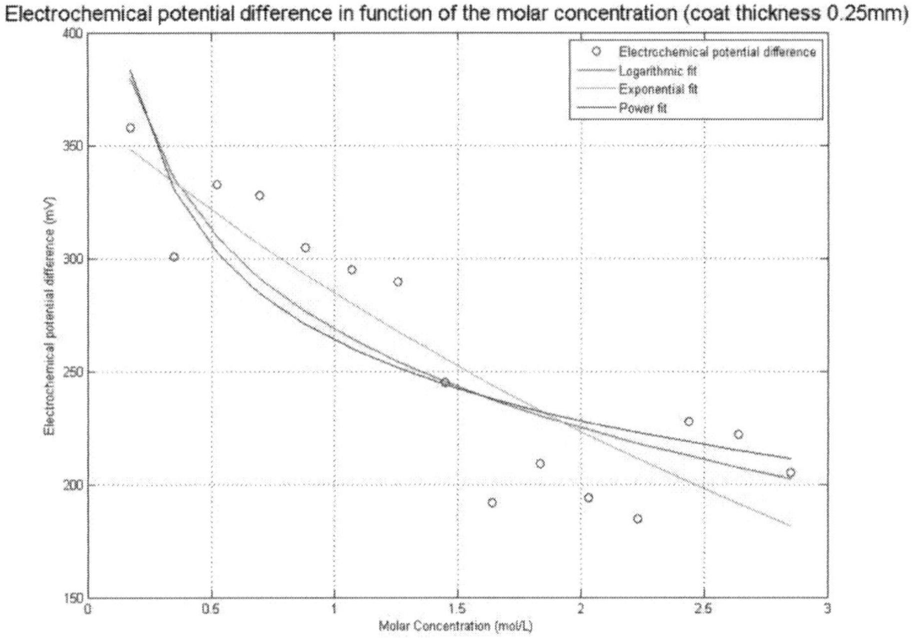

Figure 8. Electrochemical potential difference in function of the molar concentration (coating thickness 0,25 mm).

The experimental points were used in order to find a law that could relate the electrochemical potential to the concentration of the NaCl solution. There were three types of functions proposed for this purpose: exponential, logarithmic and power. The last one seems to fit the best the experimental points, because the coefficient of determination (R^2) is most of the time higher than or very close to the coefficients obtained for the other functions (as seen in Table 3). This shows that the phenomenon is related to the adsorption/desorption of the ions at the surface, which is also coherent with the decrease of the electrochemical potential. This also explains the important noise obtained for the measurements which is related to the lack of a mechanism that would impose a continuous flow of electrolyte in the experimental setup.

It may be also observed that the thicker coatings introduce a greater electrochemical potential difference which indicates a better protection of the substrate. Moreover, the surface saturation is reached for higher concentrations of NaCl.

Table 3. The results of the mathematical processing of experimental data.

Type of Law		Logarithmic	Exponential	Power
Coat thickness (mm)	Form / Coefficient	a·ln(x)+b	a·exp(b·x)	a·xb
0.05	a	-7.328	25.99	9.712
	b	11.27	-0.795	-0.5948
	R^2	0.7839	0.7066	0.8447
0.10	a	-30.75	243.5	203.5
	b	205.2	-0.1344	-0.1466
	R^2	0.7362	0.5744	0.7478
0.25	a	-63.08	363.2	264.1
	b	268.8	-0.2428	-0.2118
	R^2	0.7598	0.7986	0.7204

A comparison of the experimental results for the three thicknesses of the covering layer is depicted in Figure 9.

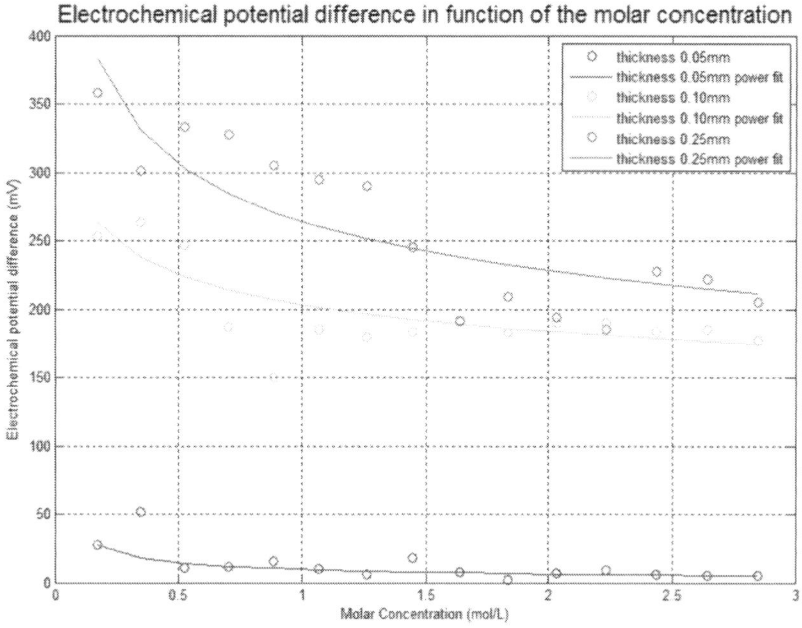

Figure 9. Electrochemical potential difference in function of the molar concentration for different thicknesses of coating.

4. Conclusions

The paper presents the principal characteristics and an original method to describe the behavior in a corrosive environment of the WC coatings deposited by plasma spraying on a stainless steel substrate; the method is based on measuring the electrochemical potential of the coating, respectively that of the substrate, immersed in a NaCl solution (corrosive agent), related to a reference calomel electrode.

The characteristics of the coatings are determined by: its chemical composition, optical microscopy, RX analysis, tensile strength, Vickers hardness, the nature and the processing degree of the substrate and the deposition conditions.

The obtained experimental results show that the coating/substrate electrochemical potential difference is a sensible parameter, which reflects the protection provided by the coating and may be used in quantifying coatings protection activity, in constant conditions concerning the nature of the couple coating/substrate and the spraying parameters.

The bigger the thickness of the coat, the greater the difference between the electrochemical potential of the coat and that of the substrate and consequently the coat protects the substrate better.

The values of the potential differences between the coating and the substrate are decreasing when raising the concentration of the NaCl solution, which shows a decrease of the protection provided by the coating.

The mathematical processing of the experimental results shows that for all the concentrations of the used NaCl solutions, the dependence between the electrochemical potential difference is very well described by some curves having equations.

References

[1] Kasai S, Yanagisawa A, Ichihara A, Shimoyama Y, Ochiai K and Onishi H 1988 *An investigation of oxide cermet coatings for hearth roll aplicatin in a continous amelling fournace,* 1st Plasma Technik Symposium,Vol. 1, Lucerne, Switzerland, May 18-20, pp 205-215

[2] Smith R W and Mutasim Z Z 1992 Reactive plasma spraying of wear – resistant coatings, *Journal of Thermal Spray Technology* **1**(1) 57-63

[3] Kvernes I, Lugscheider E and Norholm O 1988 *Potential of thermal and wear resistant coatings in combustion engines,* 1st Plasma Technik Symposium, Vol. 3, Lucerne, Switzerland, May 18-20, pp 41-53

[4] Scrivani A, Groppetti R, Bardi U, Lavacchi A, Niccolai F and Rizzi G 2001 A Comparative Study On HVOF, Vacuum Plasma Spray and Air Plasma Spray for CoNiCrAlY alloy deposition, document placed on line on June 2001, Available online http://www.unifi.it/unifi/surfchem/solid/bardi/tbcs/mcraly/CoNiCrAlYart12jun01.html

[5] ***Commission On Engineering And Technical Systems, National Research Council 1996 *Coatings for High-Temperature Structural Materials: trends and opportunities*, National Academy Press, Washington D C, USA

[6] Suryanarayanan R 1993 *Plasma Spraying-Theory and Applications*, World Scientific, Singapore

[7] Toma D, Brandl W and Mărginean G 2001 Wear and Corrosion Behavior of Thermally Sprayed Cermet Coatings, *Surface and Coatings Technology* **138**(2-3) 149-158

[8] Benea M L and Maksay S 2002 The corrosive resistance determination through potentiostatical methods at the tungsten carbide coatings on metallic support achieved through the plasma jet pulverisation, 6th International Symposium Interdisciplinary Regional Research Hungary-Romania-Yugoslavia, Novi Sad, Serbia, October 3-4, pp 221-226

[9] Benea M L and Maksay S 2002 Interpolarea matematica-metoda de interpretare a rezultatelor determinarilor de rezistenta chimica, *Analele Facultății de Inginerie din Hunedoara***IV**(1) 129-134

[10] Chapman B A, DeFord H D, Wirtz G P, Brown S D 1987 *Technology of Glass, Ceramic, or Glass-Ceramic to Metal Sealing*, Moddeman W E, Merten C W and Kramer D P (Eds), American Society of Mechanical Engineers, New York, USA

[11] Benea M L, Maksay S, Lazău R I and Becherescu D 2003Using the potentiostatic method in order to purchase the corrosion behavior of the WC coatings deposited by plasma spraying. I: The influence of the NaCl solution concentration, respectively of the temperature, upon the coating/substrate electrochemical potential difference, *Material engineering* **14**(2) 127-140

[12] Benea M L, Maksay S, Lazău R I, Becherescu D 2003 Using the potentiostatic method in order to purchase the corrosion behavior of the WC coatings deposited by plasma spraying. II: The combined influence of NaCl solution concentration and temperature upon the coating/ substrate electrochemical potential difference, *Materials engineering* **14**(2) 141-155

IOP Conf. Series: Materials Science and Engineering **85** (2015) 012005 doi:10.1088/1757-899X/85/1/012005

Laser beam treatment effect on AMPCO M4 bronze cavitation erosion resistance

I Bordeasu[1], M O Popoviciu[2], L M Micu[1], O V Oanca[1], D Bordeasu[3], A Pugna[1] and C Bordeasu[1]

[1] Politehnica University of Timisoara, 2 Piata Victoriei, 300006 Timisoara, Romania
[2] Academy of Romanian Scientists, 1 Bvd. Mihai Viteazul, Timisoara, Romania
[3] Via Universisty College, Horsens, Denmark

E-mail: ilarica59@gmail.com

Abstract. Ship propellers must resist simultaneously to ocean water corrosion and cavitation erosion. Until now, the best material used is the bronze with great copper content. These materials are expensive and there is the tendency to reduce the copper content while maintaining good properties. Such a material is AMPCO M4 used for manufacturing details for aircraft retractable landing assemblies. As a consequence we undertake cavitation erosion tests upon this bronze. In natural state (cast or even extruded) the cavitation resistance is not acceptable so, we improved the specimens by treating them with laser beams at three different impulse powers (160, 180 and 220 W). The cavitation erosion resistance was tested in the Cavitation Laboratory of Timisoara "Politehnica" University using a vibratory device respecting the conditions imposed by ASTM G32-2010 Standard. The comparisons with the genuine material (without any treatments) show that the applied procedure increased the hardness of the melted layer as well as the cavitation erosion behavior.

1. Introduction

Finding cheep materials with excellent cavitation erosion is till now an unsolved problem neither for hydraulic machineries nor for ship propellers. Researches upon new materials and new manufacturing procedures are undertaken in present, to find good and economic solution.

Statistic data show that for ship propellers the best materials are bronzes with high copper content [1], [2]. Even for such materials there occurs cavitation erosion on the propeller blades, especially for high velocity ships. In the present paper it is analyzed the increase of cavitation resistance by melting the superficial blade layer using laser beams with controlled parameters.

For laboratory test there have been chosen samples from extruded AMPCO M4 alloys. This material with good mechanical and corrosion properties is recommended for heavy-duty, high-loaded mechanical and corrosive applications involving abrasive wear, friction and chemical corrosion [3].

The cavitation researches were undertaken in the Cavitation Laboratory of the Timisoara "Politehnica" University using a vibratory device respecting the recommendations of the ASTM G32-2010 Standard [4].

The results are encouraging because the cavitation erosion resistance was improved.

2. Researched Material and Applied Treatments

The researched material is a bronze with approximate 75% Cu, 5% Ni, and 10% Al was initially developed to manufacture pieces for aircraft retractable landing assemblies [3]. Taking into account the cost and the corrosion properties of this material we proposed it for repair works or even manufacturing ship propellers.

The detailed chemical composition of this bronze is: 10.5 % Al, 5% Ni, 4.8 % Fe, 1.5 % Mn, others max. 0.5, the rest is copper. The mechanical properties with significance for the cavitation erosion resistance are: Tensile strength R_m = 1000 MPa, Yield strength $R_{p0.5}$ = 793 MPa, Brinell hardness = 286 HB30, Fatigue ($100x10^6$ cycles) =352 MPa, density ρ = 7.45 g/cm^3 [3].

The specimen surface exposed to cavitation was treated with laser beams with the following parameters:

- pulse duration: 5 ms;
- repetition frequency of pulses: 10 Hz;
- working velocity: 4.07 mm/s;
- impulse power: 160 W, 180 W, 220 W.

Figure 1 presents the aspect of the surfaces after laser treatments. It can be observed that regardless of power value the surfaces have identical aspects, with typical undulations, uniformly distributed.

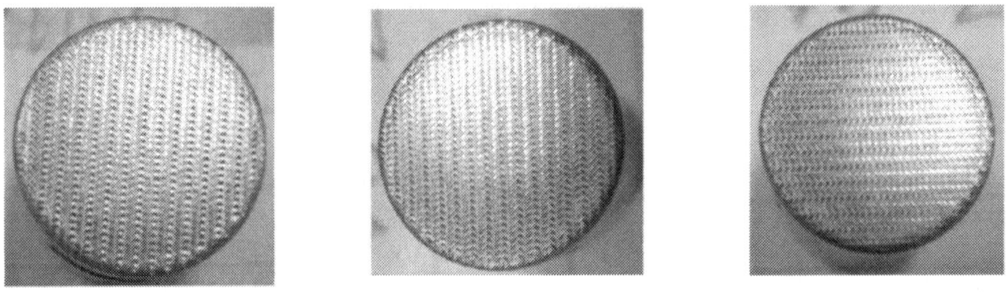

Impulse power **160 W** Impulse power **220 W** Impulse power **220 W**

Figure 1. Surface aspects (before cavitation attack – minute 0)

The melting with laser beams determined a great increase of the micro hardness HV0.5 of the surface (see tables 1, 2, 3).

Table 1. Micro hardness HV0.5 - Impulse power **160** W

Measured zone	Depth of the measured point μm	HV0.5	Mean value of HV0.5
Treated with laser beam	53	330	321
	53	330	
	54	318	
	54	318	
	54	318	
	54	318	
	54,5	312	
Basic material	57,5	280	269
	58	276	
	58	276	
	58,5	271	
	59	266	
	59	266	
	61	249	

Table 2. Micro hardness HV0.5 - Impulse power **180 W**

Measured zone	Depth of the measured oint µm	HV0.5	Mean value of HV0.5
Treated with laser beam	48,5	394	356
	50	371	
	50	371	
	51,5	350	
	52	343	
	52,5	336	
	53,5	324	
Basic material	56	285	275
	57	281	
	58	276	
	58	276	
	58,5	274	
	59	268	
	60	265	

Table 3. Micro hardness HV0.5 - Impulse power **220 W**

Measured zone	Depth of the measured point µm	HV0.5	Mean value of HV0.5
Treated with laser beam	47	420	376
	47,5	411	
	48	402	
	48	402	
	51	357	
	53	330	
	55,5	301	
Basic material	56,5	290	282
	56,5	290	
	57	285	
	57	285	
	58	276	
	58	276	
	58,5	271	

The procedure for measuring micro hardness was: after 165 minutes of cavitation exposure when all the investigations upon the eroded surface were finished, the specimens were cut in two equal parts along a longitudinal axe. At the beginning, the hardness was measured in the basic layer. For all depth chosen in the basic part, it was taken a beginning line outside of both the melted zone and the thermal influenced one. The obtained data (tables 1, 2, 3) show important micro hardness increases of the layers treated with laser beams, in comparison with untreated layers (16 % for 160 W, 29.5 % for 180 W and 33.3 % for 220 W of power impulse).

The micro hardness remains approximate unchanged for the same depth (either of the untreated material or for the treated one). As the depth of the measured point increases, the micro hardness decreases (untreated: from 290 HV0.5/56,5 µm to 271 HV0.5/58,5 µm, treated: at 220 W, from 420 HV0.5/47 µm to 301 HV0.5/55.5 µm).

3. Experimental results

Respecting ASTM G32 Standard, three specimens were tested for each material. The regression curve was obtained for the arithmetic mean. To put into evidence both the material homogeneity and the constancy of the produced cavitation there were computed the tolerances intervals for 99% probability see Figures 2-4, and was determined the Standard Error of Estimation $S_{y/x}$ see table 4.

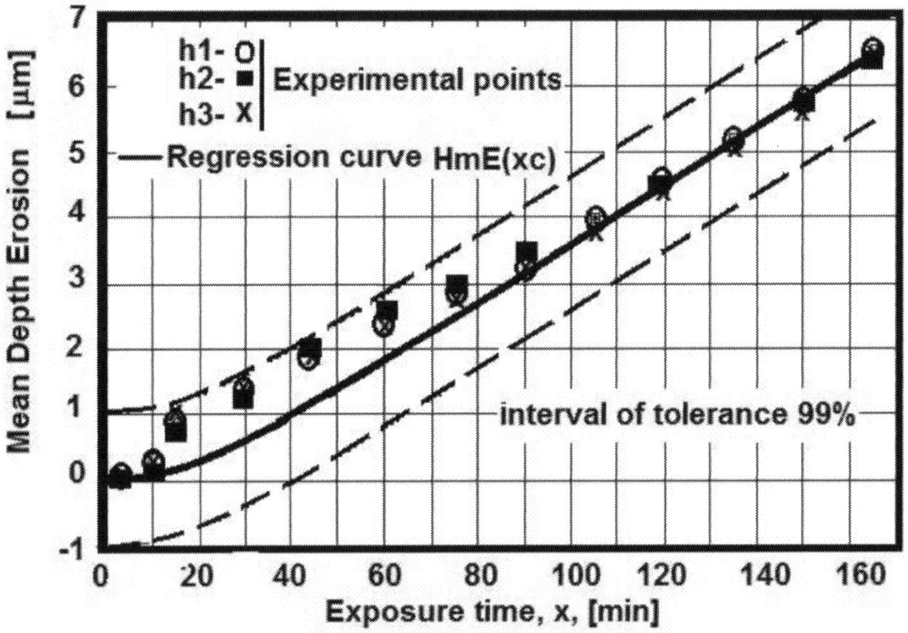

Figure 2. Measured points, regression curves and tolerance interval – Laser beam impulse power **160 W**

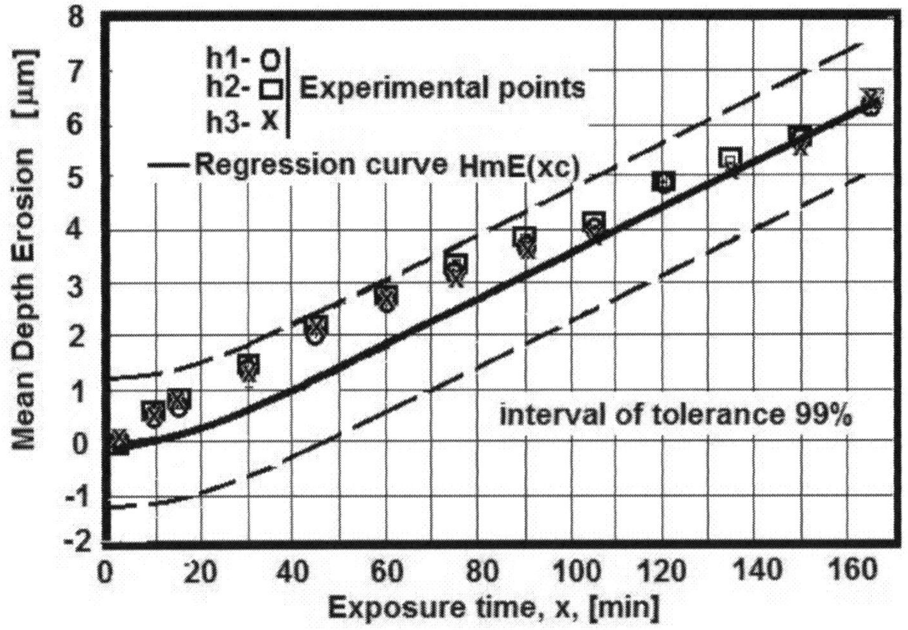

Figure 3. Measured points, regression curves and tolerance interval - Laser beam impulse power **180 W**

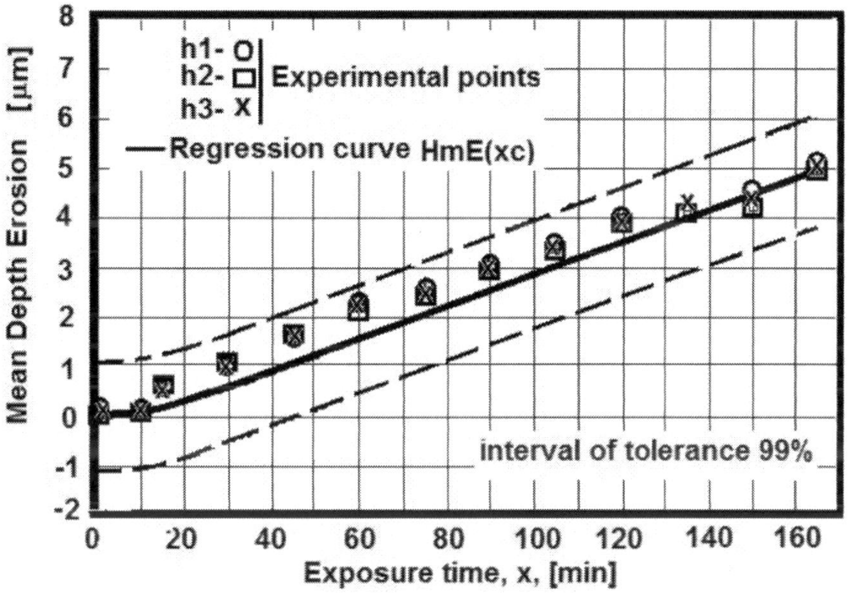

Figure 4. Measured points, regression curves and tolerance interval – Laser beam impulse power **220 W**

Even for this great probability we obtained a narrow interval of tolerance attesting both good homogeneity of the tested material after the laser beam treatment and reliable results of the test [5], [6], [7], [8], [9].

Table 4. Values of important parameter values <μm>

Laser beam power	**160 W**	**180 W**	**220 W**
Measured final mean depth erosion (165 minutes)	6,511	6,346	4,984
Final mean depth erosion from the tolerance interval (maximum value)	7,843	7,477	5,875
Final mean depth erosion from the tolerance interval (minimum value)	5,179	5,215	4,093
Standard error of estimation ($s_{y/x}$) for 99% probability	0,444	0,377	0,297

With the increase of the impulse power, the experimental measured point's scatter decreases, see Figures 2-4. For 160 W impulse powers, in the first period appear the greatest deviation from the regression curve (till approximate 0.444 μm). With the increase of the power impulse the scatter became smaller (remaining under 0.297 μm for 220 W). Because the parameters of cavitation remain constant, the causes of differences must be explained by the material behavior. The surface subjected to cavitation has in all situations an undulated shape. When, on the crests of these undulations there

IOP Conf. Series: Materials Science and Engineering **85** (2015) 012005 doi:10.1088/1757-899X/85/1/012005

are weak points they are easy removed and the mass loss is great. It result the conclusion that great power impulse reduce those weak points, probably by melting and hardening the exposed area. When these initially weak points are removed, the material shows its real behavior towards the resistance to cavitation erosion and the scatter around the regression curves is reduced.

Figures 5-7 present the obtained experimental data for the cumulative mean depth of erosion and the evolution of men depth erosion rate. In the figures "a" are presented photographs showing the aspect of the attacked surface at three different exposure times. It is to be remarked that, from the beginning, almost the entire surface is subjected to erosions (corresponding to the requirements of ASTM G32-2010).

The experimental data are approximated with a regression curves taking into account only the arithmetical mean [10], [12], [13].

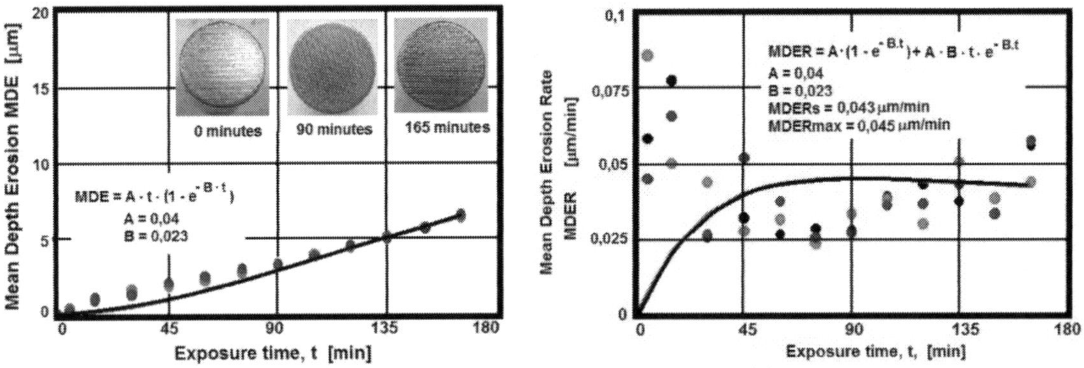

a) b)

Figure 5. Cavitation erosion characteristic curves (**160 W**):
a) Mean depth erosion against time
b) Mean depth erosion rate against time

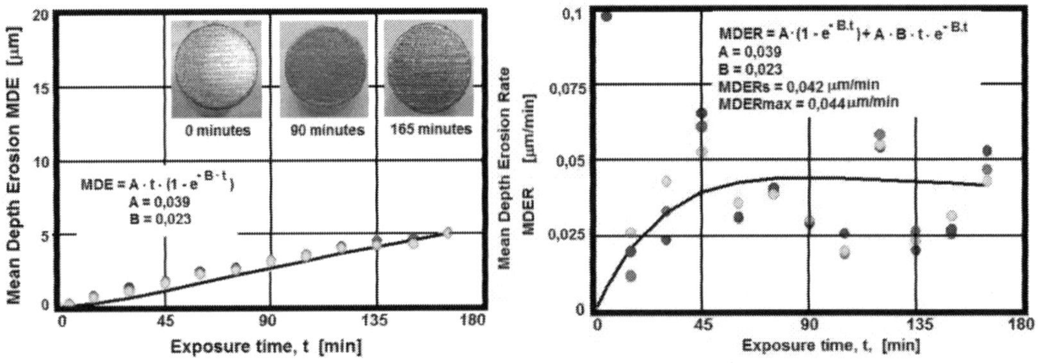

a) b)

Figure 6. Cavitation erosion characteristic curves (**180 W**):
a) Mean depth erosion against time
b) Mean depth erosion rate against time

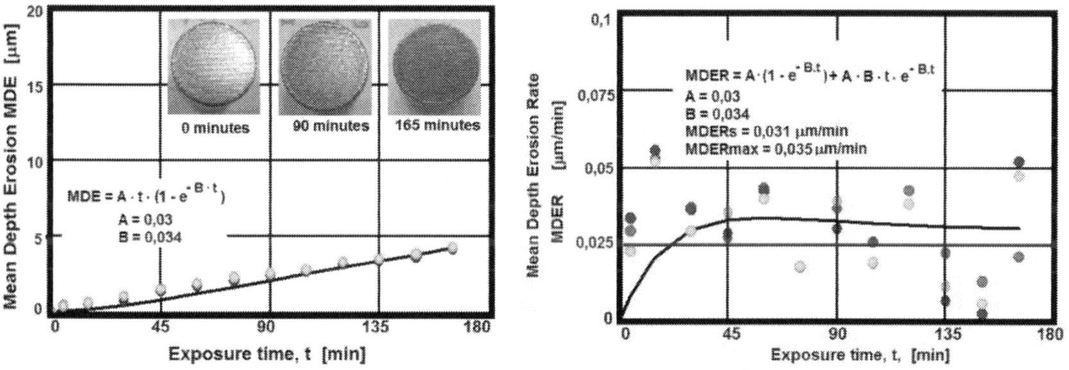

a) b)

Figure 7. Cavitation erosion characteristic curves (220 W):
a) Mean depth erosion against time
b) Mean depth erosion rate against time

For the specimens treated with 160 W laser beam, the approximation of the measured data on the curve MDE(t) is very good in the intervals of 90-165 minutes; for the MDER(t) the same approximation is very good in the interval 120-165 minutes. The stabilization value for MDER(t) is 0.043 μm/min.

For 180 W impulse power, the MDE(t) curve approximate very good the experimental data in the interval of 75-165 minutes and the MDER(t) curve in the interval 15-165 minutes; the stable velocity value is 0.042 μm/min.

For an impulse power of 220 W the whole regression curve MDE(t) approximate very good the experimental data and the stable value of the velocity is 0.031 μ/min.

In the range of 60-75 minutes, all specimens treated with laser beams show a maximum value in the velocity curves. This maximum differ with the impulse power (160 W /0.045 μm/min; 180 W/0.044 μm/min; 220 W/ 0.035 μm/min).

The small differences between the maximum values of the parameter MDER(t) and the stabilization value show that the tested materials have good cavitation erosion behavior [5], [10], [11] and testify about the increase of cavitation erosion resistance through laser beam treatments.

In Figure 8 there are given comparisons of the laser treated bronzes both with untreated bronzes and with a material (CuNiAl III RNR) which in the past was characterized with very good or excellent resistance both in laboratory and in field running. The final conclusion is that the laser beam treatment is beneficial for the cavitation erosion resistance [5], [11]. The best results were obtained for laser beam treatments with 220 W impulse powers.

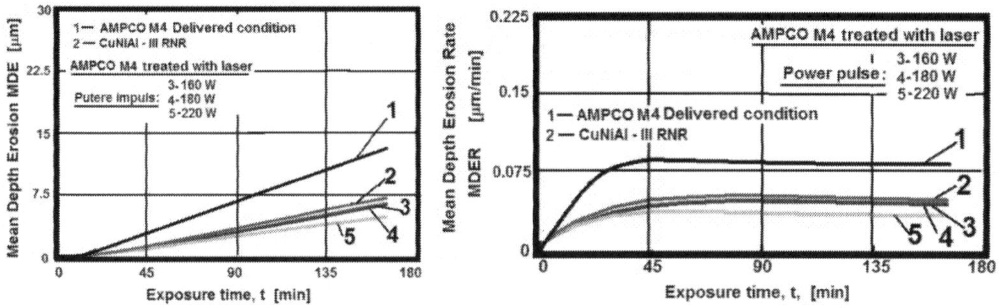

a) b)

Figure 8. Comparisons of Cavitation erosion characteristic curves
a) Mean depth erosion against time
b) Mean depth erosion rate against time

IOP Conf. Series: Materials Science and Engineering **85** (2015) 012005 doi:10.1088/1757-899X/85/1/012005

The increase of cavitation erosion resistance for an impulse power of 220 W is 38.7% in comparison with those for 160 W impulse powers and is 35.48 % greater than for 180 W. This amelioration can be explained by the hardness increase of the exposed surface (220 W gives an increase with 17%, in comparison with 160 W and 5.6 % in comparison with 180 W.

Comparing the tested bronzes, in natural states (curve 1) and after laser beam treatment (curves 3, 4, 5) with the high resistance bronze CuNiAl – III RNR (curve 2) it can be seen that the applied treatment increase the cavitation erosion behavior, regardless of the impulse power beam. Using small power values (160 W and 180 W) we have obtained only small increases (1.07 and 1.1 times). For a power impulse of 220 W the obtained resistance was 1.48 times greater than that of the non treated sample. The final conclusion of figure 8 is the possibility to increase the cavitation erosion resistance by laser beam treatments. Upon our opinion, the existent technology for laser beam treatment allows improving only small surfaces of the blades; it means that the procedure is easy to be applied for the repair works of previously eroded areas.

4. Eroded microstructure investigation

The micrographic images presented in Figures 9 and 10 confirm the favorable results obtained by the cavitation tests and show the formation of a fine martensitic structure which gives a significant increase of the cavitation erosion resistance. On the interface "treated layer/basic layer" do not appear porosities, micro cracks or other defections and the thermal influenced zone has a small extension.

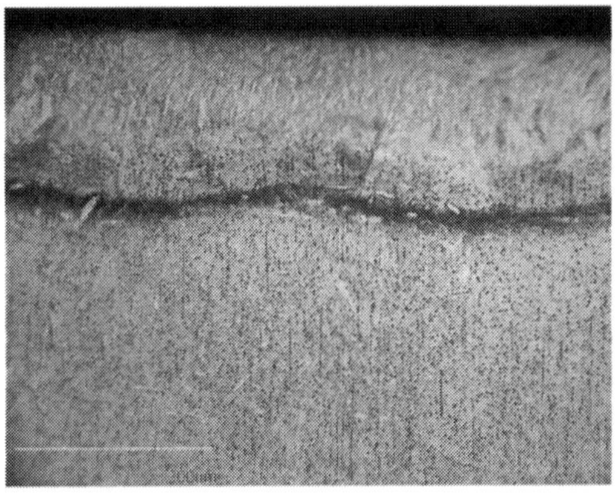

a) 160 W

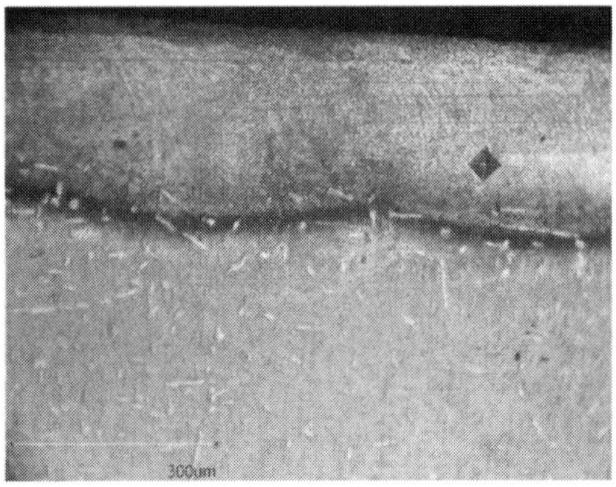

b) 180 W

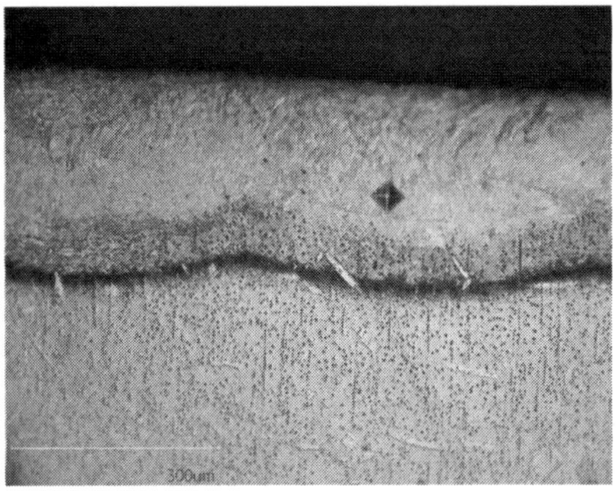

c) 220 W

Figure 9. Micro photographs of melted layer and ground layer, digital microscopy HIROX 1300

IOP Conf. Series: Materials Science and Engineering **85** (2015) 012005 doi:10.1088/1757-899X/85/1/012005

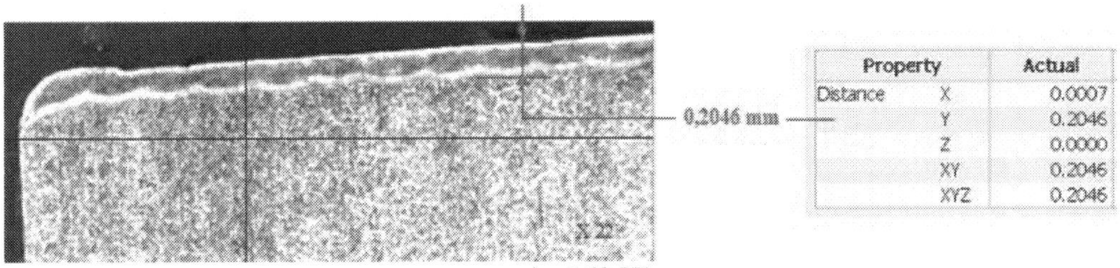

a) **160 W**

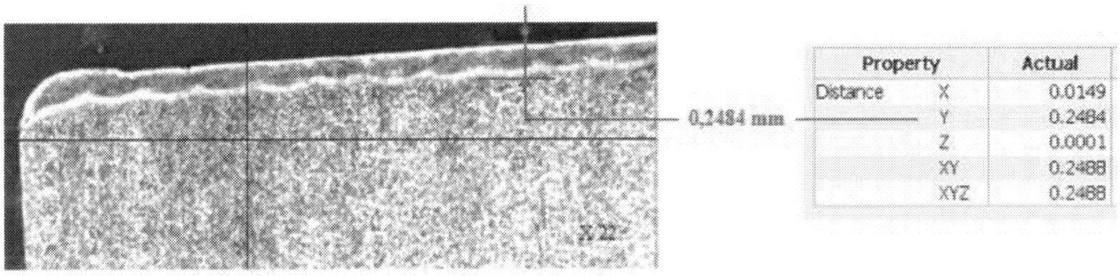

b) **180 W**

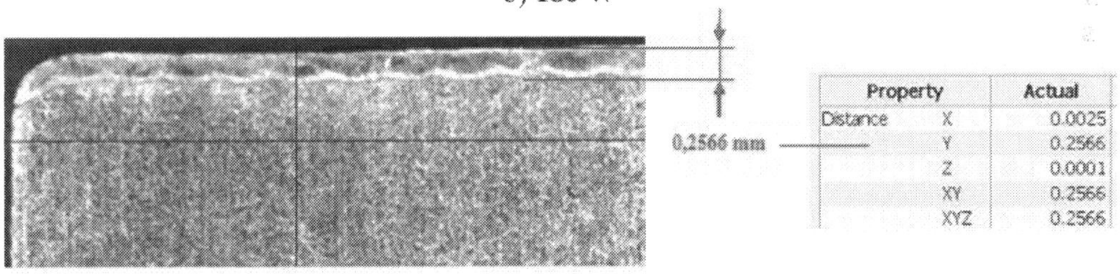

c) **220 W**

Figure 10. Scanned structure after final exposure time (165 minutes) Laser MICRO VU- Excel 664-UM

5. Conclusions

1. The reduced extension of the tolerance interval for a probability of 0.99, for all three specimens tested and for all treatments, confirms the great homogeneity of the tested material and the uniformity of the produced cavitation phenomena.
2. The surface subjected to cavitation has in all situations an undulated shape. When on the crests of these undulations there exist weak points those are easy removed and the mass loss is great. It result the conclusion that great power impulse reduce such weak points, probably by better melting and hardening the exposed area. When these initially weak points are removed the material shows its real behavior towards the resistance to cavitation erosion and the scatter around the regression curves is reduced.
3. The laser beam treatment increases the hardness and by this way the cavitation resistance is highly improved. The micro hardness values increases with the impulse power of the beam when the other parameters (pulse duration, repetition frequency, and manufacturing speed) remain constants.
4. The shape and the position of the regression curves for laser beam treated bronzes recommend the procedure for increasing the cavitation erosion resistance.
5. The researches must be continued in order to obtain the best value for the power impulse of the laser beam.
6. Using the procedure of laser beam treatment of the bronzes M4 determines a great increase to cavitation erosion resistance in comparison with natural state of the material. The resistance

became greater than that of the excellent CuNiAl III RNR bronze; in consequence the method is recommended for repair works of ship propellers.

7. The metallographic investigations confirm the formation of a fine martensitic structure determining a substantial increase of the cavitation erosion resistance.
8. The interface between treated layer/basic layers is free of porosities or micro cracks and the depth of thermal influenced zone is relatively small.

References

[1] Anton I 1985 *Cavitation,* Romanian Academy Publishing House, Bucharest, Romania

[2] Manzana M E 2012 *Studies and Experimental Researches Regarding Structural Modifications Produced by Cavitation Erosions upon Metallic Materials*, Politehnica University of Bucharest, Romania, Doctoral Thesis

[3] *** http://www.ampcometal.com/common/datasheets/en/AM4_EX_E.pdf

[4] *** 2010 *Standard method of vibratory cavitation erosion test*, ASTM, Standard G32-10

[5] Oanca O V, Bordeasu I, Mitelea I and Craciunescu C 2013 *Phenomenology of Degradation by Cavitation for Heat Treated CuNiAlFe Bronzes*, 22th International Conference on Metallurgy and Materials, Brno, Czech Republic, May 15-17, pp 1561-1566

[6] Bordeasu I and Popoviciu M O 2012 Improving cavitation erosion resistance through surface and structural hardening, *Machine Designe* **4** 171-176

[7] Frank J P and Michel J M 2004 *Fundamentals of cavitation*, Kluwer Academic Publisher, Boston, USA

[8] Hammitt F G 1980 *Cavitation and Multiphase Flow Phenomena*, McGraw Hill International Book Company, Atlanta, USA

[9] Steller J K 1992 *International cavitation test – sumary of results*, Proceedings of 3-rd International Conference on Cavitation, Cambridge, UK, December, pp 121-132

[10] Bordeasu I, Popoviciu M O, Balasoiu V and Patrascoiu C 2004 An Analytical Model for the Cavitation Erosion Characteristic Curves, *Scientific Bulletin "Politehnica" University of Timişoara, Transaction of Mechanics* **49**(63) 253-258

[11] Bordeasu I 2006 *Cavitation Erosion of Materials*, Timisoara Polytechnic Publishing House, Timisoara, Romania

[12] Bordeasu I, Popoviciu M O, Mitelea I, Ghiban B, Ghiban N, Sava M, Duma S T and Badarau R 2014 Correlations between mechanical properties and cavitation erosion resistance for stainless steels with 12% Chromium and variable contents of Nickel, *IOP Conference Series: Materials Science and Engineering* **57** 012006

[13] Bordeasu I, Popoviciu M O, Mitelea I, Ghiban B and Ghiban N 2013 Cavitation erosion resistance of two steels with the same percentage of Chromium and Nickel but different Carbon content, *IOP Conference Series: Materials Science and Engineering* **57** 012007

IOP Conf. Series: Materials Science and Engineering **85** (2015) 012006 doi:10.1088/1757-899X/85/1/012006

Experimental research on the characteristics of softening and melting of iron ores as significant factor of influence on gas permeability of blast furnace charge

E Branescu[1], A O Blajan[1] and N Constantin[1]

[1] Politehnica University of Bucharest, The Faculty of Material Science and Engineering, Splaiul Independentei no. 313, 060042, sector 6, Bucharest, Romania

E-mail: elena_branescu@yahoo.com

Abstract. It is widely accepted as a cohesive zone is directly influenced by softening and melting properties of iron ores, preparations (crowded, pellets, which represents about 90%, of the loads with metal furnace intake), or uncooked (raw ores ranked). Important results can be obtained through the study of behavior of ferrous materials at temperatures above 1000 ° C. Starting from research methods presented in the literature, this paper presents itself in carrying out their own laboratory experiments, conducted with the aim of analysing the softening and melting properties of sinter iron cores.

1. Introduction

Cohesive zone of blast furnace interior is by definition located in thebely of furnace is the area composed of pastos iron oresof softening and melting solid, coke, acid slag drops that run upward in the gaseous phase reducing valves with countercurrent.

Gas permeability of this area is very low, only one running windows gas coke solidly placed between layers of ore in the process of softening and melting. The temperature at which softening begins ferrous raw materials is called the start temperature of the cohesive zone and write with Tzc, the temperature at which melting the ends of ferrous raw materials is called melting temperature and note with Ttop. DT = difference between them Ttop-Tzc actually determine the thickness of cohesive zone.

The cohesive zone is thicker the upward flow of gaseous phase is more difficult. Softening, melting temperatures as well as the spread between them is in close correlation with the physical and chemical properties of ferrous materials with consideration of load capacity, i.e. to sinter ferrous which represents 90% of the loads of minerals. Study on the influence of chemical composition of sinter of Ib=Cao/SiO_2 and GR (reductibility degree) its effects on the cohesive zone is of great importance because of the shape and its position depends on the development of processes in the furnace.

2. Methodology

Crowded samples which have been subject to reduction under load have been specially trained in industrial agglomeration of machine at the ICEM Bucharest.

Crowded samples were taken from the mixtures listed in Table 1.

Tabel 1. The raw materials used in making crowded samples.

No sample	The raw materials %		Additions %		Initial mixture %				Sintered sample volume m^3
	Ore homogenized	Teliuc concentrate	Chalk	Coke	Ore	Teliuc	Chalk	Coke	
1	100	-	2999	8.2	60	-	18	4.9	0008
2	95	5	2908	8	57	3	17.4	4.8	00285
3	90	10	2815	8	54	6	16.9	4.8	00225
4	85	15	2724	8	51	9	16.4	4.8	00225
5	80	20	2630	7.5	32	8	10.5	3	0008
6	75	25	2540	7.5	30	10	10.2	3	0008
7	70	30	2445	7.5	28	12	9.8	3	0008
8	65	35	2350	7	26	14	9.4	2.8	0008
9	60	40	2255	7	24	16	9.0	2.8	0008
10	55	45	2165	7	22	18	8.6	2.8	0008
11	50	50	2070	6.9	20	20	8.2	2.6	0008

The tests we made shown that the higher is the reduction degree the higher is the crushing degree. Because pre-reduced materials obtained by reducing the iron ores (crude or prepared) are used in the blast furnace or the electric arc furnace, replacing the iron scrap, it is very important to know the behaviour of these materials during the heating – the correlation between the reducing degree and their melting and softening temperatures.

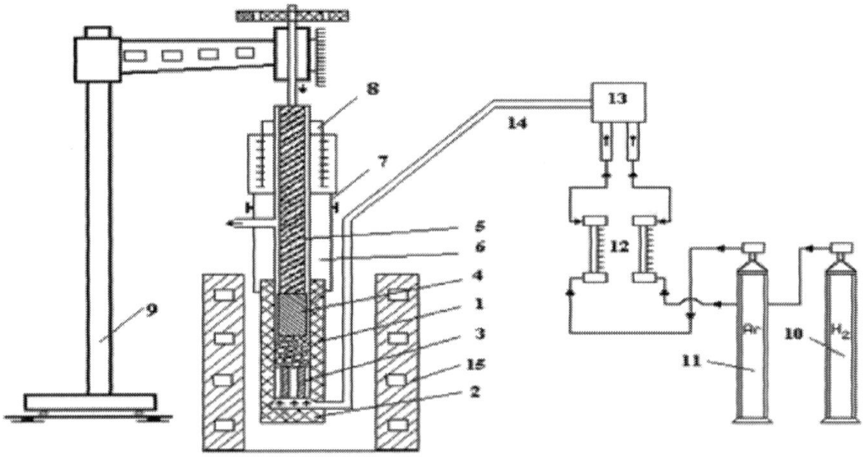

Figure 1. The experiment plant

1- graphite crucible; 2- gas distributor; 3- intermediary plate; 4 - piston; 5 - rod; 6 - graphite support; 7 - seal device; 8 - piston support; 9 - fixed support; 10 - H_2 cylinder; 11 - reducing gas cylinder; 12 - measuring device; 13 - gas mixer; 14 - gas pipe; 15 - induction electric furnace

That is why we tested the softening of these materials. The plant we used to run these tests is presented in Figure 1.

The way of the experiments was the following:

-crushed and punctured in the material has been chosen fraction of 3-5 mm of which for experimentation have used amounts ranging between 240-310g, selected quantity must fill the graphite crucible in which reduction reaction occurs;

-heating of the sample was made with the speed of 50-60 degrees/min;

-starting at 750° C is inserted into the sample It with the flow of 6.5-7.5 l/min;

-from 900° C is introduced with H_2 flow rate of 5 l/min up to 8 l/min so that the samples have a degree of reduction between 50-78%;

-the temperature is measured at 3 minute and interval from the time the piston movement at a distance of approximative 2 minutes;

-total of the piston stroke corresponding to a reduction of height equal to half of that of the sample and represents the beginning of melting characteristic (T_{top}) and travel of the piston is at the beginning of the characteristic started softening (T_{zc});

-after cooling and evacuation from the Crucible, the sample weighs in again and analyzes its layout.

3. Results

Experimental results are presented in Table 2.

Tabel 2- Experimental data on the characteristics of the softening and melting of sinters samples.

Nr of samples	Chemical composition							The degree of reduction (%)	Characteristic points		
	FeO (%)	SiO_2 (%)	CaO (%)	MgO (%)	Al_2O_3 (%)	Fe (%)	Ib		T_{zc} (°C)	T_{top} (°C)	ΔT (°C)
1	1,26	13,72	25,53	3,28	4,11	39,53	1,86	58,44	1010	1230	220
2	1,53	13,90	24,74	1,86	1,40	42,03	1,78	62,50	765	1240	475
3	3,11	13,75	24,71	1,29	2,16	41,18	1,79	70	1080	1320	240
4	6,59	10,28	26,16	3,72	0,28	43,65	2,54	68	1010	1230	220
5	8,52	10,60	24,73	3,40	1,76	43,13	2,33	69,80	1000	1330	330
6	7,40	11,05	23,56	5,45	1,96	43,18	2,13	70,35	980	1375	395
7	4,51	12,96	24,63	0,72	1,56	47,49	1,90	69,80	980	1210	230
8	10,5 6	12,64	24,11	2,80	3,06	44,26	1,90	53,60	1025	1240	215
9	3,64	11,10	24,37	4,74	1,89	44,94	2,1	54,64	1100	1270	170
10	3,94	10,08	24,72	2,49	0,21	44,98	2,45	58,50	1100	1250	150
11	8,60	10,38	20,35	1,57	0,31	46,81	1,95	56,51	1020	1240	220

The experimental data were processed mathematically statistical, obtaining based on their charts in Figures 2-16.

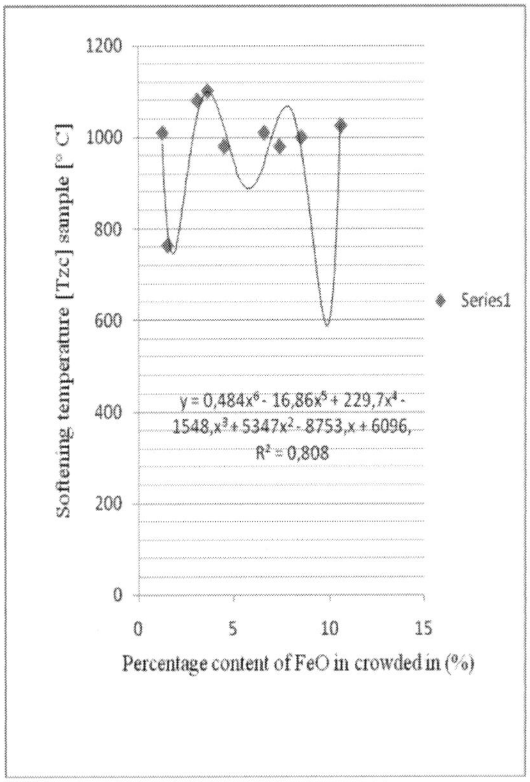

Figure 2. Graphical representation of the correlation between the softening temperature of iron agglomerate Tzc and percentage content of FeO.

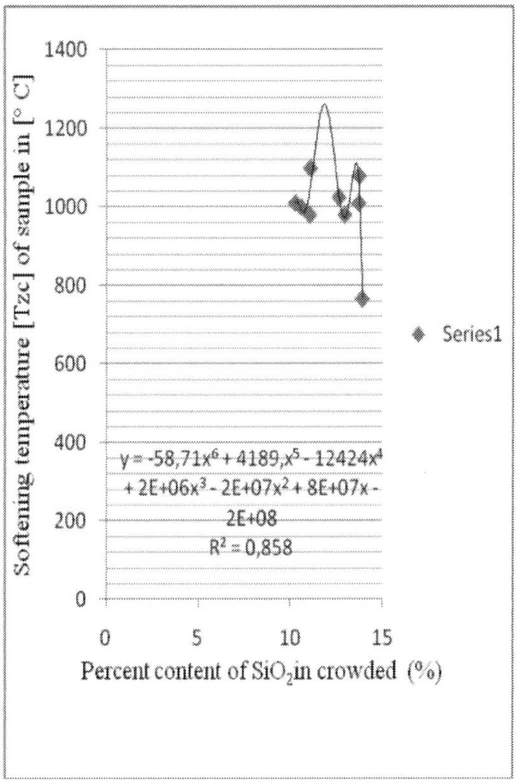

Figure 3. Graphical representation of the correlation between the softening temperature of iron agglomerate Tzc and percentage content of SiO2.

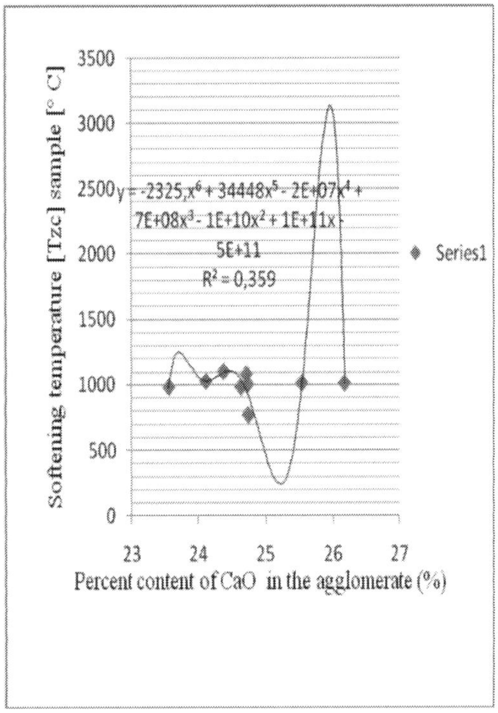

Figure 4. Graphical representation of the correlation between the softening temperature of iron agglomerate Tzc and percentage content of CaO.

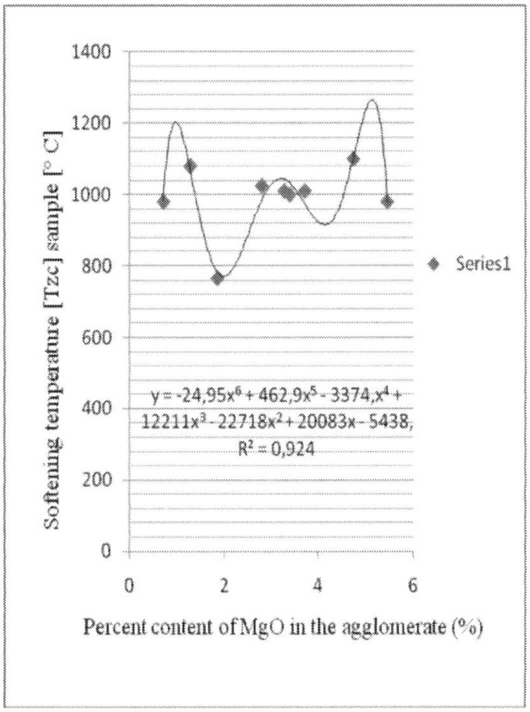

Figure 5. Graphical representation of the correlation between the softening temperature of iron agglomerate Tzc and percentage content of MgO.

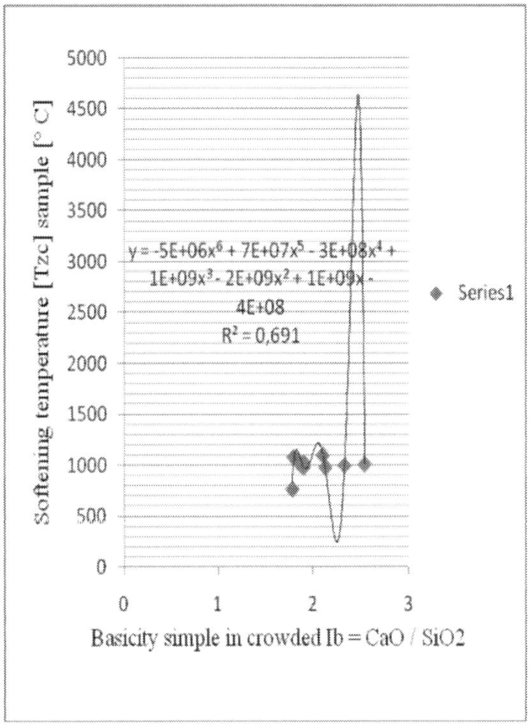

Figure 6. Graphical representation of the correlation between the softening temperature of iron and basicity Tzc simple Ib=CaO/SiO2.

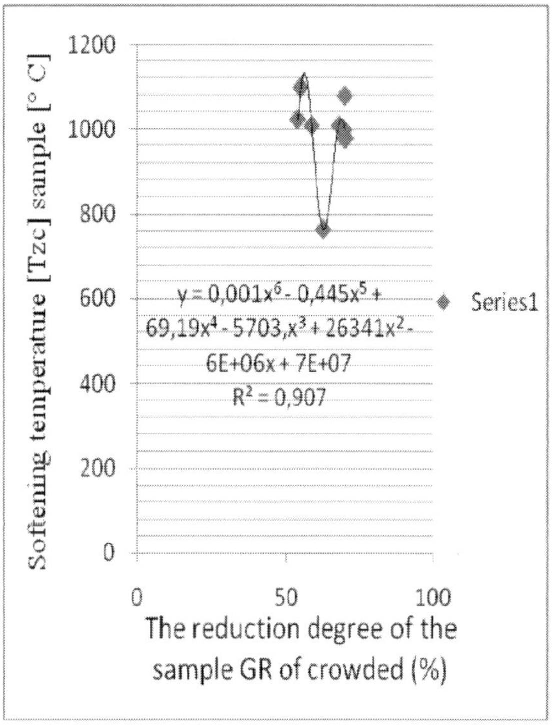

Figure 7. Graphical representation of the correlation between softening temperature of sinter ferrous Tzc and the reduction degree GR.

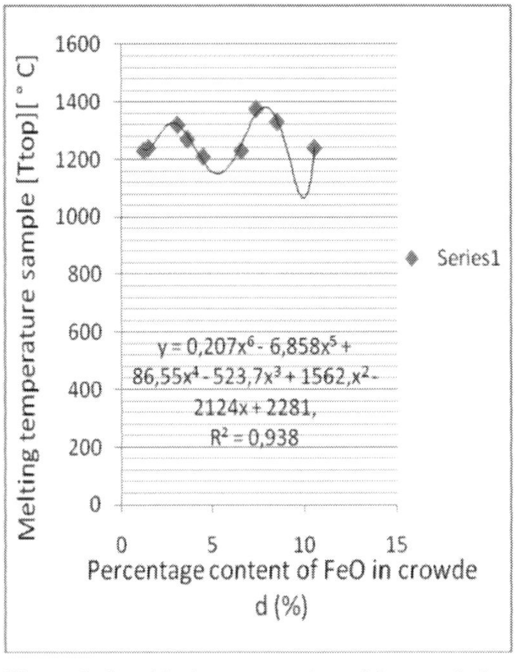

Figure 8. Graphical representation of the correlation between melting temperature of sinter ferrous Ttop and percentage content of FeO.

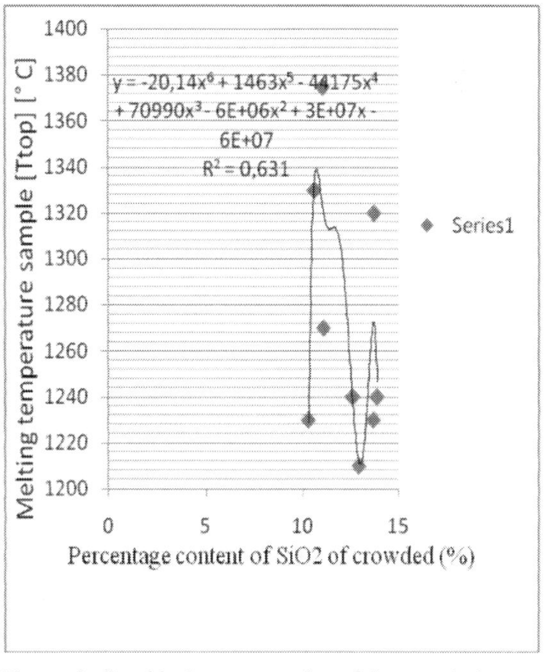

Figure 9. Graphical representation of the correlation between melting temperature of sinter ferrous Ttop and percentage content of SiO2.

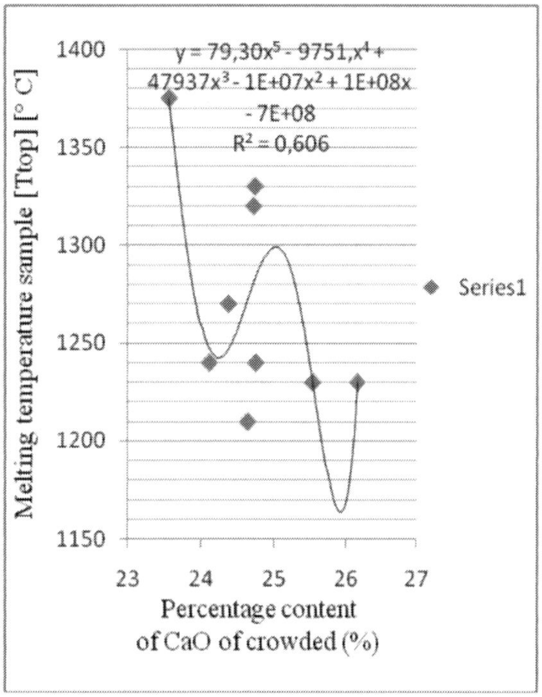

Figure 10. Graphical representation of the correlation between melting temperature of sinter ferrous Ttop and percentage content of CaO.

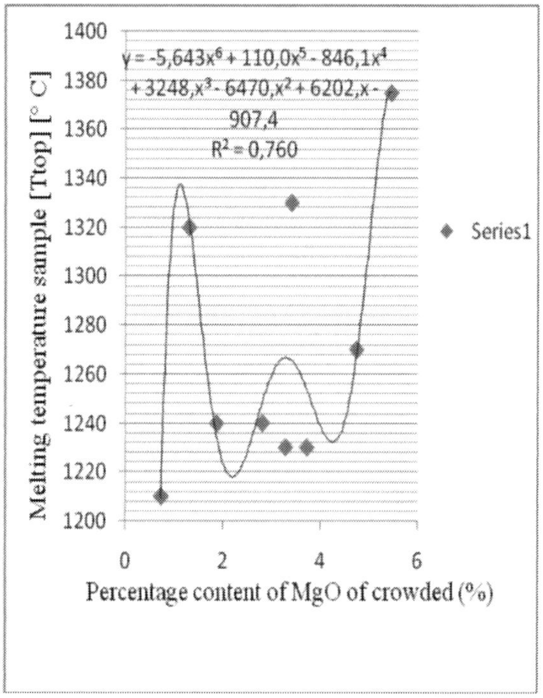

Figure 11. Graphical representation of the correlation between melting temperature of sinter ferrous Ttop and percentage content of MgO.

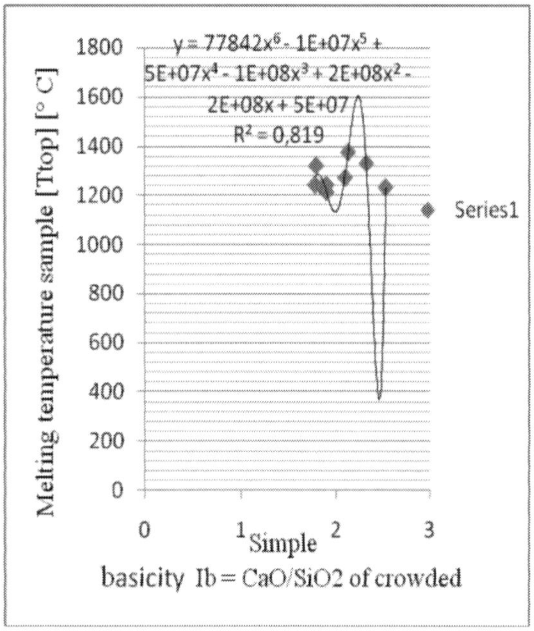

Figure 12. Graphical representation of the correlation between melting temperature of sinter ferrous Ttop and simple basicity Ib=CaOSiO2.

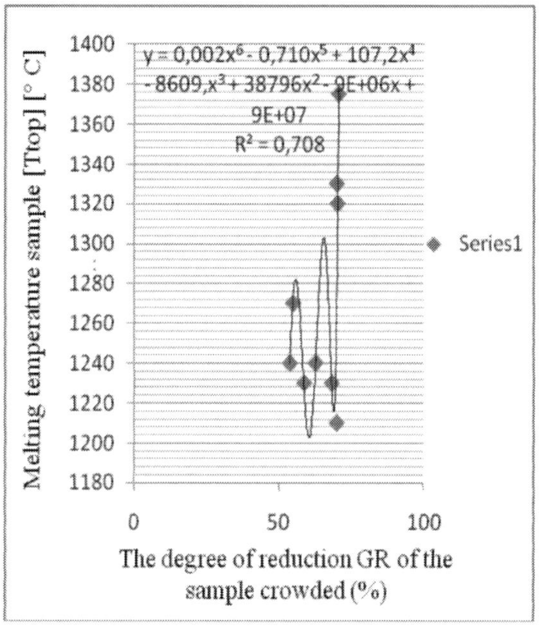

Figure 13. Graphical representation of the correlation between melting temperature of sinter ferrous Ttop and degree of reduction GR.

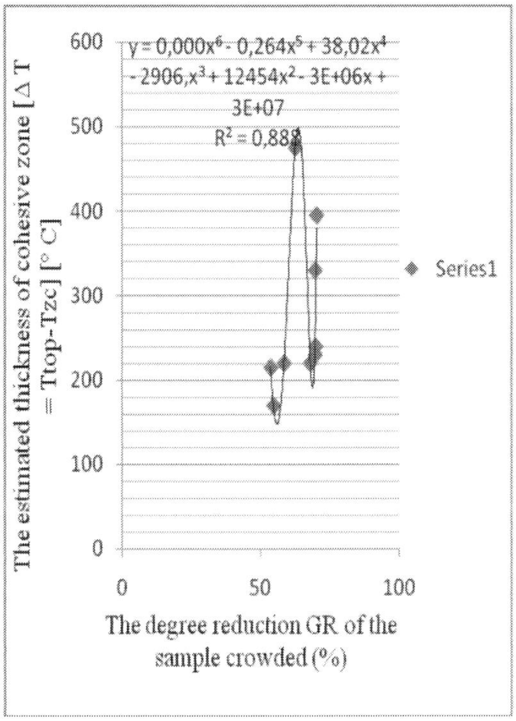

Figure 14. Graphical representation of the correlation between the estimated thickness of cohesive zone ΔT=Ttop-Tzc and the degree reduction GR of the sample crowded

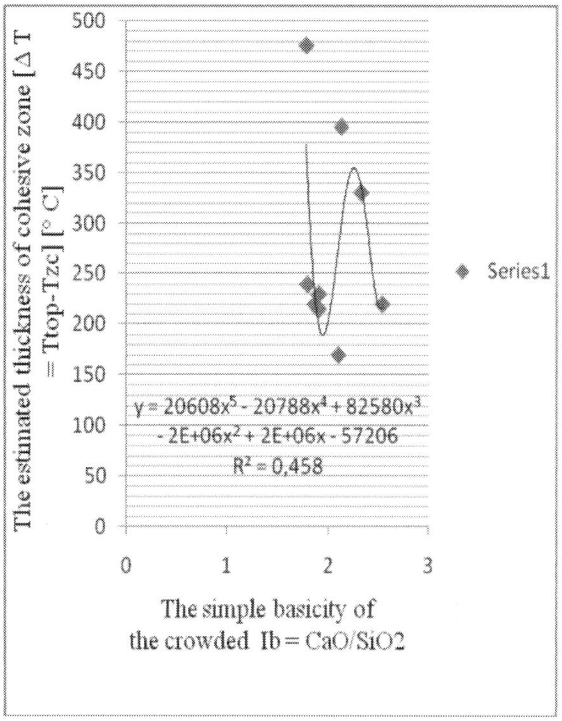

Figure 15. Graphical representation of the correlation between the estimated thickness of cohesive zone ΔT=Ttop-Tzc and the simple basicity of the crowded Ib=CaOSiO2.

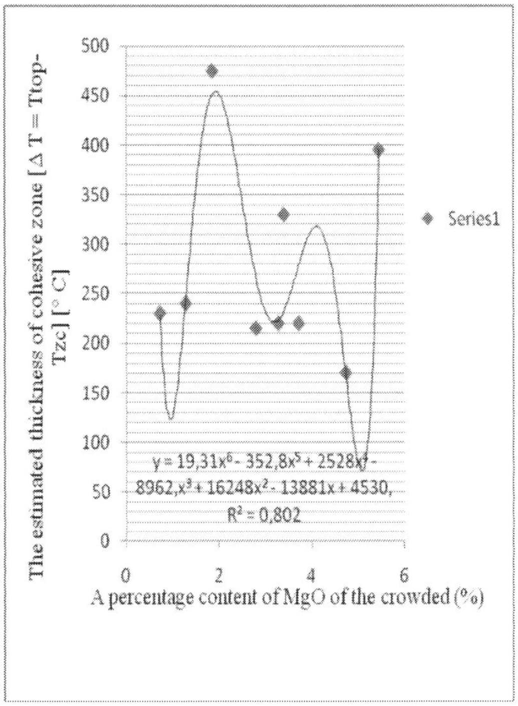

Figure 16. Graphical representation of the correlation between the estimated thickness of cohesive zone ΔT=Ttop-Tzc and percentage content of MgO of the sample crowded.

4. Conclusions

Conclusions on softening and melting characteristics of the studied agglomerates can be summarized as follows:

-the increase in waste rock MgO to increase sinter characteristic T_{zc}, T_{top}, and $\Delta T = T_{top}-T_{zc}$ (samples No. 5, 6);

-further provided with tools as random parameter studied Al_2O_3 have not submitted acceptable correlation coefficients, however the growth of ΔT, which means a premature softening layers, a compactizare their faster, causing a decrease in permeability cohesive zone;

-further provided with tools drawn with Cao/SiO_2 report variables confirmed the trends of growth and melting point of softening point with increasing basicity;

-characteristic points are not influenced by the iron content variations;

-agglomerates with higher MgO contents melted harder and longer intervals, with approx 1080°C T_{zc} and T_{top} 1280-1410°C;

-it was noted that the growth of FeO in crowded place a sharp Tzc and Ttop, what can be explained by the formation of compounds of the type fayalit, which leads to less permeable cohesive areas;

-the most powerful influence on the behavior of ferrous materials in the softening and melting, it is the degree of reduction of iron oxides, forming metallic iron support structures, which leads to ensure permeability layer plastic crowded condition.

This very important conclusion has led in the direction of a research study of reducibility and mechanical strength of components of cargo around the country, furnaces especially crowded samples.

References

[1] Constantin N 2002 *Engineering production of pig iron in blast furnace*, Printech Publishing House, Bucharest, Romania

[2] Constantin N 1994 *The intensification of the blast furnace gas job by improving circulation to increase productivity and a metallurgical coke Economy and Energy*, Politehnica University of Bucharest, Doctor Thesis

[3] Constantin N, Predescu C and Jura M 1995 *Research on mathematical model of the blast furnace cohesive zone,* International conference, Hunedoara, Romania, October 12-13, pp 18-26

[4] Constantin N, Cristian Dobrescu C and Buzduga M 2008 Research regarding the physical and chemical characteristics of pre-reduced iron ores and the analysis of the possibilities of their use in the iron and steel elaborating process, *Metalurgia International* **5** 15-21

IOP Conf. Series: Materials Science and Engineering **85** (2015) 012007 doi:10.1088/1757-899X/85/1/012007

Determination of reliability for Dacia 1304 1,9 D utility vehicle

A Budiul Berghian[1] and T Vasiu[1]

[1] Politehnica University of Timisoara, Department of Engineering and Management, 5 Revolution Street, 331128 Hunedoara, Romania

E-mail: adina.budiul@upt.ro

Abstract. The study analyses running and failure of Dacia 1304, 1,9D utility vehicle. The study comprises plotting the Pareto diagram in accordance with the data collected by observation of utility vehicle running/failure and determination of its reliability using the Weibull++8 specialized software of the company ReliaSoft, conclusions being presented in graphic form within this study.

1. Introduction

In the system of economic efficiency, which points all aspects and components of the production activity, products quality and its continuous improvement have a significant place.

Reliability is one of the prominent elements of product quality and according to SR-EN-ISO 8402-1991 represents: *Ability of a product to accomplish a specified function in fixed conditions and for a certain period of time* [1], [2].

From a certain point of view, quality can be considered as a "static" feature to satisfy some requirements at a certain moment, while reliability is "dynamic", as manifestation in time of the quality.

2. Determination of reliability for Dacia 1304 1,9D utility vehicle

Information on reliability for Dacia 1304 1,9 utility vehicles have been obtained observing the running of a lot of vehicles of the same generation and fabrication, on a period of a year. This control programme comprises data on the number of kilometers from which intervene changes in status or behavior of the utility vehicle, moment of down time, moment of re-starting up after removal of failure causes.

After processing the data obtained from running of Dacia 1304 utility vehicle using the Weibull ++8 specialized software, have been obtained the following results presented in graphic form in figures 4, 5, and 6.

IOP Conf. Series: Materials Science and Engineering **85** (2015) 012007 doi:10.1088/1757-899X/85/1/012007

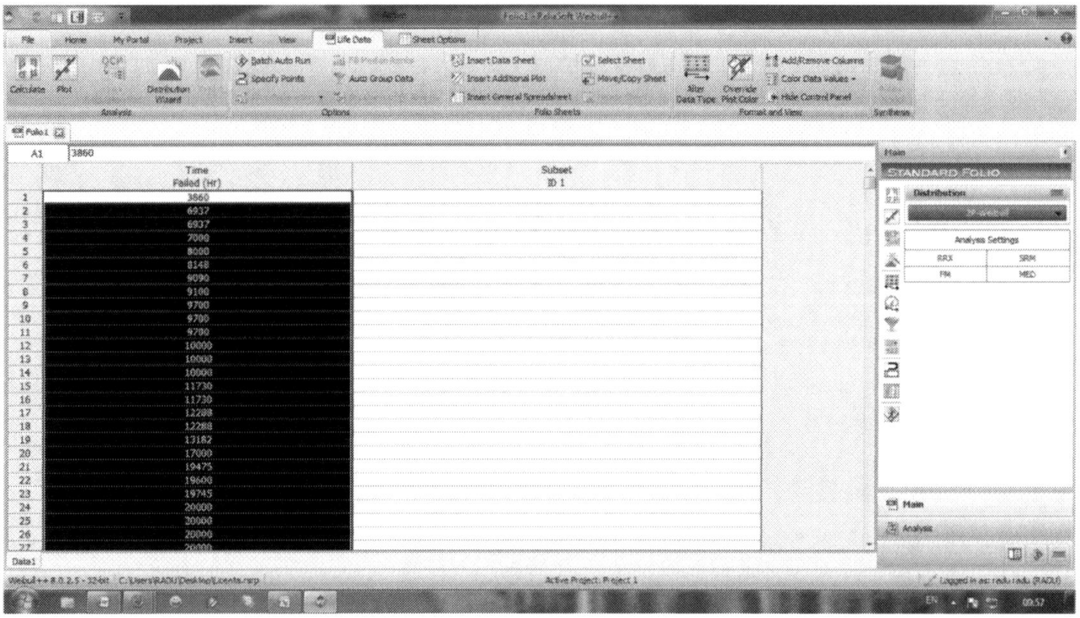

Figure 1. Interface used to determine parameters of Weibull distribution.

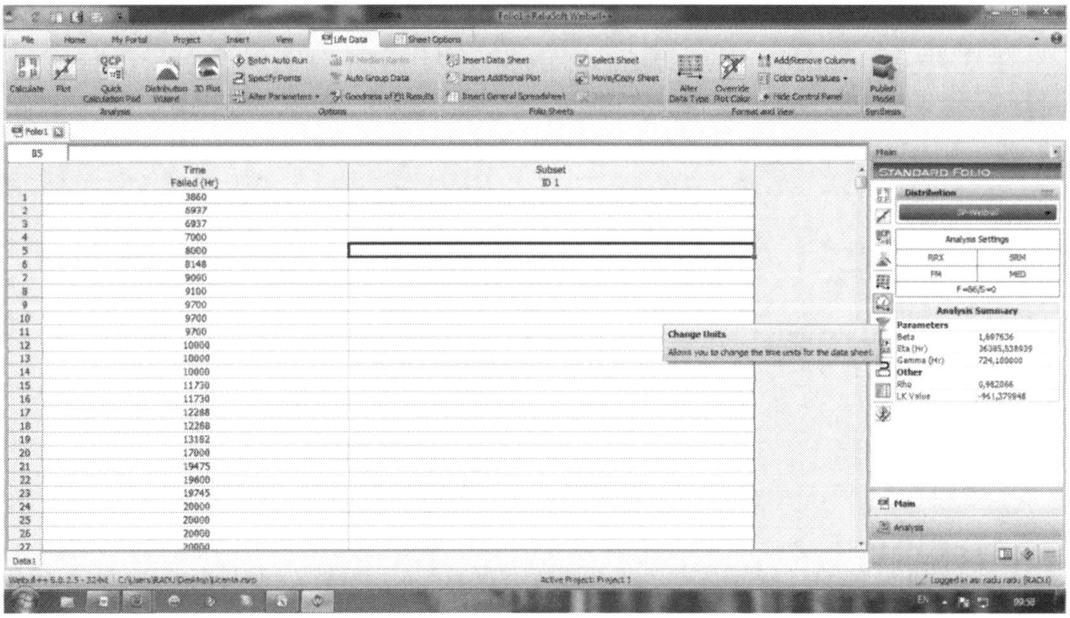

Figure 2. Interface Processing the results using Weibull ++8 software.

Certification on real reliability of the studied utility vehicle complies with Weibull distribution law can be seen in Figure 4, where is shown the Alain Plait diagram whose double logarithmic coordinates show the time dependence of the relative cumulated frequency.[3] Calculated values of the Weibull law parameters are, with correlation coefficient ρ=0,982066 : β=1,697636, η=36385,538 and

IOP Conf. Series: Materials Science and Engineering **85** (2015) 012007 doi:10.1088/1757-899X/85/1/012007

γ=724,100.[4] Thus, mathematical expression of the reliability distribution law for Dacia 1304, 1,9D is:

$$R(t) = e^{-\left(\frac{t-\gamma}{\eta}\right)^{\beta}} = e^{-\left(\frac{t-724{,}100}{36385{,}538}\right)^{1{,}697636}} \tag{1}$$

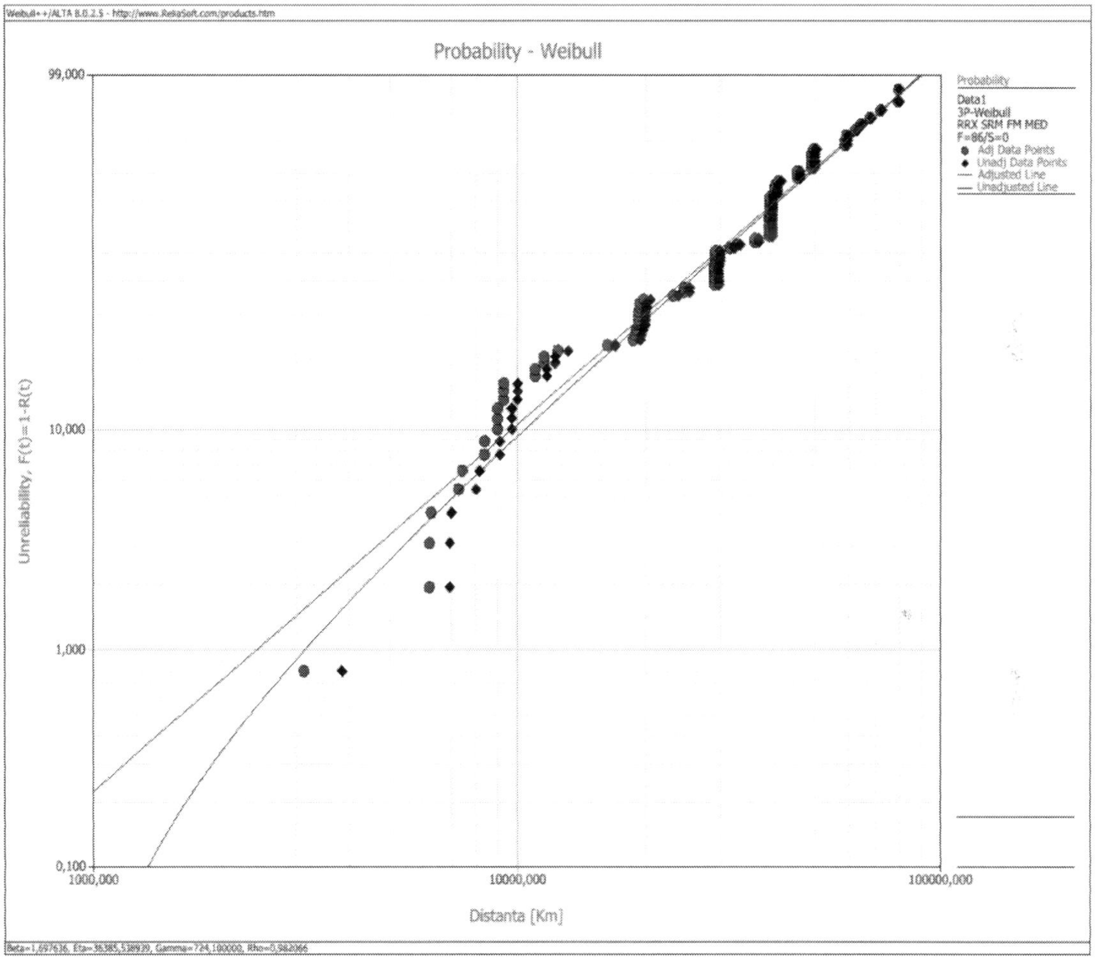

Figure 3. Alain Plait diagram.

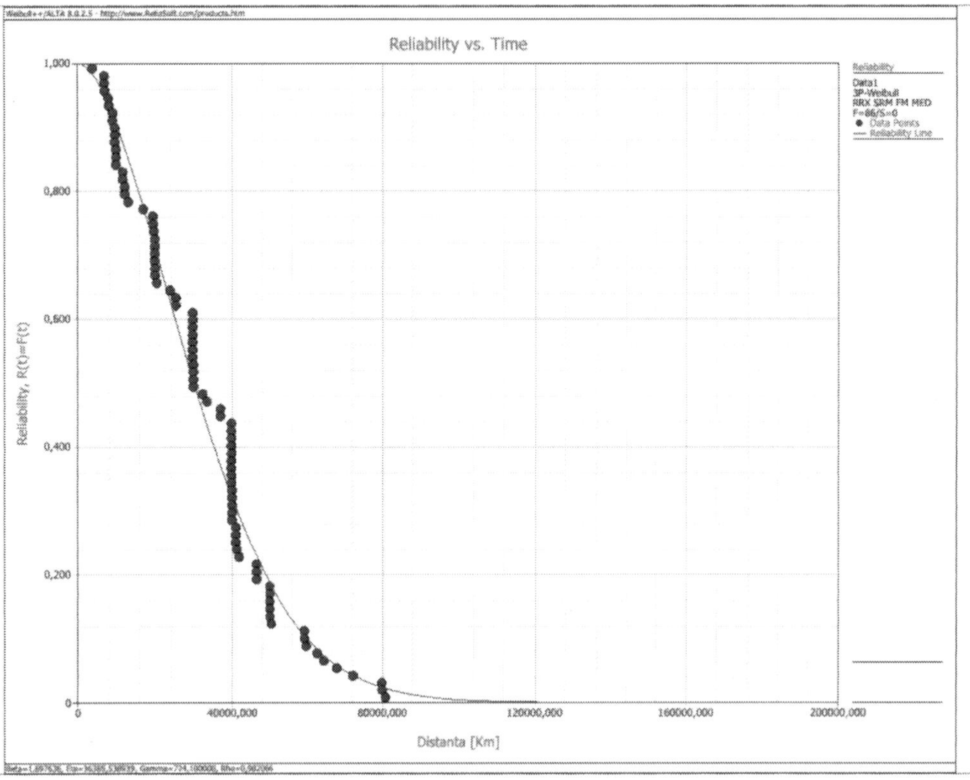

Figure 4. Graphic of reliability depending on run number of km.

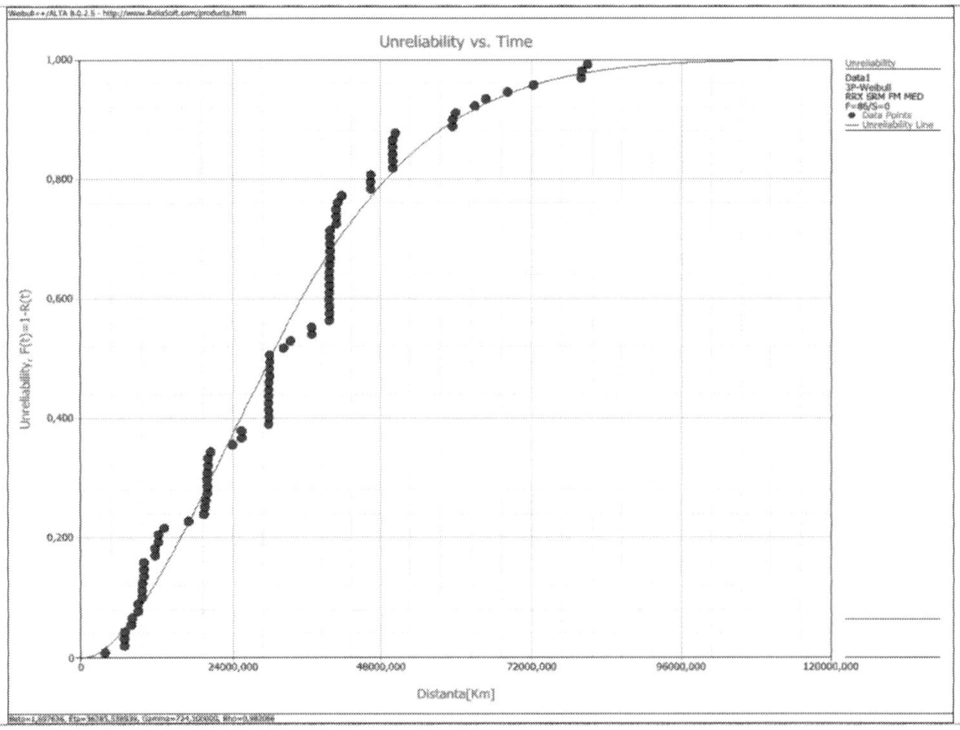

Figure 5. Graphic of non-reliability depending on run number of km.

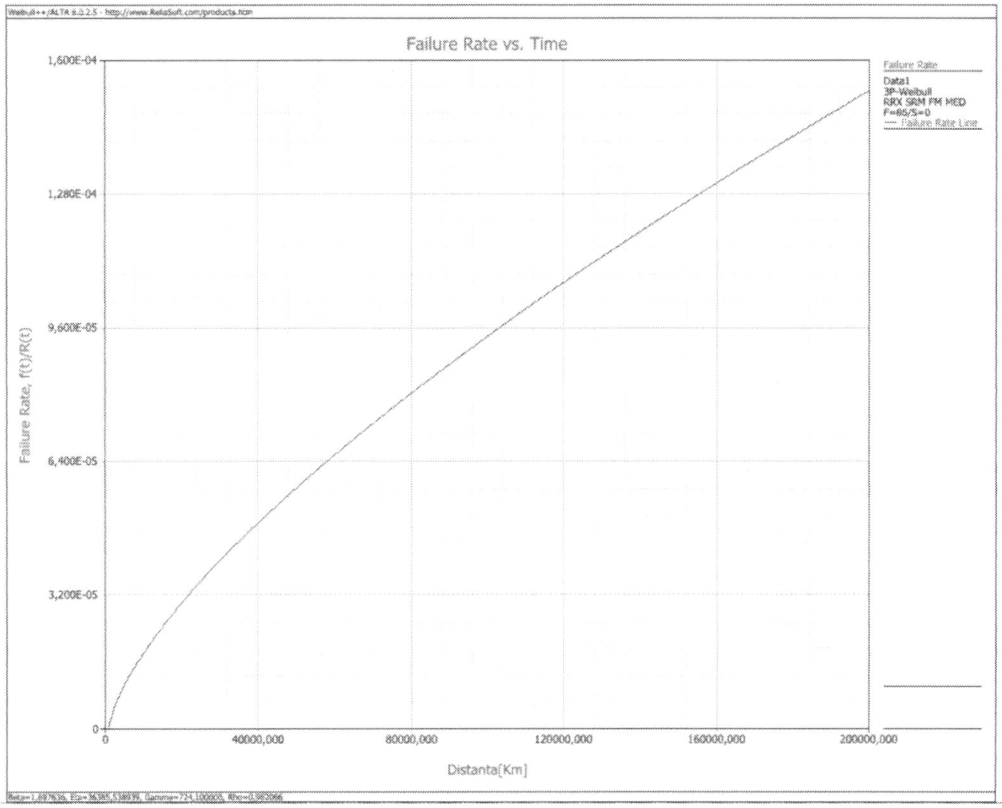

Figure 6. Graphic of down-time rate depending on run number of km.

3. Conclusions

Analyzing the Pareto diagram has been ascertained that the highest percentage failure is given by the pre-heating sparking plug of the Diesel motor, 23,26%. Furthermore, analyzing the reliability of these type utility vehicles emerges that their running follows Weibull distribution law (Figure 3) with a correlation coefficient ρ=0,982066 and the value of β indicates that the studied lot of vehicles is within the normal running time. According to the graphic of Figure 4, reliability is maintained at values relatively high up to about 40 000km, i.e. relatively less.

References

[1] ***STAS 8174/1-77, Reliability. Terminology
[2] ***STAS ISO 9001/1991, Quality Systems. Model of Quality Insurance in Project /Development, Production, Assemblage and Service
[3] Baron T 1976 *Quality and reliability of products,*Publishing House: Editura Didactică şi Pedagogică, Bucharest, Romania
[4] Budiul Berghian A and Vasiu T 2008 *Reliability and maintainability of industrial entities. Theory and applications,* Publishing House: Editura "INFOMIN", Deva, Romania

IOP Conf. Series: Materials Science and Engineering **85** (2015) 012008 doi:10.1088/1757-899X/85/1/012008

Research on uses plant ashes in processing powders for classical moulding of steel

R V Buzduga[1], N Constantin[1] and E A Lazar[1]

[1] Politehnica University of Bucharest, The Faculty of Material Science and Engineering, Splaiul Independentei no. 313, 060042, sector 6, Bucuresti, Romania

E-mail: radu_buzduga@yahoo.com

Abstract. In this work we examine various possibilities of using plant ashes in processing powders for moulding steel, in order to reduce the consumption of steel. Thus, there have been studied plant ashes which can be used to process powders and then some moulding powder compositions have been analysed.

1. Introduction

The shaping of the rough bar blister is determined by a number of factors, which depend both on the process of elaboration of steel, and the process of casting in the cast iron mold regarding: the chemical composition of the steel, the degree of overheating of the steel, the molding rate, the time of solidification of steel to the environment temperature, and the geometry of the cast iron mold.

The research in the field allows us to draw the conclusion that through the use of thermo-insulating materials (anti-blister) both as shrink-head casing and as covering materials for the surface of the steel, the discard, varying on the type of steel, can be reduced with 8-10% for the revert conical ingots, and respective 9-11% for the direct conical ones. In order to get an insulating product capable to fulfill the above mentioned conditions, the following base factors of the blister formation need to be analyzed:

- the size of the volume of contraction which depends firstly on the chemical composition of steel (especially the carbon content) and the solidification temperature;
- the unfolding in time of the solidification, especially the time for the volume of steel from the head of the ingot must be kept liquid, thus it can flow without friction within the empty contraction space, time which should be longer than the time of solidification of the ingot body;
- the heat loss within the ingot head which is the main cause in forming of the blister.

Through actions upon the head of the ingot the heat must be thus directed that the steel from the heaser to solidify later. First a heating of the heaser is induced, and then, after the molding is finished, a slow cooling is due. We are speaking thus about non-stationary heat transmission through convection.[1]

Measurements performed upon the way the heat loss is distributed from the hease (Figure 1) have shown the following repartition [2]:

- 5-8% to the ingot body;
- 24-27% through the anti-blister powder;
- 68% through the storage in the heaser walls and turning to the exterior.

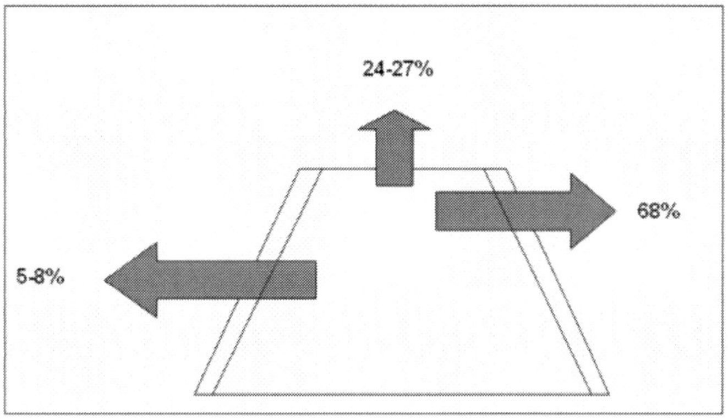

Figure 1. Heat loss through the ingot.

To draw a conclusion, the factors which influence the heat loss through radiation, through the steel glass are:

- the geometry of the ingot cast;
- the mass of the ingot;
- the specific surface of the steel glass;
- the solidification temperature;
- the amount of anti-blister materials used.

In case of large sections of the ingot, with wider solidification range of temperature, the metal volume needed to cover the heat loss is rather big, and it is difficult to reduce the heat loss from the head of the ingot, just by insulation to keep the size of the blister equal to the size of the discard.

In order to do that, we need a further contribution of heat in the head of the ingot, through the means of exothermal mass to work according to the alumina-thermal principle, and which use the oxidation heat of different metals.

For this type of thermal treatment in the hease, we need to observe a certain range of time after the exothermal reaction takes place, depending on the solidification time.

Following an empirical formula, this range should be [3]:

$$T = 0{,}0124 \cdot d \quad (1)$$

Where d is the diameter of the heaser in mm.

From the processed data we can draw a prototype model of anti-blister powder. This should meet the following characteristics:

- to have a minimum specific heat;
- to have the lowest possible thermal conductivity factor;
- to have a low volumetric mass and as high as possible thermal insulation;
- the powder layer should have optimal thickness determined by the size of the ingot.

2. Industrial experiments

We presumed the idea of getting a thermo-insulating powder with a reduced volumetric weight in order to have a better insulating performance, and, implicitly to reduce the blister and thus the specific metal consumption. This reduced volumetric weight can be accomplished by using vegetable ashes.[4]

From a chemical point of view the casting powders are part of the area of metallurgic slags, specifically the system (Figure 2).

IOP Conf. Series: Materials Science and Engineering **85** (2015) 012008 doi:10.1088/1757-899X/85/1/012008

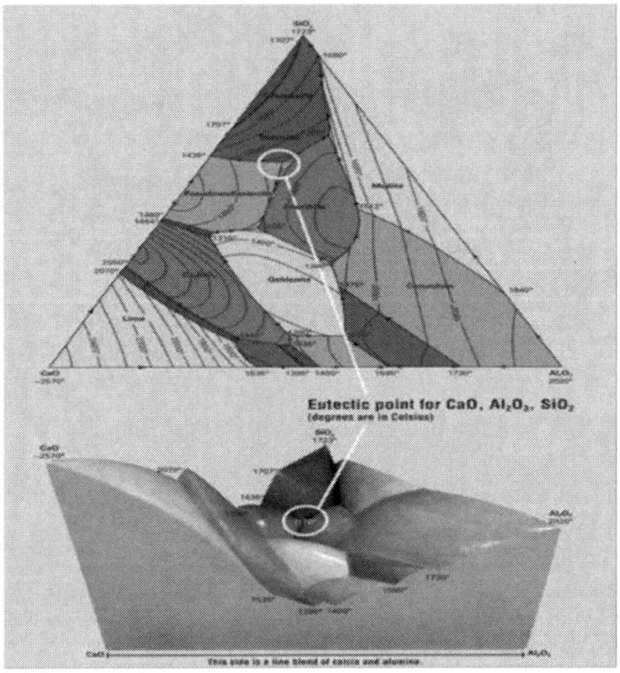

Figure 2. SiO_2-CaO-Al_2O_3 Thermal balance diagram.

The top diagram is birds-eye view of the three-dimensional model above. Each sides represents a line blend of the components of the corners, with the corners being 100% of the component and 0% of the others. The entire diagram therefore is similar to a triaxial blend. The white circles mark the lowest-melting point of calcia, alumina and silica.[3] The optimum area for the existence of casting powders in the system is the hachured one in Figure 3.

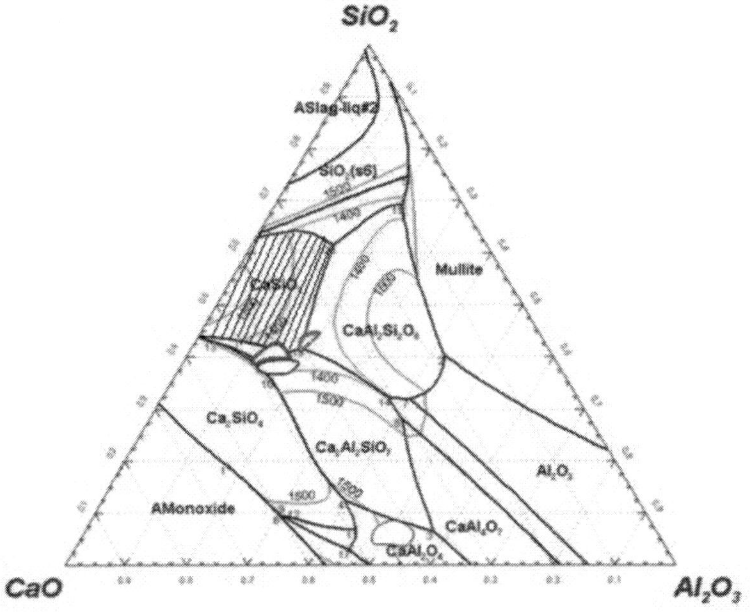

Figure 3. SiO_2-CaO-Al_2O_3 System.

IOP Conf. Series: Materials Science and Engineering **85** (2015) 012008 doi:10.1088/1757-899X/85/1/012008

Except for these basic oxides, SiO_2, CaO, Al_2O_3 the casting powders may include MgO, Na_2O, CaF_2, MnO, Fe_2O_3, C, etc. The presence of these components modify the characteristics derived from the SiO_2-CaO-Al_2O_3 system, affecting all the other characteristics of the slag, each in its particular way.

The ashes from different plant can be classified in acid ashes, alkalescent ashes and mixed ashes. The present paper takes into consideration just the acid ashes.

Thus, we have analyzed a number of 11 ashes from different plant which have been obtained through calcination at 600°C in the furnace. The results are given below, in Table 1.

Table 1. Plant compositions

Plant Ashes	Chemical analysis, [%]								Volumetric weight, [g/cm^3]
	SiO_2	Al_2O_3	Fe_2O_3	P_2O_5	CaO	MgO	Na_2O+K_2O	PC	
Wheat thatch	74,84	1,89	0,58	2,32	7,27	2,93	4,87	5,36	0,09
Wheat chaff	67,03	4,14	4,12	2,03	7,06	1,94	7,85	5,82	0,12
Barley	59,77	1,31	0,51	4,32	9,18	2,76	17,69	4,30	0,10
Oat	49,85	1,52	0,61	4,92	10,32	2,88	25,48	4,25	0.13
Rice skin	43,29	21,98	5,23	2,40	10,62	6,52	4,48	5,30	0,12
Rice thatch	78,71	2,58	0,69	-	5,12	1,89	5,63	5,24	0,06
Thatch waste	57,64	1,79	0,95	3,13	7,56	2,32	21,67	4,78	0.15
Corn	72,68	3,91	0,79	2,56	5,03	2,64	6,03	6,20	0.11
Cane	77,48	1,00	0,20	-	6,98	1,50	7,12	5,70	0.12
Sedge	54,42	5,25	1,13	4,14	17,86	3,85	6,73	6,50	0.10
Fern	50,26	0,14	0,43	-	17,86	9,51	16,78	4,80	0.10

To secure an optimum composition for the thermo-insulating powder, capable to guarantee the chemical and mineralogical composition, but also the minimum volumetric weight possible, thus a higher capacity on thermal insulation, the following ashes have been selected:

- rice thatch ashes;
- wheat thatch ashes.

A number of 8 compositions of thermo-insulating powders have been developed, for which vegetable ashes have been used, and as carbon contribution – the amorphous graphite. The compositions developed are listed in Table 2.

Using the vegetable ashes might give us the following advantages:

- the ashes obtained by calcination at 600°C is a frail powder which does not need grinding to ensure a grinding gauge of 80% under 90 μm;
- the technology of fabrication is simple, because there are but two components within the recipe, thus the errors of dosage are minimized, and the fabrication flow is not impure;
- the amorphous graphite is used in its basic form, not needing other preparations.

Table 2. Compositions

Components	Recipe componence, [%]							
	P1	P2	P3	P4	P5	P6	P7	P8
Rice thatch ashes	50	60	70	80	0	0	0	0
Wheat thatch ashes	0	0	0	0	50	60	70	80
Amorphous graphite	50	40	30	20	50	40	30	20

The compositions prepared are characterized from physical, chemical and thermic point of view the results of the analysis being presented in Table 3. The sintering, melting and flow temperatures were established with the high temperatures microscope, and the viscosity was determined theoretically with a soft. The covering thermo steel powder, produced by Stolberg, was considered the standard.[4]

Table 3. Mass characterization

Analysis	P1	P2	P3	P4	P5	P6	P7	P8	Standard
SiO_2, [%]	40.71	48.31	55.91	63.51	38.77	45.98	53.20	60.41	69.20
Al_2O_3, [%]	1.29	1.55	1.81	2.06	0.95	1.13	1.32	1.51	1.09
Fe_2O_3, [%]	0.35	0.41	0.48	0.55	0.29	0.35	0.41	0.46	-
CaO, [%]	2.56	3.07	3.58	4.10	3.64	4.36	5.09	5.82	-
MgO, [%]	0.95	1.13	1.32	1.51	1.47	1.76	2.05	2.34	0.75
Na_2O+ K_2O, [%]	2.82	3.38	3.94	4.50	2.44	2.92	3.41	3.90	0.87
PC, [%]	51.27	42.06	32.86	23.65	51.33	42.14	32.94	23.75	25.68
CaO/SiO_2	0.0628	0.0635	0.0640	0.0645	0.0629	0.0635	0.0956	0.0963	-
Volumetric weight, g/cm^3	0.33	0.27	0.22	0.16	0.34	0.29	0.24	0.19	0.48
Granulation,% under 0.06 mm	82	83	80	82	81	82	83	82	80
Humidity,%	0.83	0.86	0.83	0.85	0.82	0.83	0.82	0.85	0.86
High temperatures microscope analysis									
T_s, [°C]	1200	1230	1250	1260	1200	1220	1230	1250	>1300
T_t, [°C]	1260	1250	1260	1280	1250	1260	1290	1280	
T_c, [°C]	1300	1300	1310	1290	1300	1310	1310	1290	
Viscosity, Ns/m^2, Pas (calculated)									
T.1=T.c	2.2393	3.1029	4.5153	4.8578	2.0456	2.9341	4.2632	4.1713	5.4738
T.2 = Tc +50	1.6304	2.2911	3.3891	3.7046	1.4302	2.1206	3.1041	3.8246	4.2165
T.3= Tc +100	1.1246	1.5823	2.3677	2.6545	1.0025	1.3541	2.1034	2.7271	3.4025
T.4= Tc +150	0.7218	0.9766	1.4494	1.7074	0.6527	0.8032	0.6812	1.8025	1.9734
T.5= Tc +200	0.4220	0.4739	0.6341	0.8634	0.3836	0.4153	0.4821	0.8733	0.9887
Superficial tension, [mJ/m^2]	481.172	516.324	540.117	569.192	478.371	512.898	536.112	565.11	574.103

3. Results, discussions

Following the analysis of the data from the estimations performed on the prepared compositions there have resulted that:

- indigenous materials were used, materials which are easy to secure, in adequate amounts, at a relatively low cost;
- several compositions of casting powders have been drawn: 4 coverage powders compositions from rice thatch ashes P1 – P4 used for traditional steel casting; 4 coverage powders from wheat thatch ashes P5 – P8 used for traditional steel casting.
- the raw material used has been prepared according the laboratory flows. We can retain the preparation of rice thatch or wheat thatch ashes with additional SiO_2 which gives the product higher qualities;
- the dosing of the components has been made gravimetrically according to the drawn recipes;
- homogenizing the raw is done in electric agate grinding mortar, for 15 minutes, period needed to homogenize the components;
- the physico-chemical characterization of the components is done for the purpose of the best selection of the compositions.

The analysis of the data has given the following results:

- the values of the volumetric weights are affected by the volumetric weight of the raw material used and by the percentage of it in the recipe;
- the values of humidity are below 1%;
- the values of the granulation are min. 80% under a sieve of 0,06 mm;
- the values of the chemical composition are affected by the type of raw material used;
- the values of the temperatures of sintering, melting, flowing, are affected by the composition of raw material and its granulation;
- the recipes P4 and P8 have the closest characteristics to the standard powder, so they have been retained for further thorough study.

4. Conclusions

From the analysis of the presented data, we have the following results:

- from the point of view of the chemical composition, the vegetable ashes can be used to produce steel casting powders;
- regarding the chemical composition of the ashes, through their process, we can get powders with acid nature, their composition being determined by the proportion of the components;
- as a result of low volumetric weight, these powders have a high insulating capacity;
- we consider that continuing this research is more than beneficial, and we must have in view more complex recipes in order to secure powders from 3 – 4 sorts of vegetable ashes, focusing on rice and oat ashes, due to their high content of Na_2O şi K_2O with an important significance on the viscosity and superficial tension.

Acknowledgement

The work has been funded by the Sectoral Operational Programme Human Resources Development 2007-2013 of the Ministry of European Funds through the Financial Agreement POSDRU/159/1.5/S/132397.

References

[1] Comsa A 1987 *Research on improving the ingots quality by using a new types of lubricating powder*, CCPR SA, Brasov, Romania
[2] Comsa A 1988 *Research on uses rice straw to manufacture powders and insulation boards,* CCPR SA, Brasov, Romania
[3] Solacu S 1968 *Physical chemistry of technical silicates*, Tehnica Press co., Bucharest, Romania
[4] Teoreanu I, Ciontea N 1985, *Refractory products technology*, Tehnica Press co., Bucharest, Romania

International Conference on Applied Sciences 2014 (ICAS2014) IOP Publishing
IOP Conf. Series: Materials Science and Engineering **85** (2015) 012009 doi:10.1088/1757-899X/85/1/012009

Using microcontrollers for orienting a set of mirrors to focus the light beam

C Silaşi[1], C Cunţan[2] and I Baciu[2]

[1] "Key Safety Romania", Ribita Plant
[2] Politehnica University of Timisoara, Department of Electrical Engineering and Industrial Informatics, 5 Revolution Street, 331128 Hunedoara, Romania

E-mail: corina.cuntan@upt.ro

Abstract. This paper presents an independent control system for two mirrors tracking the position of the sun and allowing the focalisation of sunlight to increase the efficiency of the energy conversion process.
The application is designed to operate on the principle of a master that controls slave devices. The mirrors are designed to be independently manoeuvrable, this being possible by using microcontrollers to handle the management of their local orientation. For orienting a surface in order to capture or reflect the solar radiation, it is necessary to know the sunlight direction with a very high accuracy.

1. Introduction

The microcontrollers are VLSI type integrated circuits (VLSI = Very-large-scale integration), incorporating most of the components required to build a computer microsystem: central processing unit (CPU), program memory, data memory, interrupting system, digital input/output ports, analogue to digital converters, digital to analogue converters, counters and timers, serial network interfaces, and others. In many cases, a single circuit solves all the monitoring and control functions of an application [1].

Besides the microcontroller, an application requires a program to execute, and other few elements that constitute a logical interface with the stabilizing elements [2].

The programming can be done in several languages, such as: Assembler, C and Basic, which are the mostly used programming languages. The *Assembler* belongs to the low-level languages that are slowly programmed, but use the smallest memory space and give the best results in terms of program execution speed.

The programs in *C* language are easier to write, easier to understand, but are slower in execution than the programs in *Assembler* language. *Basic* is the easiest to learn and its instructions are the closest to human thinking, but, just like the *C* programming language, it is slower than *Assembler*.

After the program is entered into the microcontroller, this one must be connected to the application module. Based on its clock, the microcontroller executes the program instructions. Once connected, the microcontroller executes the recognition instructions from the beginning of the program, and then starts the actual operation phase. The circuit operation depends on many parameters, of which the most important are the ability of the hardware developer and the programmer experience in obtaining the best results when using the designed program [3].

A microcontroller-based system allows obtaining a maximum power at a given time, by using a solar orientation system of certain systems of reflection and focalization of the solar radiation.

The solar energy is a green energy source, free and inexhaustible. With the help of solar radiation collection systems, the solar energy can be converted into electricity or heat, which can be used for own use, or injected into the public network. The technological development and, at the same time, the growing price of energy in general, justifies the interest of an increasingly number of users to purchase energy production / capture systems.

2. Presentation of the paper

This paper deals with the design and implementation of a solar radiation focalisation system, as basis for a range of household and industrial applications.

The use of microcontrollers in industrial processes has a high dynamics in the current period, because of their advantages over the other data processing & capture solutions.

For these reasons, we describe in this paper the use of microcontrollers in the management and control of a focalisation system used for the solar radiation incident on a central point, for further applications of the focused energy.

This can be done by using measuring devices able to direct the sunlight, or algorithms that calculate the direction of solar radiation based on various models developed over time.

For finding the direction of solar radiation, we chose the method described in "Solar Position Algorithm for Solar Radiation Applications", because it allows the calculation with uncertainties of ±0.0003°, in the period from the year -2000 to 6000 [4].

The determination of the vector that defines the direction of solar radiation, hereinafter referred to as the *sun vector*, allows us to calculate the orientation of the reflective surface as being perpendicular on the bisector of the angle formed by the sun vector and the vector connecting the centre of the reflective surface and the centre of the outbreak.

In this application, the PC program of interfacing with the central microcontroller sends the following data to the microcontroller: year, month, day, time, minute, second, time zone (negative in relation to Greenwich meridian), longitude, latitude, height at which the system is located, atmospheric pressure and ambient temperature. Based on these data, the microcontroller calculates the solar radiation vector and sends it to all the network microcontrollers, along with the outbreak coordinates. The slave microcontrollers calculate the mirror-outbreak direction and place the mirror perpendicular to the bisector of the angle between the two vectors.

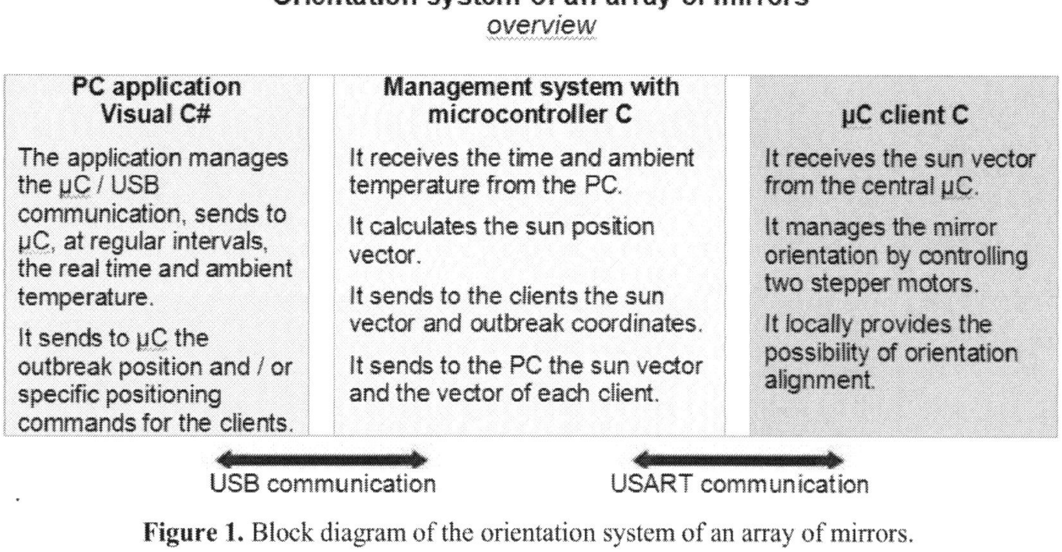

Figure 1. Block diagram of the orientation system of an array of mirrors.

The application has three main components (Figure.1):

The PC application for system monitoring and control is carried out in C#, and is intended to provide the system's interface with the user. The application runs on a normal PC and communicates with the central microcontroller by USB. The main role of the application is to keep the main system configuration data and to provide them to the central microcontroller at the user's command or at regular intervals, according to the settings.

The application running on the central system (central microcontroller) has the following functions:

- to manage the communication with each client and the PC application;
- to calculate the sun vector to be sent to the clients;
- to store and send the outbreak position to the clients.

The application running on the client system (slave microcontroller) has the role to calculate the mirror orientation based on the data received from the central microcontroller and to effectively manage the mirror rotation with two stepper motors.

Regarding the practical implementation solution, we chose as master the ATXMEGA256 A3 microcontroller[5]. (Figure 2)

This microcontroller has a number of features that make it suitable for this application. It is an advanced microcontroller, has enough memory (256k) to run such an application, has sufficient communication resources to be able to manage the communication with a large number of clients and, in the same time, it has peripherals that allow the acquisition of local information, as follows: temperature, atmospheric pressure, GPS coordinates from specialized devices or sensors, if required.
Also, we chose to use a development board, built around the AtMEGA256, of ATxmega AL-XSLED_EXT type, due to the fact that this board implements a standard solution for using the microcontroller, in terms of energy supply, and has also a solution for the USB communication, i.e. a mini USB connector for communication with the application installed on the PC, and two programming interfaces, JTAG and ISP.

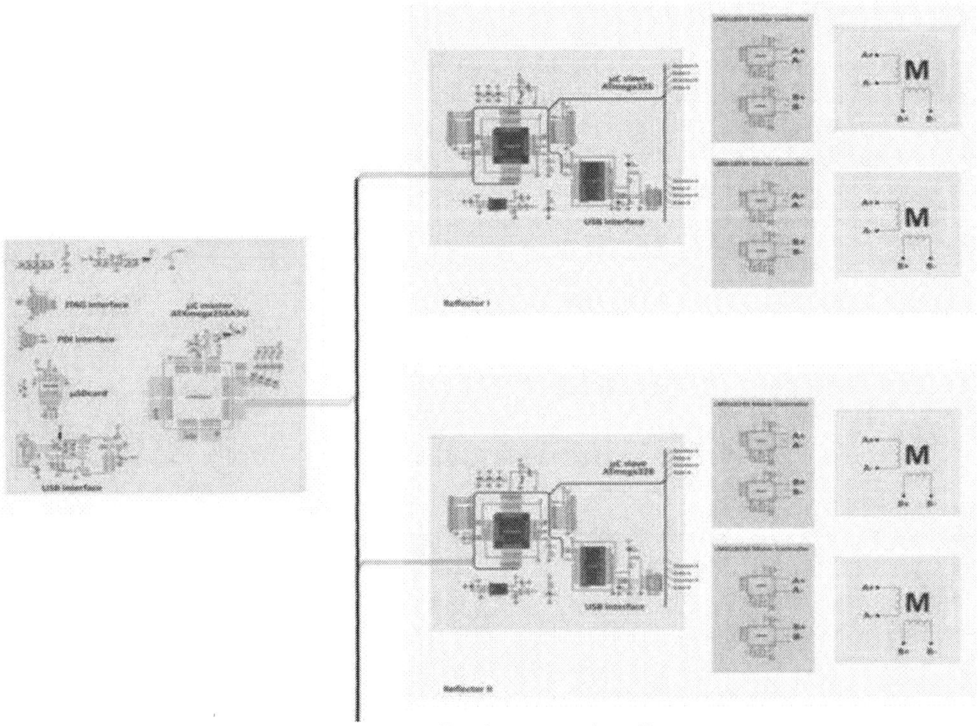

Figure 2. The application operation diagram.

Another advantage of using a development board is its solid construction, electrically and mechanically tested. All these features make the chosen solution to be suitable for the intended use.

The master board has the port USARTC1 configured as communication port with the clients. The port USARTD1 is connected to the USB interface by default, reason why it is used as a connection with the PC application.

The port USARTC1 uses the pins 6 & 7 and RX & TX, respectively. The pins TX and RX of the clients will be connected to those pins.

The USART communication is conceived as point to point bidirectional communication between two participants in communication.

This method requires the creation and implementation of a protocol to provide each client with its own identification number, which is going to be used as a key for communication with that client.

Also, the master keys can be configured so that, if such a key is received, all the clients react to the received information; these are the so-called broadcast messages. In this configuration, the communication will be made only from the central microcontroller to the clients.

The chosen slave microcontroller is the model AtMega328. This microcontroller has a storage space of 32k, sufficient to host the application to be run, it has input-output ports that can be used to control the two stepper motors required for the reflector orientation, and serial communication resources for integration into the network containing the central microcontroller [3], [6].

In a similar way as the master microcontroller, we can use a development board built around the AtMega328 microcontroller, of MINI-AT type, board that implements a standard solution for using the microcontroller in the circuit. The board has a USB interface for programming and communication with a PC, and a number of connectors placed on the sides of the board to which all the pins of the controller are connected, so that the device can be easily used on prototyping boards.

We use stepper motors having the following features: number of steps per revolution: 200, winding current: 2.6A, and supply voltage: 5V. The motors develop sufficient torque to be able to rotate the reflector safely (Figure 2).

The actual stepper motor control is carried out with motor drivers LMD18245 that can withstand a current of 3A and a voltage of 55V. This type of driver has inputs that require an input voltage level of 5V, which allows us to use it with a microcontroller that has an output voltage level of 5V.

As regards to the driver, 5 output ports are required for each motor. Therefore, we need 6 output ports to control two motors.

The driver inputs are controlled by the outputs of the microcontroller, as follows:

- The port D2 controls the current direction in the winding A of the motor 1;
- The port D3 controls the current direction in the winding B of the motor 1;
- The port D4 controls the braking input, which stops the current passing through the motor windings 1;
- The port C0 controls the current direction in the winding A of the motor 2;
- The port C1 controls the current direction in the winding B of the motor 2;
- The port C2 controls the braking input, which stops the current passing through the motor windings 2.

In addition to these ports, the drivers need a reference voltage and a 4-bit codification for the voltage to set the current through the windings.

The reference voltage is 5V, and can be taken from the source. I chose to statically treat the determination of the current through the windings, based on experimental tests, by connecting the bits to the 5V supply voltage.

2.1. Programming

The programming has three distinct components:

The program running on the PC and provides the user interface;

The program running on the master controller;

The program running on the slave controller.

The program running on the PC is made using C# in the programming environment Microsoft Visual Studio 2010, and consists of a graphic interface, whose main elements are: an area that includes the serial communication port settings, an area that includes the reflectors array settings, and an area of buttons for commands (Figure 3)[7].

As may be noticed, in the "SPA setting" area you can enter the data required for the calculation algorithm for the solar radiation direction vector.

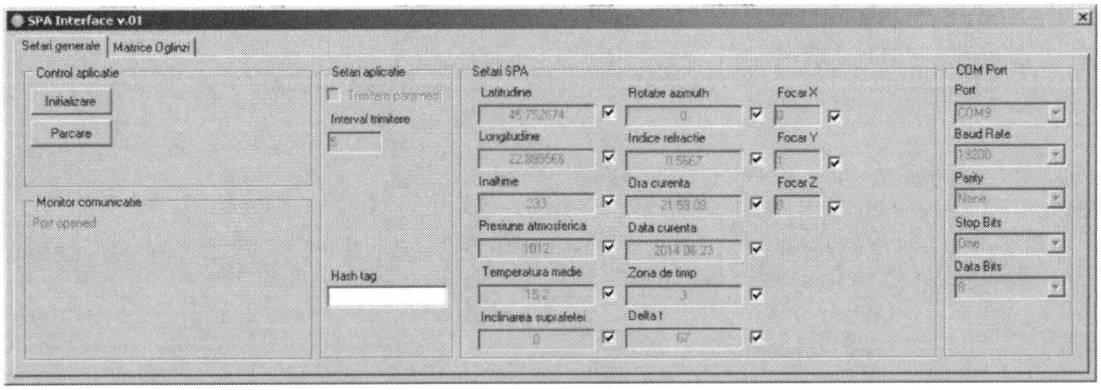

Figure 3. Graphical user interface for the program running on PC.

These data are read at startup from a "config.cfg" file (Figure 4) placed in the same location as the executable program, given that most data required by the algorithm are specific to the geographical location and can be considered constants.

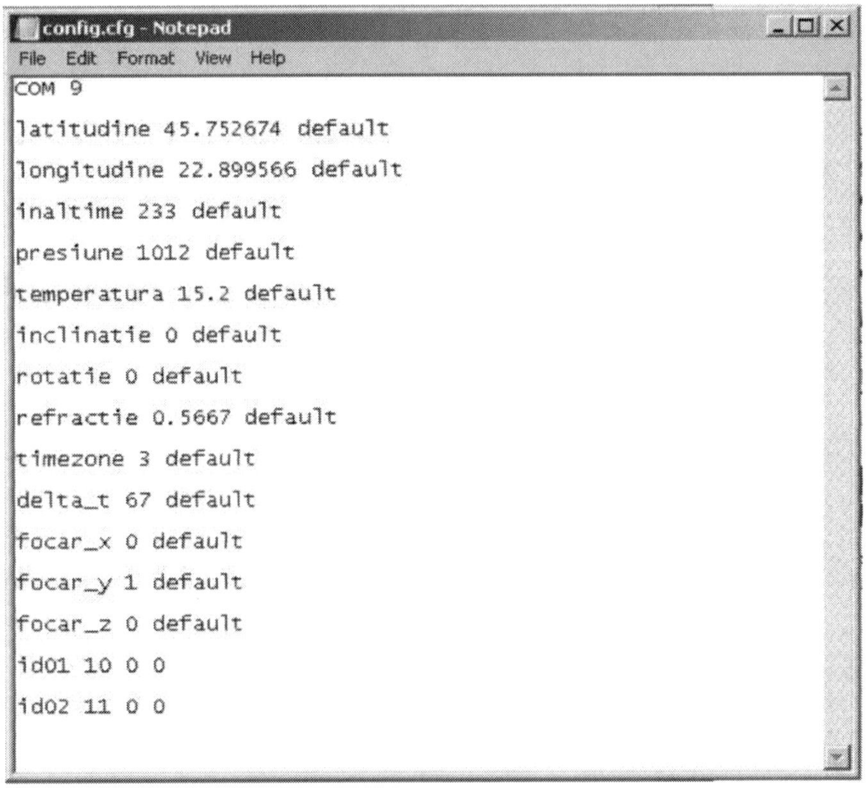

Figure 4. Data required.

When closing the program, the data are saved in the file.

The exceptions to the above are the fields containing the current date and time, these ones being obtained at regular intervals from the system on which the application is running.

These settings can be changed if unticking the checkbox on the right of each field. By ticking the check-box, the default value of the field is changed for the next loading of the program (Figure 3).

These data are sent only once to the central microcontroller, or every time when a parameter is changed, by using the Setup (Initialization) button.

In the second page of the application, we define the positions of all the clients relative to the same reference system as the outbreak. To initialise the network of reflectors, we must send these data to the central microcontroller, which, in its turn, sends to the clients the location of each one.

This information is sent only once, at the initialization of the system; then, it can be sent on request whenever any changes occur.

After initialization, the application starts to send, at regular intervals, the date and time to the microcontroller, these being the only data still required by the microcontroller.

The program of the central microcontroller is written in the C programming language. For its development, we used the AtmelStudio application, which is the integrated development environment provided by the microcontroller manufacturer Atmel.

Basically, the application running on the microcontroller calculates the solar radiation direction vector (based on the data received from the PC) and the current date and time, and sends this vector, along with the outbreak location, to the clients.

The microcontroller program is written in the C programming language; for its development, we have also used the AtmelStudio application, which has the role to receive data from the central controller, the solar radiation direction vector, the outbreak position and its own coordinates. Taking into account all these data and the current orientation of the mirror, it shall act accordingly.

The communication protocol involves the use of simple rules ensuring the transmission of information from the PC to the central microcontroller, and further to the client microcontrollers.

For easing this, the messages are organised as strings with a maximum length of 20 characters. A message is considered to be complete when it has 20 characters, or when the character ":" is added to its end (if it has less than 20 characters) (Figure 5).

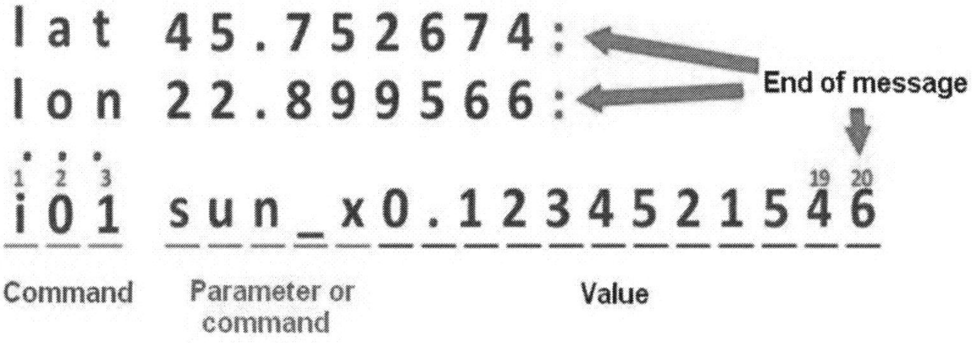

Figure 5.

The commands have three characters in length for the central microcontroller and five characters for the clients, and are interpreted by a program that initially converts the three / five characters into an integer using a string hash function, to be able to use the SWITCH-CASE instruction for interpretation.

The mechanical part of the application (Figure 6) consists of a footed stand to which the azimuth axle motor is attached. The motor drives a gear wheel on which the zenith axle is mounted on two stands, axle which is driven by another motor. The reflector is fixed on this axle, by means of two stands.

The design is made in Solid Works, a CAD application for industrial design, and the parts are cut from 10 mm thick polyamide, using a CNC water jet cutting machine, the metal shaft is made of steel machined on a lathe, and the bearing housings are made of textolite sheet, machined on rotary cutter mill and lathe.

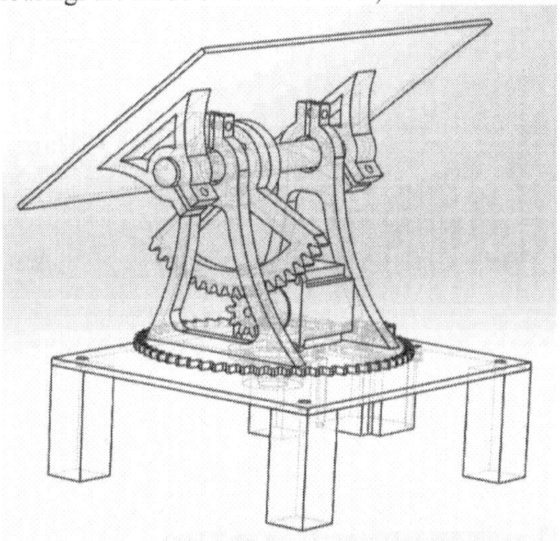

Figure 6. Computer-aided design (CAD) of the mirror orientation system.

3. Conclusions

For the practical implementation of the system, we need to create the PC programs, to build the central microcontroller and client microcontrollers, to assemble the mechanical parts and electronic circuits, and to check the application operation. (Figure 7)

For programming, we firstly realise the motor command, meaning the connection of the drivers to the control board, followed by the operational testing. The next step is to develop the communication part and to test its operation.

The string of characters received has been experimentally obtained after a series of trials, using a PC serial communication utility that sends commands to the microcontroller, followed by setting the microcontroller to send back to the PC the decomposed string, for checking it.

Figure 7. Assembling and connecting of the electronic components.

The sent string can be seen at the bottom of the window, and the decomposed response at the top of the window. (Figure 8)

To improve the communication between the clients and the central microcontroller, we can realise a bidirectional network to enabling the central microcontroller to take information from the clients and make the right decision.

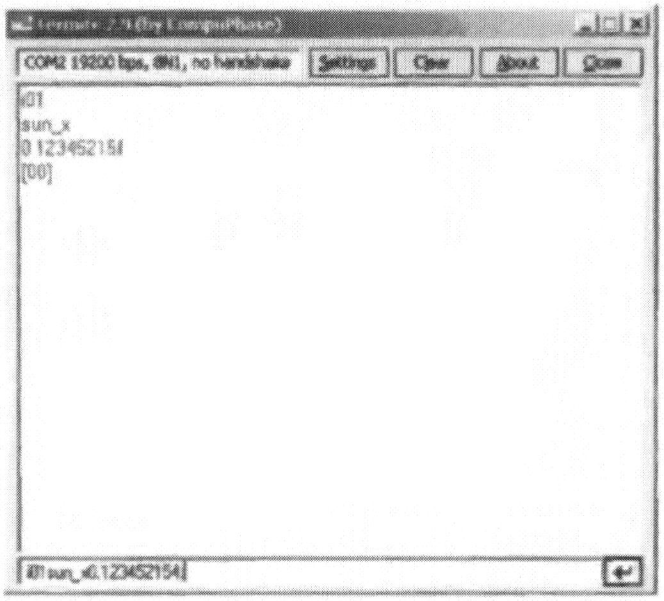

Figure 8. Testing the communication with the client.

References

[1] *** http://www.unitbv.ro/faculties/biblio/interfete_specializate/curs.pdf

[2] Sebestyen Gh 2006 *Informatică Industrială* [Industrial Informatics], Publisher: Editura Albastră, Cluj Napoca, Romania

[3] Susnea I and Mitrescu M 2005 *Microcontrollers in Practice (Springer Series in Advanced Microelectronics)*, Springer, Berlin, Heidelberg, Germany

[4] ***Solar position algorithm for solar radiation applications by Ibrahim Reda and Afshin Andreas, available on the U.S. Department of Energy website http://www.nrel.gov, National Renewable Energy Laboratory of the Office of Energy Efficiency and Renewable Energy (EERE)

[5] Xavier F C 2000 *An introduction to the industrial applications of microcontrollers*, 3rd Conference On Physics And Industrial Development: Bridging The Gap (Copid 2000), September 4-7, Durban, South Africa

[6] ***http://retele.elth.ucv.ro/Bratu%20Cristian/MAP/002%20-%20Curs%20002%20-%20MAP%20-%20Microcontrolerul%202.pdf

[7] Zachariadou K, Yiasemides K and Trougkakos N 2012 A low-cost computer-controlled Arduino-based educational laboratory system for teaching the fundamentals of photovoltaic cells, *European Journal of Physics* **33**(6) 1599

Dual stator winding variable speed asynchronous generator: optimal design and experiments

L N Tutelea[1], S I Deaconu[2] and G N Popa[2]

[1] Politehnica University of Timisoara, Electrical Engineering Department, 1-2 Vasile Parvan Street, 300223 Timisoara, Romania
[2] Politehnica University of Timisoara, Department of Electrical Engineering and Industrial Informatics, 5 Revolutiei Street, 331128 Hunedoara, Romania

E-mail: sorin.deaconu@fih.upt.ro

Abstract. In the present paper is carried out a theoretical and experimental study of dual stator winding squirrel cage asynchronous generator (DSWA) behavior in the presence of saturation regime (non-sinusoidal) due to the variable speed operation. The main aims are the determination of the relations of calculating the equivalent parameters of the machine windings to optimal design using a Matlab code. Issue is limited to three phase range of double stator winding cage-induction generator of small sized powers, the most currently used in the small adjustable speed wind or hydro power plants. The tests were carried out using three-phase asynchronous generator having rated power of 6 [kVA].

1. Introduction

In the context of rapid technical progress, the electrical drives have to ensure an operation with high demands on change and adjustment of speed, starting, braking and reversing, respectively a movements correlation of working mechanisms of the same production unit. All these technical requirements have created prospects for the development of complex drive systems, using power electronic converters based on semiconductor components that provide automatic management of production processes, with a low energy, turning to computer and microprocessor [1], [2], [3].

Despite its simple and robust construction, the motion control for this type of generators should take into account the complexity of the dynamic model which is nonlinear and variable in time and that the physical parameters of the machine are not always known with great precision. Under these conditions the motion control means controlling the speed and/or the position, respectively torque control. As on obtain a faster torque response as the motion control is more efficient. Mainly there are two control strategies: scalar control and vector control [4], [5]. Usually, the electric machines are designed to be supplied in sinusoidal regime. If the generator is connected to the grid through a static frequency converter, because of the higher harmonics (non-sinusoidal regime), both its parameters and functional characteristics values will be more or less different from the case of sinusoidal power supply. The presence of these higher harmonics will have as result the appearance of a deforming and saturation regime in the machine, with adverse effects in general in its operation [4]. The appearance of the deforming regime in the machine is inevitable because any static frequency converter based on semiconductor technique produces voltages or currents, which contain, in addition to fundamental and harmonic, higher odd time harmonics.

The introduction of distributed generation through renewable sources of energy has opened a challenging area for power engineers. As these sources are intermittent in nature, variable speed electric generators are employed for harnessing electrical energy from these sources. However, power electronic control is required to connect these sources to the existing grid [6]. A wind turbine can be designed for a constant speed or variable speed operation. Variable speed wind turbines can produce 8% to 15% more energy output as compared to their constant speed counterparts, however, they necessitate power electronic converters to provide a fixed frequency and fixed voltage power to their loads. The major components of a typical wind energy conversion system include a wind turbine, generator, interconnection apparatus and control systems. Most turbine manufacturers have opted for reduction gears between the low speed turbine rotor and the high speed three-phase generators. Direct drive configuration, where a generator is coupled to the rotor of a wind turbine directly, offers high reliability, low maintenance, and possibly low cost for certain turbines [6], [7], [8], [9], [10], [11].

For small to medium power wind or hydro turbines, permanent magnet generators and squirrel cage induction generators are often used because of their reliability and cost advantages. Interconnection apparatuses are devices to achieve power control, soft start and interconnection functions. Very often, power electronic converters are used as such devices. Most modern turbine inverters are forced commutated PWM inverters to provide a fixed voltage and fixed frequency output with a high power quality. For certain high power wind or hydro turbines, effective power control can be achieved with double PWM (pulse width modulation) converters which provide a bidirectional power flow between the turbine generator and the utility grid [12], [13], [14], [5], [16], [17].

2. Optimal design

Many unknown parameters are involved in the design of the Dual Stator Winding Asynchronous Generator (DSWA). As a result, it is necessary to assign some description to these parameters. They will be further explored in the design equations. Table 1 gives a list of the parameters used in the design approach and their values.

Table 1. Preliminary Design Parameters.

Symbol	Description	Value
p	number of poles	8
$\mathbf{D_{si}}$	inner stator diameter	180 mm
$\mathbf{D_{so}}$	outer stator diameter	267.8 mm
$\mathbf{N_{ss}}$	stator slots number	72
$\mathbf{N_{rs}}$	rotor slots number	58
$\mathbf{l_i}$	ideal core lenght	230 mm
$\mathbf{h_{ag}}$	air-gap height	0.35 mm
$\mathbf{k_c}$	Carter factor for air-gap	1.22
$\mathbf{h_{ss}}$	total height of stator slot	30.5 mm
μ_0	permeability of the air-gap	$12.56\ 10^{-7}$ H/m
$\mathbf{h_{rs}}$	total height of rotor slot	31.5 mm
mwp	main winding position	up
layersm	number of layers of main winding	1
layerse	number of layers of excitation winding	2
ncme	number of elementary conductors on main winding	24
ncee	number of elementary conductors on excitation winding	24
$\mathbf{R_m}$	main winding phase resistance	2.4 Ω
$\mathbf{R_e}$	excitation winding phase resistance	3.78 Ω
$\mathbf{R_r}$	rotor phase resistance	2.36 Ω
$\mathbf{L_{\sigma m}}$	main winding leakage phase inductance	14.5 mH
$\mathbf{L_{\sigma e}}$	excitation winding leakage phase inductance	0.6 mH

Symbol	Description	Value
$\mathbf{L_{me}}$	mutual inductance between main and excitation winding	14 mH
$\mathbf{L_{\sigma r}}$	rotor winding leakage phase inductance	15.3mH
$\mathbf{S_{copper}}$	copper section	S_{1net} = 129 mm^2
	phase connection	Y
$\mathbf{I_s}$	stator current	I_1 = 10,38/6 A
$\mathbf{U_f}$	phase voltage	U_1 = 139/240 V
f	frequency	15 Hz
$\mathbf{P_n}$	rated power	6 kVA
n	speed	415 rpm
$\mathbf{U_{DC}}$	DC voltage	460 V
$\mathbf{k_e}$	main winding to excitation winding voltage ratio	24/31

The general optimization problem could be divided in three quasi-independent sub-problems: choosing the objective function, choosing the optimization variables and the machine model, and then solving the problem. The optimal design of the DSWA generator is subject to multi-objective criteria and constructive constrains such as: reducing the initial cost, reducing the generator size and weight, improving the efficiency, and limit the components temperature to a feasible level. The multi-objective criteria should be aggregated somehow in a single objective function if the design objective is to obtain a unique solution. This could be a total cost function including penalty for unmet constrains. The objective function, C_t becomes [19]:

$$C_t = C_i + C_E + C_a + C_p \,, \tag{1}$$

where C_i is the initial cost, C_E the cost of the lost energy, C_a additional cost to consider the impact of the machine size and finally C_p is the penalty cost. Sixteen optimization variables are used to control the machine main size dimensions as well the slots details. The probability to reach the global optimum using Hooke Jeeves (HJ) algorithm could be increased by starting the algorithm several times from different points of the optimization variable space [20]. A few sample results for the HJ optimization evolution to the best design are shown in figures 1, 2, 3 and 4.

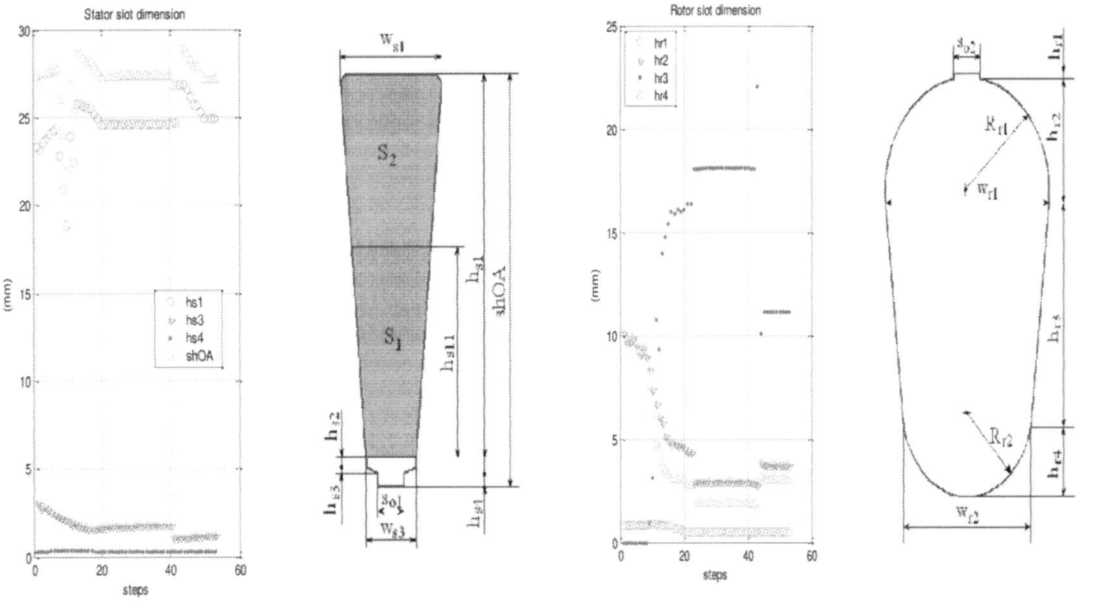

a) **b)**

Figure 1. a) Optimal stator slot dimensions; b) optimal rotor slot dimensions.

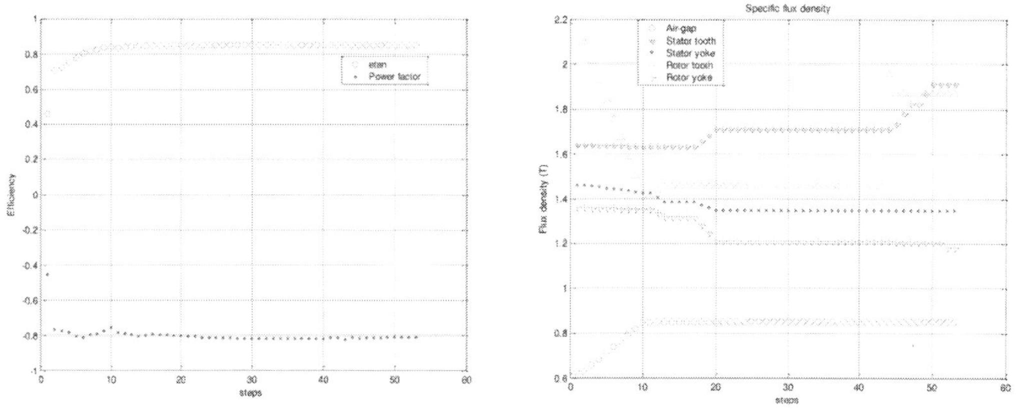

a) **b)**

Figure 2. a) Efficiency and power factor; b) specific flux density.

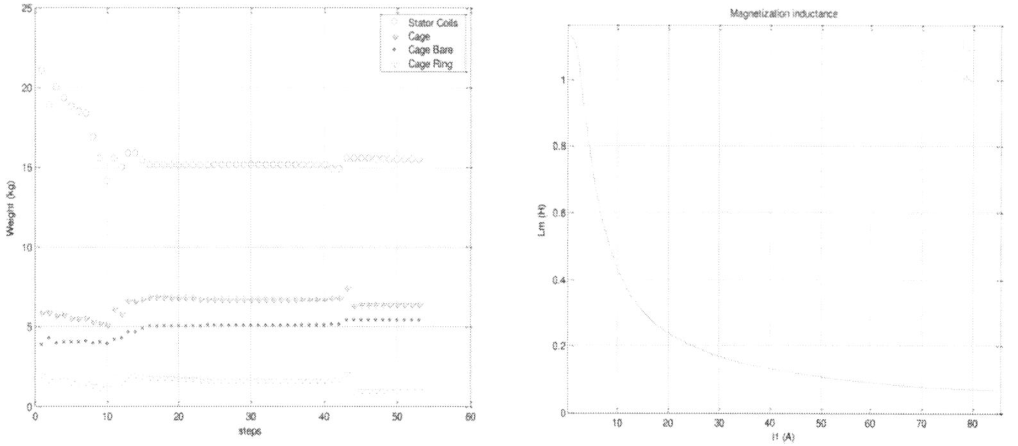

a) **b)**

Figure 3. a) Weight; b) magnetization inductance.

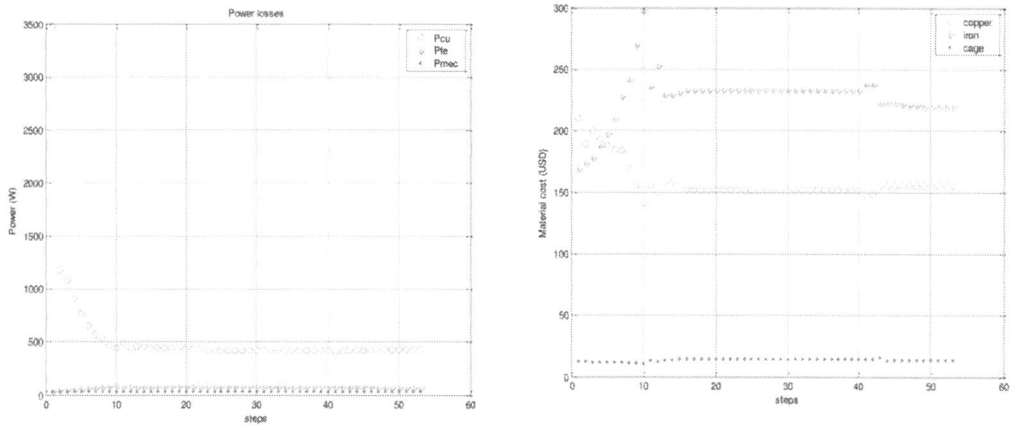

a) **b)**

Figure 4. a) power losses; b) material cost.

3. Experimental results

To prove the above design methods, a 6kVA 400V/415 rpm prototype of the DSWA generating system has been developed (Figure 5). The prime mover is simulated by a three-phase cage-type induction machine driven by an inverter of ABB ACS 800-11. The parameters of the prototype are given in Table 1. In experiments, the value of the auxiliary excitation capacitors is 35 μF, and the value of the filter inductances is 23 mH.

Figure 5. Generator DSWA coupled with prime mover induction motor under experimental tests.

Through the experimental tryouts it is desired, in a first step, the computation of the DSWA parameters and characteristics for the stationary regime and their comparison with the values obtained through optimal design and finite element analysis. In figure 6 the equivalent phase scheme of the machine is presented. Resistances R_m and R_e are measured in DC current and the inductances will be computed through no load (real and ideal) and short circuit methods.

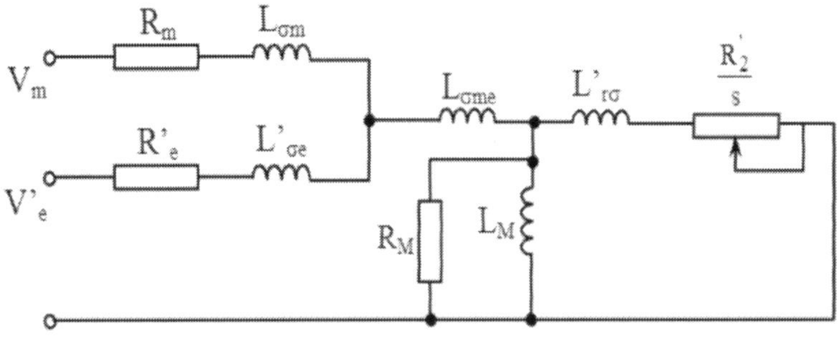

Figure 6. Equivalent phase scheme of the DSWA.

The results of the measurements are presented in tables 2 and 3. In Figures 7, 8, 9 and 10 are presented some experimental characteristics and parameters.

Table 2. Ideal no load probe with the main winding supplied from the AT and open circuit excitation.

Crt. no.	V_{Am} [V]	V_{Bm} [V]	V_{Cm} [V]	I_{Am} [A]	I_{Bm} [A]	I_{Cm} [A]	P [kW]	Q [kVar]	V_{Ae} [V]	V_{Be} [V]	V_{Ce} [V]	$I_{motor\ antrenare}$ [A]	n [rot/min]	Medie P [%]
1	271,53	270,08	269,56	2,29	2,29	2,36	0,21	1,85	256,9	258,0	256,6	11,6	750,1	0,52
2	253,74	253,13	251,78	2,12	2,12	2,18	0,21	1,62	240	240,7	239,8	11,65	750,2	1,045
3	231,05	228,92	229,96	1,92	1,91	1,94	0,12	1,32	218	219	218,5	11,6	749,9	1,405
4	213,68	212,41	212,06	1,74	1,74	1,77	0,06	1,12	202	203	202	11,65	749,9	1,585
5	190,81	190,52	190,12	1,55	1,56	1,57	0,03	0,89	181	182	180	11,65	750	1,865
6	173,32	172,22	172,45	1,40	1,41	1,43	0,02	0,73	164,5	165,6	164,5	11,6	749,8	1,48
7	156,58	154,61	155,02	1,28	1,26	1,25	0,03	0,59	147,3	149	148	11,55	749,5	1,57
8	138,21	136,14	136,31	1,11	1,11	1,09	0,03	0,45	129,5	131	130	11,65	749,6	1,32
9	115,76	113,68	114,14	0,93	0,92	0,89	0,04	0,31	108,5	109,8	108,5	11,6	749,8	1,375
10	95,20	93,18	93,59	0,78	0,78	0,74	0,01	0,22	89	90,4	89,3	11,65	749,6	1,385
11	74,59	73,15	73,78	0,63	0,59	0,59	0,01	0,13	69,8	70,8	70,2	11,6	749,5	1,355
12	57,79	56,52	57,21	0,49	0,46	0,46	0,01	0,08	54	55	54,5	11,6	749,8	1,37
13	43,58	42,49	43,18	0,38	0,35	0,35	0,01	0,05	40,5	41,4	41	11,6	749,8	1,225
14	29,38	28,75	29,21	0,28	0,25	0,26	0,00	0,02	27,3	27,8	27,6	11,6	749,7	1,32
15	15,18	14,55	15,01	0,16	0,14	0,14	0,00	0,01	13,8	14,4	14,1	11,6	749,6	1,285
16	6,23	5,95	6,52	0,08	0,06	0,06	0,00	0,00	5,7	5,9	6	11,65	749,6	1,285

Table 3. Short circuit probe for the main winding, which is supplied from the AT in two phases and open circuit excitation.

Crt. no.	V_{Am} [V]	V_{Bm} [V]	V_{Cm} [V]	I_{Am} [A]	I_{Bm} [A]	I_{Cm} [A]	P [W]	Q [Var]	V_{Ae} [V]	V_{Be} [V]	V_{Ce} [V]
1	7,7	0,1	7,7	0,52	0	0,52	4	8	4,18	8,08	4,13
2	13,4	0,2	13,5	1	0	0,98	12	24	7,28	13,92	7,25
3	19,7	0,2	19,8	1,52	0	1,52	30	52	10,57	20,16	10,55
4	25,6	0,2	25,8	2,04	0	2,02	65	90	13,65	26,05	13,66
5	31,0	0,3	31,2	2,5	0	2,5	80	134	16,41	31,42	16,47
6	36,4	0,3	36,7	3,	0	2,98	112	188	19,17	36,78	19,27
7	42,3	0,4	42,6	3,52	0	3,5	156	254	22,15	42,55	22,28
8	48,1	0,4	48,5	4,04	0	4,02	204	330	25,05	48,25	25,22
9	53,7	0,4	54,1	4,52	0	4,52	260	414	27,85	53,76	28,15
10	59	0,5	59,4	5	0	4,98	316	500	30,41	58,95	30,78
11	64,6	0,5	65,1	5,5	0	5,48	384	598	33,22	64,5	33,63
12	70,5	0,6	71	6,06	0	6,02	464	718	36,13	70,15	36,6
13	75,8	0,6	76,4	6,52	0	6,48	540	830	38,8	75,36	39,26
14	81,1	0,6	81,7	7,02	0	6,98	624	952	41,6	80,65	42,02
15	86,4	0,7	87,1	7,5	0	7,46	712	1084	44	85,8	44,7

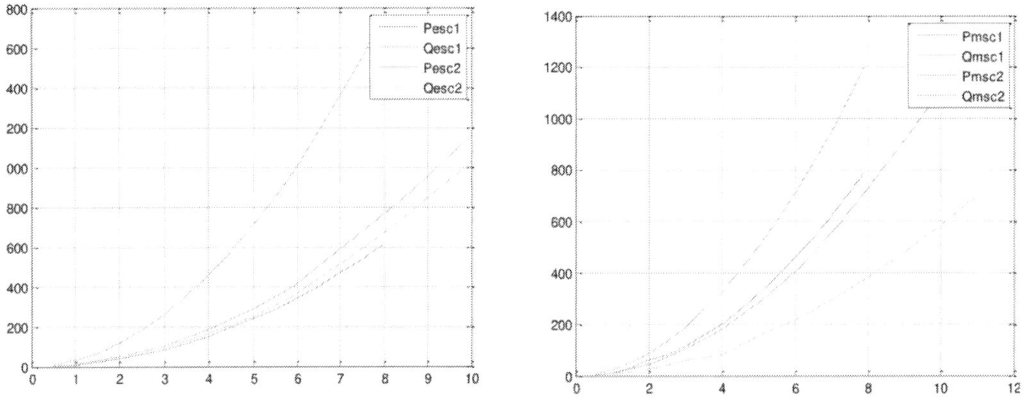

a) **b)**

Figure 7. Short-circuit active and reactive power versus current:
a) excitation winding; b) main winding.

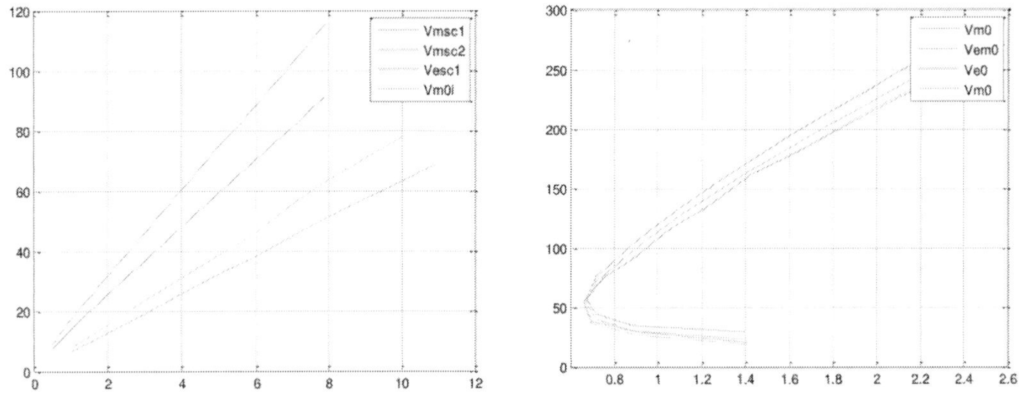

a) **b)**

Figure 8. Voltages versus current: a) short-circuit; b) no load.

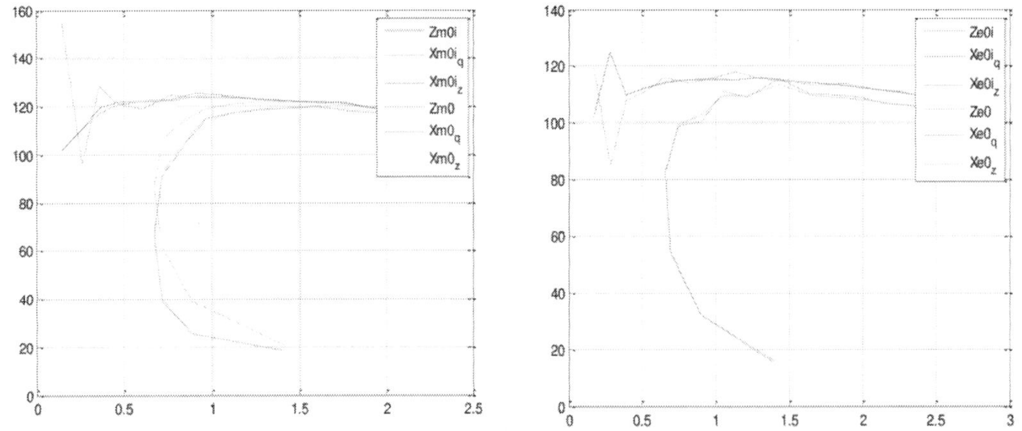

a) **b)**

Figure 9. Impedances and reactances versus current at no load:
a) main winding; b) excitation winding.

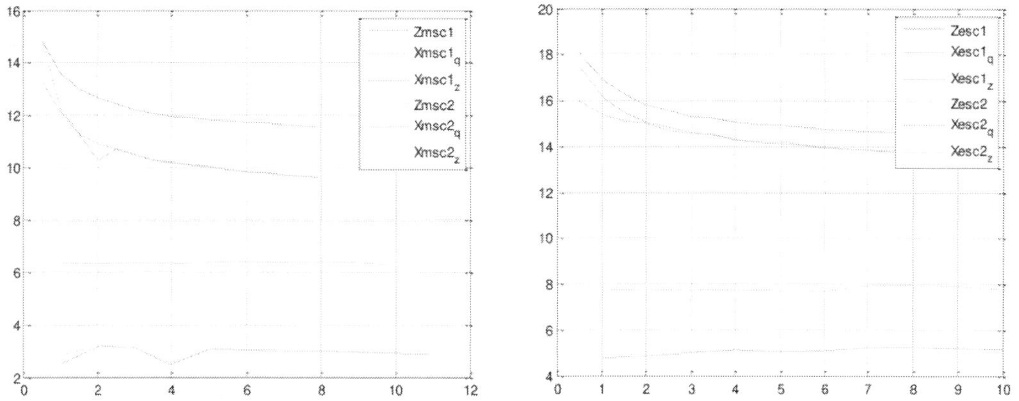

a) **b)**

Figure 10. Impedances and reactances versus current at short-circuit: a) main winding; b) excitation winding.

4. Conclusions

In this paper was presented a new type of twin stator windings induction machine operating in generator mode. The main advantage of the DSWA is its improved capability to operate at variable low speed for wind or hydro power plants.

A mathematical perfectly saturated model should be implemented in flux rotating frame coordinate and then the transients and steady state magnetization inductance could be considered in the model.

The preliminary and optimal design had shown the parameters and performance complex dependence on the number of poles. However, the 16 poles generator is considered the optimum because it has the maximum efficiency, while the active material cost is also near its minimum and the power factor (which influences the power convertor cost) is not large.

In the experimental tests are prove quantitatively and qualitatively the good correlations with simulations, and the practicality of the proposed system for low power variable speed wind or hydro power plants applications.

References

[1] Muşuroi S, Olărescu V N, Vătău D and Şorândaru C 2009 Theoretical and experimental determination of equivalent parameters of three-phase induction motor windings in case of power electronic converters supply, *WSEAS Journal of Systems* **8**(10) 1115-1124

[2] Cunţan C and Baciu I 2012 *Highlighting the Effects of an Active Filter on the Power Supply Lines of an Electric Locomotive, by Using the PSCAD Software*, 2012 IEEE International Conference on Industrial Tehnology (ICIT), Athene, Greece, March 3-5, pp 438-443

[3] Marungsri B, Meeboon N and Oonsivilai A 2006 *Dynamic Model Identification of Induction Motors using Intelligent Search Techniques with taking Core Loss into Account*, Proceedings of the 6-th WSEAS International Conference on Power Systems, Lisbon, Portugal, September 22-24, pp 108-115

[4] Muşuroi S, Şorândaru C, Olărescu V N and Svoboda M 2009 Mathematical Model Associated to Three-Phase Induction Servomotors in the Case of Scalar Control, *WSEAS Transactions on Systems* **8**(10) 1125-1134

[5] Vasudevan M, Arumugam R, Venkatesan G and Paramasivam S 2003 Estimation of stator resistance of induction motor for direct torque control, *WSEAS Transactions on Circuits and Systems* **2**(2) 327-332

[6] Mittal R, Sandhu K S and Jain D K 2009 Controlled Operation of Variable Speed Driven Permanent Magnet Synchronous Generator for Wind Energy Conversion Systems, *WSEAS Transactions on Systems* **8**(2) 189-199

[7] Singh B 1995 Induction generator-a prospective, *Electric Machines and Power Systems* **23** 163-177
[8] Zinger D S and Muljadi E 1997 Annualized Wind Energy Improvement Using Variable Speeds, *IEEE Transactions on Industry Applications* **33**(6) 1444-1447
[9] Bansal R C, Bhatti T S and Kothari D P 2003 A bibliographical survey on induction generators for application of non-conventional energy systems, *IEEE Transaction on Energy Conversion* **18**(3) 433-439
[10] Sandhu K S and Dheeraj J 2008 A Simple Approach to Estimate the Steady-State Performance of Self-Excited Induction Generator, *WSEAS Transactions on Systems and Control* **3**(3) 208-218
[11] Sharma S and Sandhu K S 2008 Role of Reactive Power Source on Power Quality of Three-phase Self-Excited Induction Generator, *WSEAS Transactions on Power Systems* 3(4) 216-225
[12] Panoiu M, Panoiu C and Sora I 2009 Modeling the electromagnetic pollution of the electric arc furnaces, *Revue Roumaine des Sciences Techniques–serie electrotechnique et energetique* **54**(2) 165-174
[13] Panoiu M, Panoiu C, Osaci M and Muscalagiu I 2007 Simulation Results for Modeling the AC Electric Arc as Nonlinear Element using PSCAD EMTDC, *WSEAS Transactions on circuits and systems* **6**(1) 149 – 156
[14] Tutelea L N, Deaconu S I, Boldea I and Popa G N 2014 Dual rotor single-stator axial air-gap PMSM motor/generator drive for high torque vehicle applications, *IOP Conference Series: Material Science and Engineering* **57** 012009
[15] Baciu I, Cuntan C, Panoiu M and Abrudean C 2013 *Highlighting the existing harmonics on locomotives with DC motors*, Recent Researches in Electric Power and Energy Systems, Chania, Crete Island, Greece, August 27-29, pp 120-125
[16] Popa G N, Abrudean C, Deaconu S I, Popa I and Vaida V 2009 *A Case Study of ESP Electrical Characteristics from a Thermal Power Station*, 44th Annual Meeting of the IEEE-Industry-Applications-Society, Houston, Texas, USA, October 4-8, pp 515-520
[17] Boldea I and Nassar S A 2001 *The Induction Machine Handbook*, CRC Press Boca Raton FL, USA
[18] Madescu G, Boldea I and Miller T J E 1996 *An Analytical Iterative Model (AIM) for Induction Motor Design*, Industry Applications Conference IAS'96, San Diego, USA, October 6-10, pp 566 – 573
[19] Tutelea L N, Deaconu S I, Budişan N and Boldea I 2013 *Double Staor Winding Induction Generator for Wind and Hydro Applications: 2D-FEM Analysis and Optimal Design*, 15th European Conference on Power Electronics and Application, EPE ECCE Europe, Lille, France, September 3-5, pp 1-10
[20] Tutelea L N, Deaconu S I, Boldea I and Budişan N 2013 *Design, control and 2D-FEM validation for an Double Stator Winding Induction Generator*, IECON 2013, Vienna, Austria, November 10-13, pp 2732-2737

IOP Conf. Series: Materials Science and Engineering **85** (2015) 012011 doi:10.1088/1757-899X/85/1/012011

Study on sources of charging lead acid batteries

C M Diniş[1], G N Popa[1] and A Iagăr[1]

[1] Politehnica University of Timisoara, Department of Electrical Engineering and Industrial Informatics, 5 Revolution Street, 331128 Hunedoara, Romania

E-mail: corina.dinis@fih.upt.ro

Abstract. The paper presents the general characteristics of lead acid batteries and two charging methods of these batteries. For charging of lead batteries was used an intelligent power source K 8012 (from Velleman). The power source allows fixing the level of the battery voltage and battery capacity. The intelligent power source uses the joint method (at constant current and, then, at constant voltage) and warning that indicates different situations in the charging process. Other method of charging presented in the paper is at constant voltage using a stabilized power source. In the paper experimental measurements were carried out using data acquisition card SER 10 BIT (from Conrad) for charging/ discharging of a lead acid battery 12V/9Ah (using an intelligent power source) and charging of another high capacity lead acid battery 12V/47Ah/390 A (using a stabilized power source). At the discharging of the lead acid batteries it were used automotive lamps as electric loads.

1. Introduction

Lead is usually associated with other minerals Zinc, Copper, Silver. Natural properties such as mass, malleability, low melting point, corrosion resistance, electrical properties and prolonged resistance, making it indispensable in world industry. It has the highest rate of recycling of all materials of the world [1], [2].

A battery stores electrical energy for later use. It produces power from a chemical reaction produced between two different materials (positive and negative plates) that are placed into the electrolyte. Electrodes are immersed in a water solution of sulphuric acid. Through the so-called formatting (the procedures can differ from production to another), which consists mainly of the battery power supply, the electrodes transform, positive plates get brown coloured and covered with PbO_2, and negative plates get grey coloured and covered with blackish lead [3], [4].

In a normal lead acid battery approximate voltage is 2 V/cell; so a total voltage on a battery is 12 V (usually, 6 cells). The battery is faster discharge is an electric load between the positive and negative terminals. The electrodes are made of lead, allied with 5 ... 10% antimony which limits them to the anodic polarization corrosion. Constructive, they are grating from grill that have a larger surface area. The terminal + is made of lead with a paste of lead dioxide; the electrode can be so symbolized as PbO_2/Pb. The separator is a grill of PVC, hard rubber or cellulosic material impregnated with meant to prevent internal short-circuit between electrodes. Sulphuric acid electrolyte is 30 ... 40% (ρ = 1.21... the 1.28 g/cm^3), depending on the battery type. Usually, the casing is made of plastic and for large batteries rubber resistant.

IOP Conf. Series: Materials Science and Engineering **85** (2015) 012011 doi:10.1088/1757-899X/85/1/012011

The characteristics of the battery are: E_{cell} = 2.14 V; $U_{initially}$ = 2.1...1.9 V; U_{final} = 1.6...1.75 V; W_g = 30...40 Wh/kg, number of loading-unloading cycles: 1000...8000, the life time is between 5 to 7 years, h_E = 75...85%. Usually the batteries have 3 or 6 series cells, with nominal voltage of 6V or 12V, used for cars [5-7].

During unloading shall consume sulphuric acid and produces water electrolyte solution, so it dilutes, so the concentration (c_{H2SO4}) of the electrolyte (or its density) is a measure of the degree of charge of the battery. It is recommended to not work with batteries for which $c_{H2SO4} < 18\%$. The reaction of the electrolysis of $PbSO_4$, held to 2.2 V will be held prior to the electrolysis of water. When the electrolysis of lead sulphate is finished, start the electrolysis of water and battery "boils" i.e. hydrogen and oxygen gas on the electrodes and barbotage through the electrolyte liquid. Increase the life time of a lead battery is done through the right choice of materials and by correct exploiting [4], [7].

Although it presents some inconveniences related to especially great mass and fluid-electrolyte, lead acid battery is widely used primarily due to large voltage debited, the reversible reactions and the low secondary reactions. Continuously improved, studied and used as a reference, the battery will lead in the 3rd millennium will be an important source of energy.

2. Charging lead acid accumulators

There are the following methods for loading accumulators lead: loading with constant current, constant voltage, and mixed. Charging with constant current is achieved while maintaining the current at a fixed value throughout the loading process. The method has the advantage that it requires less time, but conducts to intense gas release in the final phase, when it starts the electrolysis of water. Charging with constant voltage is achieved while maintaining the voltage at a fixed value 2.3 V/cell. In this case the current shocks may occur at the beginning of the process, but avoid electrolyte boiling because, in the final phase, the charging current is lower. Mixed load consists of the combination of the two methods in order to eliminate their disadvantages, begins charging with constant current, and when the emission of gas starts it continues with constant voltage.

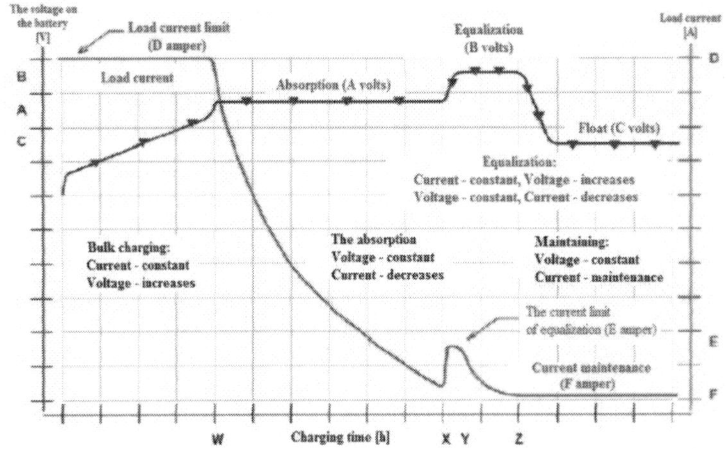

Figure 1. Loading characteristics of lead acid accumulators [8].

The charging lead acid batteries uses an algorithm based on voltage measurement algorithm, similar to the charging of lithium-ion batteries. Loading time is between 12-16 hours, or up to 36-48 hours in the case of very large capacity batteries. Using the large load currents, the load can be reduced up to 10 hours or less.

General, loading algorithm has four stages and is presented in Figure 1 [8]. Graph axis indicators represent general values for time and for charging current and voltage of the load. On the axis of the voltage, A represents the voltage value maintained at a constant value during the absorption stage; C represents; B represents the value of the voltage maintained constant charger during the equalization stage; C represents the value of the constant voltage that keeps the charger during the maintenance stage (Float).

3. Intelligent charger for lead acid batteries

For charging of lead acid batteries with higher capacity can be used intelligence charging source K 8012 from Velleman (Figure 2). The intelligent charging source allows fixing the level of the battery voltage (6 V or 12 V) and battery capacity (less than 4 Ah or greater than 4 Ah). The source K 8012 has short-circuit protection, when connecting the battery terminals wrong and has four warning lights (LEDs): one that indicate the absence of battery, other that indicates the wrong connection of the battery to the source, a LED that indicates the battery charging process and one which indicates the full charge of the battery [9].

Figure 2. Intelligent charging source K 8012 – overview.

4. Experimental measurements with intelligent charging source K 8012 and lead acid battery 12 V, 9 Ah

For testing it used a data acquisition card (DAC) with 8 uni-polar analogue channels or 4 analogue differential channels with 12 bits (Conrad Electronic). The DAC connects to the RS 232 serial interface (with 9 poles) through the COM_1 or COM_2 from a personal computer (PC). It can purchase up to 8 analogue uni-polar signals in the wide 0 ... 5 V. For a resolution of 12 bits (4096 steps) correspond a voltage step by 1.22 mV (5/4095). To acquisition the signals it is used a Basic program created for this module. The DAC can be directly supply from the PC (switch S_2 Int) or external to a 9 V D.C. power source (switch S_2 Ext; power supply between the ground (GND) and U_B extern, 9 V D.C.) [10].

Communication with the PC is made through the serial interface and is a duplex communication (while the PC configuration issues for the following measurements, simultaneously receiving the result of the previous measurement). The transmission is performed on 16 bit, 12 bit of them are used for measuring, and 4 bits are used to address the selected channel and the type of measurement (uni-polar or differential).

The software used is Ser10Bit (in Basic). Without this software module can not be put into operation. Communication is performed on hexadecimal addresses that are enabled by the INP and OUT Basic commands.

The measurements were used in the experimental setup from figure 3. It uses two simple sources of voltage type MPS-6003S, one of the sources is used for intelligent charging source K 8012 and the other power source is used to amplify voltage signals from the shunt resistance R_s (5 A/100 mV). The voltage on the R_s is applied an operational amplifier (from four available) type LM 324 configured in non-inverting summation. The resistors R_1 and R_2 were chosen so that 5 A current which would pass through the lead acid battery B results a 5 V voltage, after operational amplifier connected to channel CN_2 of DAC. The current measurement result (vertical) from graphs (Figures 4-7) is 1 A/div.

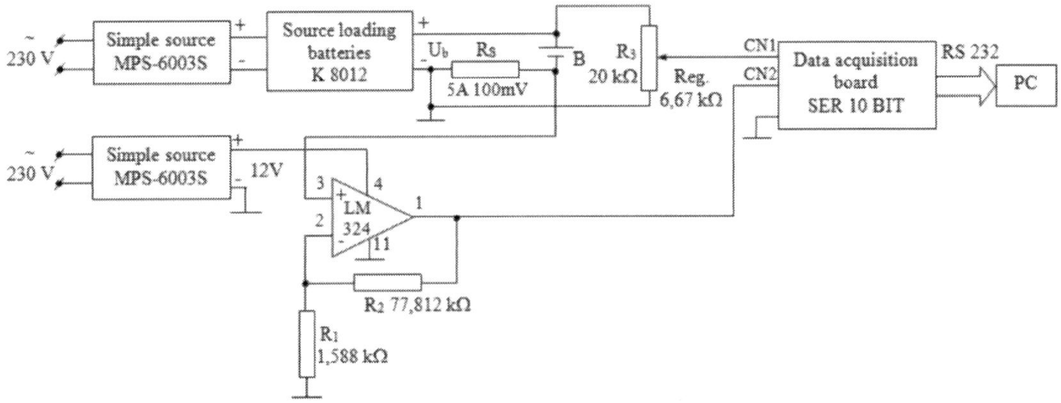

Figure 3. Experimental setup for charging lead acid battery with the source K 8012, 12 V, 9 Ah.

To measure the voltage on the battery has been used R_3 potentiometer, and the cursor was positioned at one-third of the value (from the GND). For the measurement from Figures 4-7 the voltage has 3 V/div. The cursor has been applied directly on channel 1 (CN_1) of the DAC. In this diagram setup of charging lead acid battery with intelligent source K 8012, - terminals are not common. The - terminal of the battery is not the same with - terminals of the power sources MPS-6003S. In fiure 3: CN_1 – voltage measurement channel (3V/div); CN_2 – current measurement channel (1A/div); B – lead acid battery 12 V, 9 Ah.

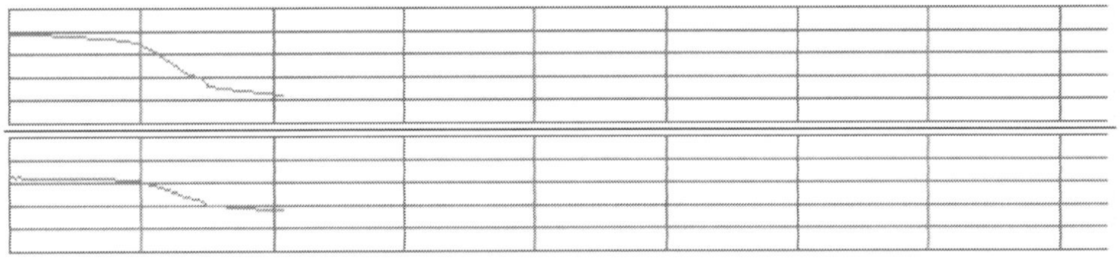

Figure 4. The voltage and current through the battery B 12 V, 9 Ah, at discharge, when it connected a 12 V/45 W lamp: up-voltage 3 V/div, down-current 1 A/div, time 30 min/div.

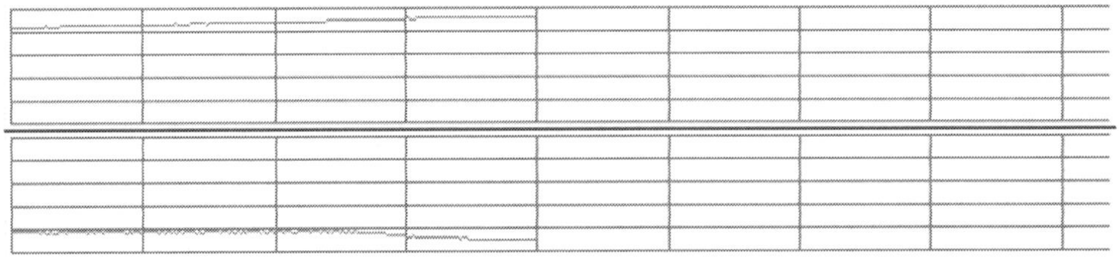

Figure 5. The voltage and current through the battery B 12 V, 9 Ah, charging through the intelligence charging source K 8012: up-voltage 3 V/div, down-current 1 A/div, time 1 hour/div.

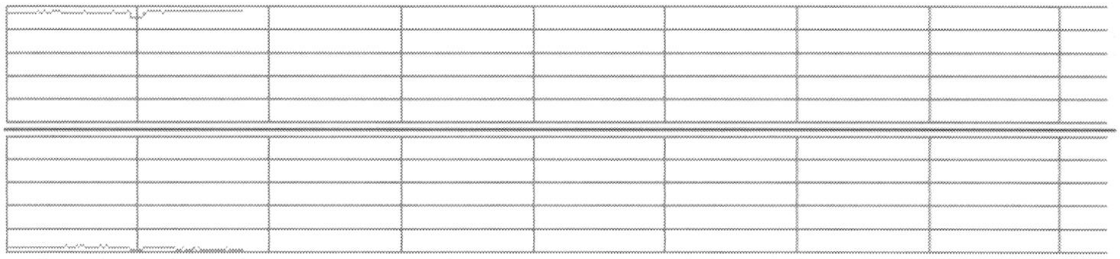

Figure 6. The voltage and current through the battery B 12 V, 9 Ah, at the beginning of charging through the intelligence charging source K 8012: up-voltage 3 V/div, down-current 1 A/div, time 1 hour/div.

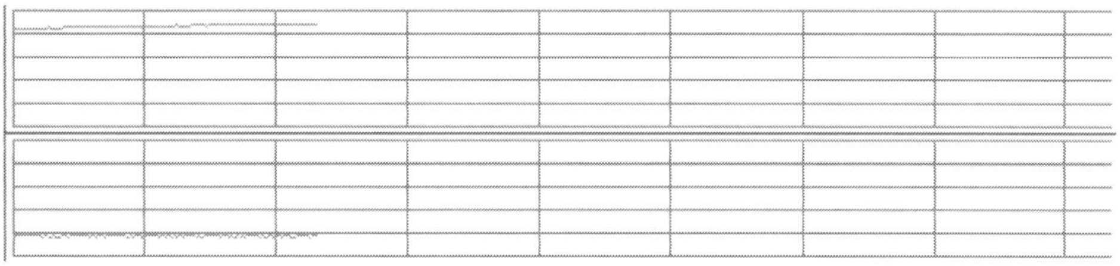

Figure 7. The voltage and current through the battery B 12 V, 9 Ah, for a short period of time, charging through the intelligence charging source K 8012: up-voltage 3 V/div, down-current 1 A/div, time 1 hour/div.

Figure 8. Overview from experimental setup with the lead acid battery (12 V, 9 Ah) charging from K 8012 source.

$$U_e = U_i\left(1 + \frac{R_2}{R_1}\right) \tag{1}$$

$$k_A = 1 + \frac{R_2}{R_1} \tag{2}$$

The intelligent power source K 8012 use the mixed charging: at current constant and voltage constant. At the charging of the battery 12 V, 9 Ah, the voltage increases slightly, and the current is maintained constant for a long period of time. Upon completion of charging the battery, the voltage has maximum value and the current is limited at almost 0.1 A. At discharging the battery through a lamp, the voltage on the lamp decreases faster than the current. Voltage and current suddenly decrease, after a period of time.

5. Experimental measurements with power source and lead acid battery 12 V, 47 Ah, 390 A

The measurements were used with experimental setup from figure 9. The method used for charging is constant voltage. It uses a double voltage source AX 3003D-3 and a simple voltage source type MPS-6003S. Double voltage source is used for charging the battery B, and the other power source is used to amplify voltage signal from the shunt resistance R_s (100 mV/5 A). The voltage on the R_s is applied an operational amplifier (from four available) type LM 324 configured in non-inverting summation. The resistors R_1 and R_2 were chosen so that 5 A current which would pass through the lead acid battery B results a 5 V voltage, after operational amplifier connected to channel CN_2 of DAC. The current measurement result (vertical) from graphs (Figures 10-11) is 1 A/div. To measure the voltage on the battery has been used R_3 potentiometer, and the cursor was positioned at one-third of the value (from the GND); thus the measurement from figures 10-11, for voltage 3 V/div. The cursor has been applied directly on channel 1 (CN_1) of the DAC. In this diagram setup of charging lead acid battery with power sources AX 3003D-3 and MPS 6003S, - terminals are common. The - terminal of the battery is not the same with - terminals of the power sources (AX 3003D-3 and MPS 6003S).

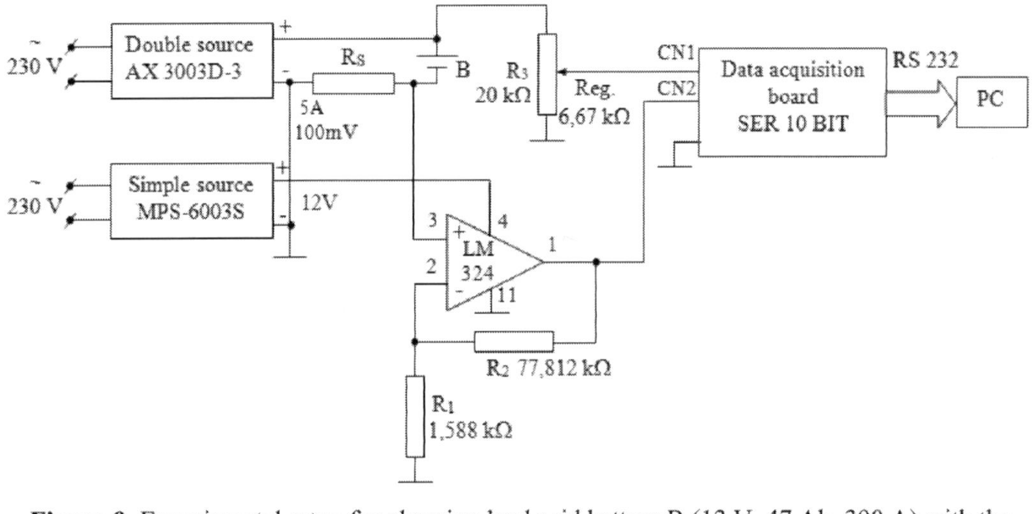

Figure 9. Experimental setup for charging lead acid battery B (12 V, 47 Ah, 390 A) with the power source AX 3003D-3.

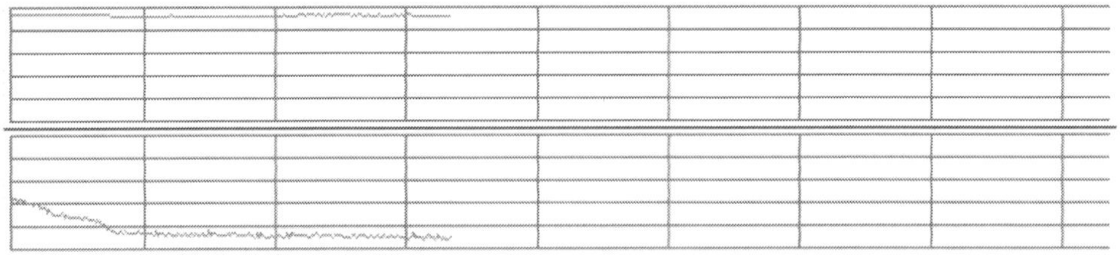

Figure 10. The voltage and current through a battery (12 V/47 Ah), discharging with constant voltage from a double source stabilized type AX 3003D-3, 13 V/6 A: up-voltage 3 V/div, down-current 1 A/div, time 1 hour/div.

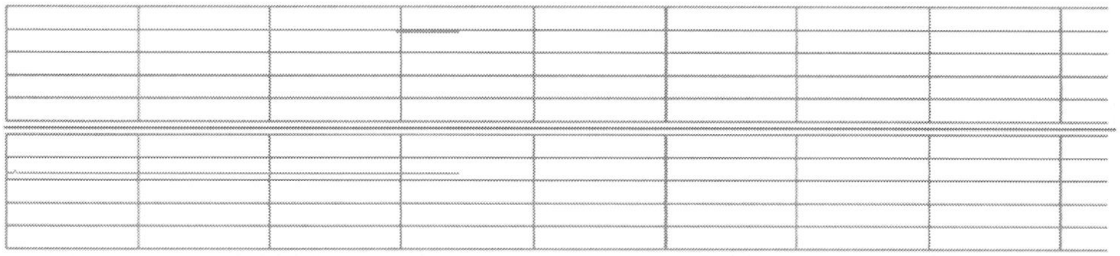

Figure 11. The voltage and current through a battery (12 V/47 Ah), charging with constant voltage from a double source stabilized type AX 3003D-3, 13 V/6 A: up-voltage 3 V/div, down-current 1 A/div, time 1 hour/div.

Figure 12. Overview from experimental setup with the lead acid battery (12 V, 47 Ah, 390 A) charging from power source AX 3003D-3.

At this experiment was used a large lead acid battery (12 V, 47 Ah) that was charging at the constant voltage. In Fig.10 when the battery is discharging through connecting two lamps of 12 V/45 W. It was found that after three and a half hours, a voltage drop occurs, due to the large capacity of the battery and the current has decreased due to the increase of the internal resistance of the battery (the battery is old). At the charging of the battery, the voltage grows more slowly, and the current is maintained constant for a long period of time.

6. Conclusions

The greatest enemies of lead acid batteries are over-charging and over-heating. For charging of lead acid batteries it can use three methods of charging: at constant current, constant voltage, or mixed (which is the best option). In this paper, were used two methods: mixed charging by using an intelligent charging source (source K 8012) and the constant voltage (stabilized power supply). Intelligence source K 8012 ensures joint charging: at constant current (a length of time), and then the constant voltage. Power supply K 8012 has protection for current limiting through battery. This allows for setting the level of the battery voltage and battery capacity. The source K 8012 has short-circuit protection, at the wrong connections of the battery terminals at the source, warning when the battery is missing and when is full charge. By using an intelligent charging source does not take place the boiling, over-charging and over-heating of the lead acid battery, because the current limiting when it is considered that the battery has charged completely.

For lead acid batteries charging can be used a stabilized source with constant voltage. The battery must be supervised constantly, not to overcharge and boil. Some sources stabilized voltage sources have over-current protections and allow this, but it must be properly adjusted to limit the inrush current of the battery.

Although dedicated sources for charging lead acid batteries are more expensive, are recommended for charging them because in this way it ensures proper charging and a long life.

References

[1] Buchmann I 2011 *Batteries in a Portable World - A Handbook on Rechargeable Batteries for Non-Engineers*, Third Edition, Cadex Electronics Inc., Richmond, Canada

[2] Hariprakash B, Gaffoor S A and Shukla A K 2009 Lead-Acid Batteries for Partial-State-of-Charge Applications, *Journal of Power Sources* **191**(1) 149–153

[3] Chen H Y, Li A J and Finlow D E 2009 The Lead and Lead-Acid Battery Industries During 2002 and 2007 in China, *Journal of Power Sources* **191**(1) 22–27

[4] Wan X F, Wu J P and Hu H L 2009 *The Smart Battery Management System*, International Conference on Test and Measurement, ICTM '09, Hong Kong, China, December 5-6, pp 29-32

[5] Linden D 2005 *Batteries and Fuel Cells, Standard Handbook of Electronic Engineering*, Fifth Edition, McGraw-Hill Book Co., New York, USA

[6] Wong Y S, Hurley W G and Wolfle W H 2008 Charge Regimes for Valve-Regulated Lead-Acid Batteries: Performance Overview Inclusive of Temperature Compensation, *Journal of Power Sources* **183**(2) 783–791

[7] *** 1989 *Battery Application Manual*, Gates Energy Products Inc., Gainesville, FL, USA

[8] Cadar D V 2011 *Contributions to the battery management systems*, Technical University Cluj-Napoca, Romania, Doctoral Thesis

[9] *** 2008 *K 8012 Lead Acid Battery Charger*, User's manual, Velleman, Belgium

[10] *** 1996 *Ser 10 Bit Data Acquisition Card*, User's manual, Conrad, Germany

CFD numerical simulation of air natural convection over a heated cylindrical surface

M Flori[1] and L Vîlceanu[1]

[1] Politehnica University of Timisoara, Engineering and Management Department, 5 Revolution Street, 331128 Hunedoara, Romania

E-mail: flori.mihaela@upt.ro

Abstract. In this study a CFD numerical simulation is used to describe the fluid flow and heat transfer in air surrounding a heated horizontal cylinder. The model is created in 2D space dimension involving a finite element solver of Navier-Stokes equations.
As natural convection phenomenon is induced by a variable fluid density field with temperature rising, the Boussinesq approximation was coupled to the model.

1. Introduction

In industrial applications it is common that the density of a process fluid varies. These variations can have a number of different sources but the most common one is the presence of an inhomogeneous temperature field [1].

The representative problem is of a fluid external flow which is in contact with a warmer surface (in many cases, horizontal circular pipes). Initially, the heat is transmitted by conduction from the cylinder surface to surrounding fluid with formation of a thermal boundary layer around the cylinder. While the fluid begins to increase in temperature, its density decreases. The buoyant force causes the less dense fluid to rise, promoting fluid motion. With time the velocity around the cylinder increase and the heat is transmitted not only by conduction but also by convection [2].

According to Newton's Law of Cooling the rate of heat transfer between a warmer cylindrical surface and surrounding fluid, depends on the heat transfer coefficient $h[W/(m^2 K)]$, cylinder diameter d[m], surface and ambient temperatures (T_s[°C] and T_∞[°C]) by equation [1]:

$$q = h \cdot \pi \cdot d \cdot (T_s - T_\infty)\,[W/m] \tag{1}$$

This study aims determining the temperature and velocity fields around a heated cylinder placed horizontally in air environment, due to natural convection. The cylinder surface has a constant temperature bigger than that of surrounding air. Therefore, a multiphysics model involving a finite element solver of Navier-Stokes equations was generated. So, both heat transfer and air flow phenomena around the horizontal cylinder are taken into account with the following assumptions: flow around the cylinder is laminar and driven in two-dimensions; initial, the fluid has a uniform temperature and is at rest, i.e. the velocity field is zero; all fluid physical properties, except density, have no significant changes with temperature.

Also, knowing the temperature field it can be determined the total normal heat flux coming from heated cylinder surface and further calculate the average heat transfer coefficient between heated horizontal cylinder and surrounding air.

IOP Conf. Series: Materials Science and Engineering **85** (2015) 012012 doi:10.1088/1757-899X/85/1/012012

2. Multiphysics model description

The CFD analysis was done by coupling the following physics involving finite element solvers of commercial Comsol Multphysics software (v. 4.2): Single-Phase Laminar Fluid Flow and Heat Transfer in Fluids. The analysis type was set as Stationary.

Concerning space dimension, in CFD analysis involving cylinders, by neglecting any effect from the geometry domain walls, the solution is constant in the longitudinal direction of cylinder [1]. So, the model was created in 2D enabling solution for half geometry along a symmetric vertical axis to reduce the modeling time.

Modeled geometry is a horizontal cylinder (having a radius of 0.06 m and constant temperature T_s of 323.15 K) placed in room-temperature air environment (T_∞=293.15 K). The air domain is considered rectangular, having a width of 0.2 m and a height of 0.8 m. As study concern natural convection around the isothermal cylinder surface, it is represented as a semicircular cut in the rectangle left wall, its center being positioned in origin (Figure 1a). As may be seen in Figure 1a, model domain boundaries were numbered from 1 to 7 as following: boundaries 1 and 2 overlap symmetry vertical axis, boundaries 3, 4 and 5 delimit the air domain, while boundaries 6 and 7 correspond to cylinder half-surface.

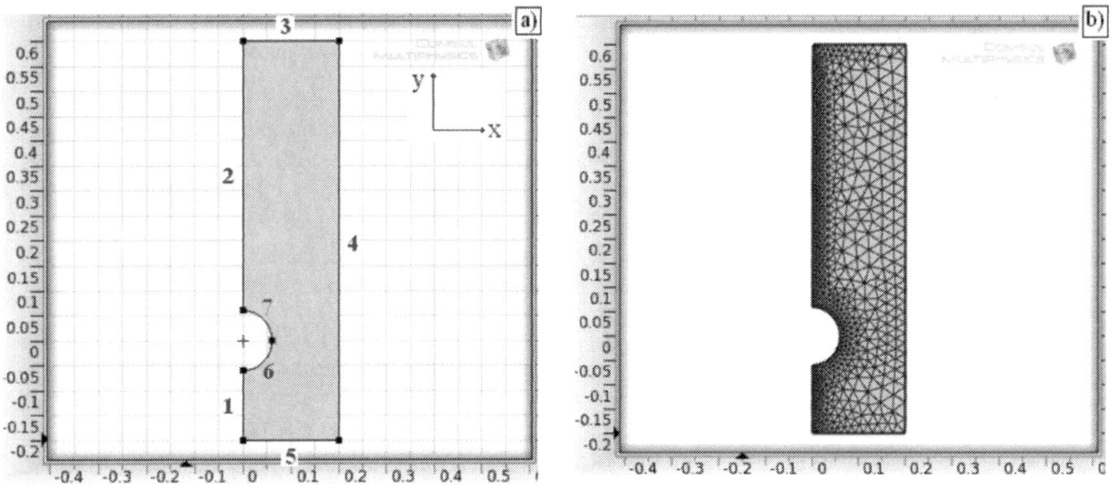

Figure 1. 2D Geometry (a) and mesh distribution (b) of model domain; dimension scale in [m].

By employing a triangular mesh with extremely fine feature at domain left boundaries and fine feature for the rest of domain, the modeled geometry was discretized in 1503 elements (Figure 1b). The refined mesh along the symmetric axis and cylinder surface was added because at these boundaries the changes in fluid flow due to natural convection are expected to be more developed.

Further, a list of initial parameters were defined in the model, i.e. air thermo-physical properties evaluated at the mean fluid temperature $(T_s+T_\infty)/2$ of 308.15 K: density $\rho_\infty = 1.139\ \mathrm{kg/m^3}$, dynamic viscosity $\mu = 18.89e^{-6}\ \mathrm{Pa \cdot s}$, thermal conductivity $k = 0.02684\mathrm{W/(m \cdot K)}$ and heat capacity at constant pressure $C_p = 1007\ \mathrm{J/(kg \cdot K)}$ [2].

Boussinesq approximation was also coupled to the model to represent air density changes induced by temperature field, and set according to the following equation [3-5]:

$$\rho = \rho_\infty[1 - \beta(T_s - T_\infty)]\ \ [\mathrm{kg/m^3}] \tag{2}$$

where $\beta = 1/T_\infty\,[\mathrm{K^{-1}}]$ is the coefficient of thermal expansion.

The Boussinesq approximation assumes that variations in density have no effect on the flow field around horizontal cylinder, except that they give rise to buoyant forces [1].

2.1. Laminar Fluid Flow physics settings

Boundary conditions specified in the model were: *Symmetry* for boundaries 1 and 2, *Open Boundary/Normal Stress* for boundary 3 and *Wall, No-Slip* for boundaries 4 to 7.

The *Symmetry* condition means that an identical process takes place to the left, outside the domain. The *Open Boundary* condition describes boundaries that are open to large volumes of fluid which can both enter and leave the domain. The *Wall* feature includes a set of boundary conditions describing the fluid flow condition at a wall. The *No Slip* condition assumes that the velocity is zero (u=0), i.e. the fluid at the wall is not moving [1].

In order to express the gravity influence on the fluid flow around the cylinder, a volume force (buoyant force) was also defined on y-direction, with the following expression [5]:

$$F_y = g \cdot (\rho_\infty - \rho)\ [N/m^3] \tag{3}$$

where $g = 9.81 m/s^2$ is gravity acceleration.

2.2. Heat Transfer physics settings

As stated above, in the model the cylinder surface (boundaries 6 and 7) is kept at constant temperature $T_s = 323.15K$, while the initial temperature of air domain is room temperature $T_\infty = 293.15K$. As air warms up, it rises upwards to upper boundary (no. 3) therefore it was set as *Open Boundary/Normal Stress*. Also, thermal radiation from cylinder surface to ambient air is negligible compared to convection, and so is neglected in the model.

2.3. Model governing equations

The Navier-Stokes equations (4 and 5) describing fluid dynamics and the heat transfer in the fluid equation (6) are coupled through the bidirectional multiphysics model, with the following expressions:

- conservation of momentum equation:

$$\rho(u \cdot \nabla)u = \nabla\left[-pI + \mu\left(\nabla u + (\nabla u)^T\right) - \frac{2}{3}\mu(\nabla \cdot u)I\right] + F \tag{4}$$

- continuity equation, which represents the conservation of mass:

$$\nabla \cdot (\rho u) = 0 \tag{5}$$

- conservation of energy equation, formulated in terms of temperature, excepting the heat source terms:

$$\rho C_p u \cdot \nabla T = \nabla \cdot (k \nabla T) \tag{6}$$

where besides air physical properties, u[m/s] is velocity, p[Pa] is pressure, T[K] is absolute temperature and F[N/m^3] is volume force (body force).

The momentum and energy equations are inherently coupled since the temperature appears in the momentum equation and the velocity in the energy equation [1].

3. Results and discussion

CFD analysis results were obtained in terms of temperature (in [K]) and velocity (in [m/s]) distributions in the fluid around the heated horizontal cylinder due to natural convection (Figures 2 and 3).

Temperature field develops around the cylinder (see magnification in Figure 2) and also above it along the symmetry line which passes through the cylinder center. At small temperature differences between warm surface and surrounding fluid, a laminar flow is expected [6].

Also, the velocity field develops in lateral direction, but has bigger intensity above cylinder surface where maximum velocity values may be identified in vertical direction along the symmetry line (surface colored in red in Figure 3).

Moreover numerical values of temperature and velocity variations with y-coordinate of computed domain were obtained and presented in Figures 4.a) and b). In these plots, the y-coordinate axis (horizontal axis) corresponds to the vertical symmetry line of computed domain where x-coordinate is

0. So, in y-direction the computed domain was plotted from $y = -0.2$ m to $y = 0.6$ m, resulting domain height of 0.8 m.

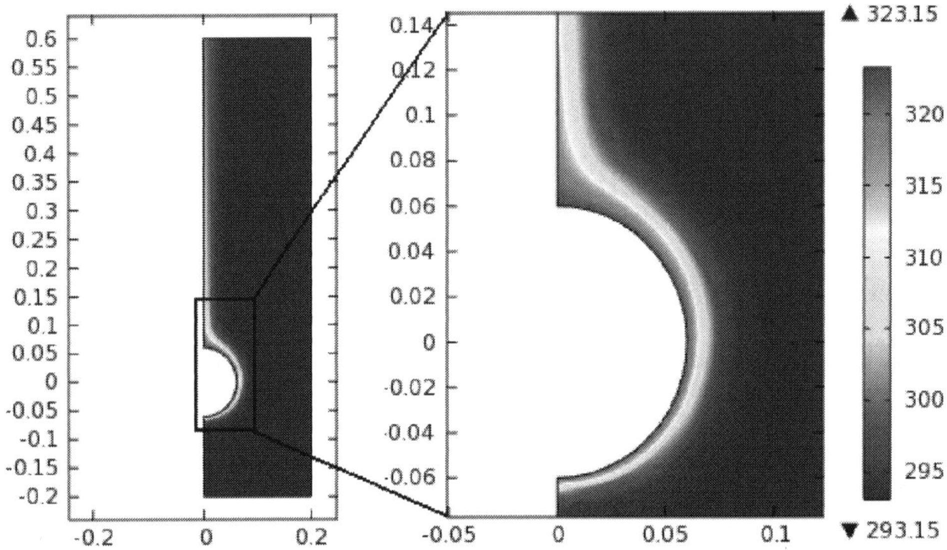

Figure 2. Computed temperature field; color scale in [K].

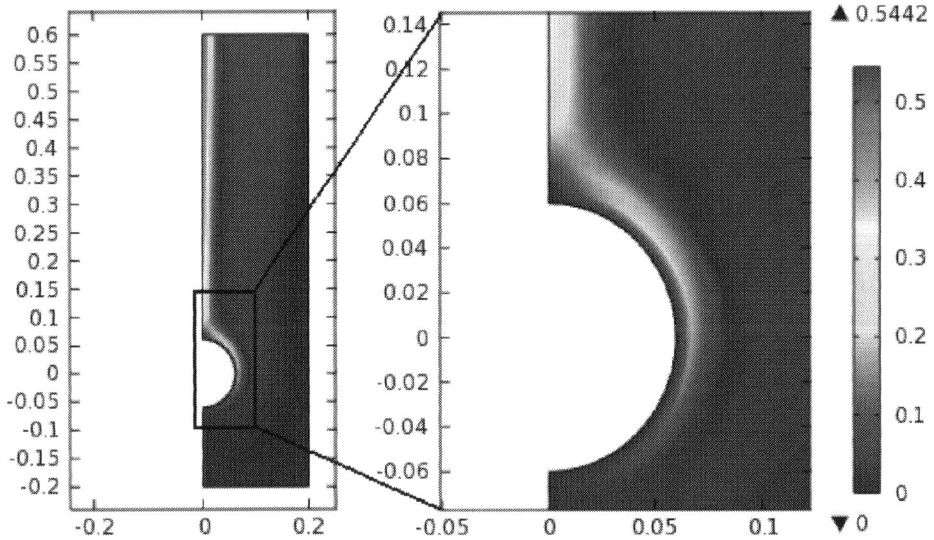

Figure 3. Computed velocity field; color scale in [m/s].

Analyzing variation from Figure 4.a one may observe that below the heated cylinder surface ($y < -0.06$ m) the air warms up only on a small distance of about 0.02 m. Above the heated cylinder surface air warms up constantly, having maximum value near cylinder surface ($y = 0.06$ m) and about 302 K at height $y = 0.6$ m.

Analyzing air velocity distribution from Figure 4.b, one may observe that below the cylinder surface the velocity is zero, indicating that no convection currents are formed here. Above cylinder surface air velocity increase rapidly attending a maximum value of 0.54 m/s at height $y = 0.6$ m.

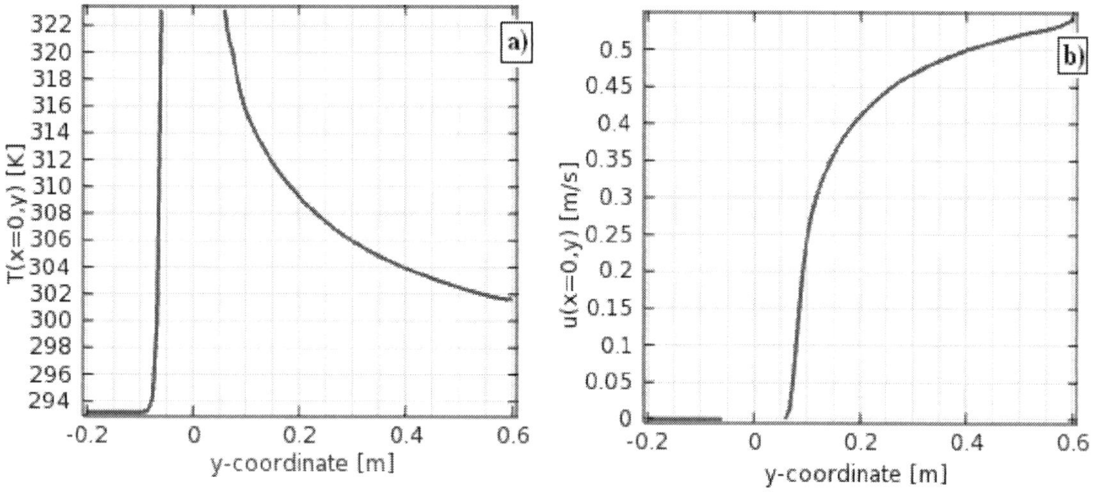

Figure 4. Variation of air temperature (a) and velocity (b) with y-coordinate.

Moreover, the computed temperature field generates a total normal heat flux coming from heated cylinder surface of $q = 57.242\ W/m$. With this value, average heat transfer coefficient between heated horizontal cylinder and surrounding air is calculated with relation (1) and obtained value is $h=5.06\ W/m^2K$.

4. Conclusions

A finite element analysis of free convection around a heated horizontal surface was done to determine temperature and velocity distributions in air surrounding environment. Besides initial and boundary conditions, the influence of gravity on air flow around the cylinder was expressed in terms of volume force, while the Boussinesq approximation was used to represent the variable density filed with temperature rising.

By computing the temperature field around horizontal cylinder due to natural convection, the total normal heat flux was determined and average heat transfer coefficient between heated horizontal cylinder and surrounding air was calculated.

References

[1] ****COMSOL Multiphysics User's Guide*, ©Copyright COMSOL AB., 2010
[2] Linenhard J H IV and Linenhard J H V 2008 *A heat transfer textbook. Third edition*, Phlogiston Press, Cambridge MA
[3] Sairamu M, Nirmalkar N and Chhabra R P 2013 Natural convection from a circular cylinder in confined Bingham plastic fluids, *International Journal of Heat and Mass Transfer* **60** 567–581
[4] Chandra A and Chhabra R P 2012 Laminar free convection from a horizontal semi-circular cylinder to power-law fluids, *International Journal of Heat and Mass Transfer* **55**(11–12) 2934-2944
[5] Pasut W and De Carli M 2012 Evaluation of various CFD modelling strategies in predicting airflow and temperature in a naturally ventilated double skin façade, *Applied Thermal Engineering* **37** 267–274
[6] Atayılmaz Ş Ö and Teke I 2009 Experimental and numerical study of the natural convection from a heated horizontal cylinder, *International Communications in Heat and Mass Transfer* **36** 731-738

Analysis the electrical parameters of a high-frequency coreless induction furnace

A Iagăr[1], G N Popa[1] and C M Diniş[1]

[1] Politehnica University of Timisoara, Department of Electrical Engineering and Industrial Informatics, 5 Revolution Street, 331128 Hunedoara, Romania

E-mail: angela.iagar@fih.upt.ro

Abstract. The paper analyzes the most important electrical parameters of a coreless induction furnace, having a capacity of 7 kg molten iron (i.e. 2.5 kg aluminum). The furnace is supplied by a high-frequency (HF) static converter ITS2 12K20, with the output frequency 5...12 kHz, 600...1200 Vac rated HF voltage, and 20 kW rated HF power. Monitoring of electrical parameters was done for an aluminum charge, using a power quality analyser CA8334. The measurement results showed that induction furnace operation causes unbalance and harmonics in three-phase currents absorbed from the distribution network. Harmonics in the line currents are caused mainly by static converter, and to a lesser extent by furnace load and interaction of eddy currents induced in the charge and the magnetic field of the inductor. To reduce the negative impact of the current harmonics on the distribution network is necessary the design and achievement of electrical filters for odd-order harmonics (5th, 7th, 11th, 13th, 17th, 19th, 23th, 25th).

1. Introduction

The Induction melting is currently the fastest electric technology in metal production [1], [2]. Medium and high-frequency coreless induction furnaces are suitable for melting, re-melting and alloying, due to the advantages related to low installation cost, low operating cost, maximum alloy flexibility (can be emptied very quickly), good heating efficiency, precise temperature control. Induction melting involves automatic stirring, resulting increases of homogeneity and quality of metal alloy. These furnaces are used for elaboration of low-carbon high alloyed steel, alloy cast iron, hard iron, bronze, nickel and precious metals [1-8].

New installations of medium and high-frequency induction melting are characterized by performant power supplies, improved furnace refractory linings, heat recovery, and overall system control [1], [2], [3].

Electronic power supplies control current and frequency for efficient induction melting, providing high-power density in the furnace charge [1].

Static converters have the following advantages: very good yield; variable and auto-adjust frequency during the melting process (eliminate the need to modify the capacitors bank for power factor correction); are compact and easy to install; low wear; maintenance is minimal due to the absence of moving parts.

Two main types of static converters are used in medium and high-frequency induction melting installations: voltage-fed power supply (with series-resonant furnace circuit) and current-fed power supply (with parallel-resonant furnace circuit) [2].

Voltage-fed power supply takes advantage of switching technology that is capable of handling high currents. Current-fed power supply takes advantage of rugged and economical switching-device technology, but has less control over the furnace current than voltage-fed power supply [2], [3], [6].

Beside of these advantages, both current and voltage-fed supplies generate harmonics and interharmonics [3], [4], [5].

Moreover, current-fed power supply generates voltage notching, that can cause tripping of other power supplies and DC drives [2].

Induction furnaces, as nonlinear loads, can produce significant distortions that affect the correct functioning of other neighboring loads, resulting interferences on the control systems, heating of rotating machines and capacitors banks, increasing electrical losses [1], [7], [9].

Further are analyzed the most important electrical parameters of a high-frequency coreless induction furnace, having a capacity of 7 kg molten iron (i.e. 2.5 kg aluminum), and power quality problems caused in distribution network.

2. Technical characteristics for electrical installation of analyzed induction furnace

The installation of furnace (figure 1) is supplied from the three phase low voltage (0.4 kV) through a general distribution board equipped with an automat circuit-breaker with thermal protection (Ir=50 A), electromagnetic protection (Iem=5It), differential protection (Id=300 mA) and overvoltage at 50 Hz protection (U>=260-280 V). The on/off switching of the installation is made with three static contactors (Solid state relay-SSR).

The furnace is supplied by a high-frequency (HF) static converter ITS2 12K20, with following electric characteristics: mains supply voltage 3x400 V, 50 Hz; maximum mains current 3x40 A; power factor 0.95; output frequency 5...12 kHz; rated HF voltage 600...1200 Vac; rated HF power 20 kW, control voltage 24 Vdc [10].

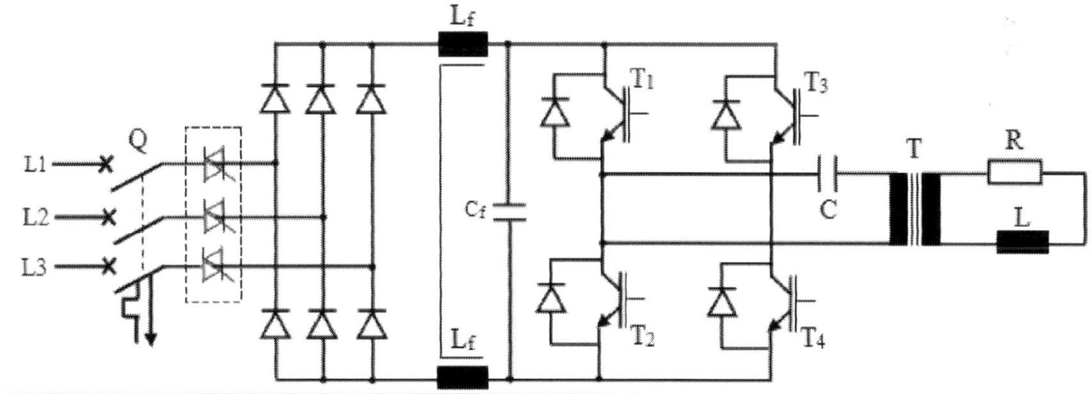

Distribution network 3x400V/ 50Hz and circuit-breaker	Solid state relay (SSR)	Diode bridge rectifier	DC link	IGBT inverter and load (load consists of the series-resonant circuit built from: isolation of transformer T, capacitor bank C and inductor with charge of furnace, having R, L equivalent parameters)

Figure 1. Electric scheme of the high-frequency coreless induction furnace.

Also, the electric scheme from Figure 1 includes a diode bridge rectifier DD B6U 85N with following electric characteristics [11]: repetitive peak reverse voltage 1600V; non-repetitive peak reverse voltage 1700V; RMS forward current (per chip) 60A; output current 85…104 A; surge forward current 550A; I^2t=1500 A^2s; forward voltage max. 1.44 V; reverse current 5 mA.

In DC link is placed a large DC choke (L_f=1.5 mH) and two parallel-connected DC capacitors (C_f=2x100 μF, 900 V) for energy storage and filtering. The DC choke serves as a current-limiting reactor against faults on the inverter side [2].

Static converter includes an inverter with IGBT transistors FF300R12KS4, having following electric characteristics [12]: collector-emitter voltage 1200 V; continuous DC collector current 370 A; repetitive peak collector current (tp=1 ms) 600 A; total power disipation 1.95 kW; gate-emitter peak voltage ± 20 V; collector-emitter saturation voltage 3.75 V.

For supplying the inductor is used a HF transformer (T) with primary voltage 1200 V, apparent power 480 kVA and frequency domain 5…12 kHz. In Figure 1, R, L represent the equivalent parameters of inductor-charge system.

3. Electrical parameters from induction furnace installation

Monitoring of electrical parameters from induction furnace installation was done for an aluminum charge, using a power quality analyser CA8334 [13]. Measurements were made on AC low voltage side, between breaker and static converter.

The following table presents informations about the heating stages of aluminum charge, power set on each stage, power absorbed by inductor, frequency, supply voltage of inductor and current absorbed on medium-frequency side.

Table 1. Heating stage of aluminum charge

Time period	$\%P_N$	$P_{inductor}$ [kW]	f [kHz]	U_{MF} [V]	I_{MF} [A]	Heating stage
9:08-9:39	5%	4,5	5	555	130	Crucible preheating
9:40-10:10	25%	14,5	6,3	1040	324	Heating of aluminum charge
10:11-10:23	100%	26,2	6,7	1300	435	Melting of aluminum charge
10:24	100%	27	7,1	1273	454	End of melting (furnace emptied)

Preheating of the crucible start at 8:30 with an active power set of $5\%P_N$. At 9:40 was set a new level of active power ($25\%P_N$) and at 10:11 the active power was set to maximum (100% P_N). At 10:24 the aluminum charge was ready for casting.

CA8334 gave an instantaneous image of the main characteristics of power quality for the analyzed furnace. The main parameters measured by the CA8334 analyser were: True RMS (TRMS) AC phase voltages, and TRMS AC line currents; peak voltage, and current; active, reactive, and apparent power per phase; harmonics for voltages and currents up to the 50th order [13].

This analyser provide numerous calculated values and processing functions in compliance with EMC standards in use (EN 50160, IEC 61000-4-15, IEC 61000-4-30, IEC 61000-4-7, IEC 61000-3-4). The values computed by the CA8334 are [13]: total harmonic distortion of voltages and currents, distortion factor of voltages and currents, K factor for current, voltage and current unbalance, power factor and displacement factor, extreme and average values for voltage and current, peak (crest) factors for current and voltage.

3.1. Crucible preheating and heating stage of aluminum charge

The following figures show the time variation of TRMS AC line currents, total harmonic distortion (THD), crest factor (CF) and unbalance of line currents, active, reactive, and apparent powers, power factor (PF) and displacement factor (DPF) recorded in crucible preheating and heating stage of aluminum charge (time period 9:08-10:10). By green is presented phase 1 (L_1), by yellow phase 2 (L_2) and by magenta phase 3 (L_3).

THD of line currents (Figure 2.b) was 138% at the crucible preheating stage, and decreased to 104% in the heating stage of aluminum charge.

During the entire period of aluminum charge heating, crest factor of currents (Acf) exceeds 1.41, characteristic value of sinusoidal waveforms (Figure 3.a). This indicates very high harmonic pollution of the line currents. In the crucible preheating stage (9:08-9:39), crest factor of currents had an average value of 2.9 (maximum 3.2). In the time interval 9:40-10:10 average value ridge drops to 2.38 (maximum 2.7), due to the reduction of current harmonics.

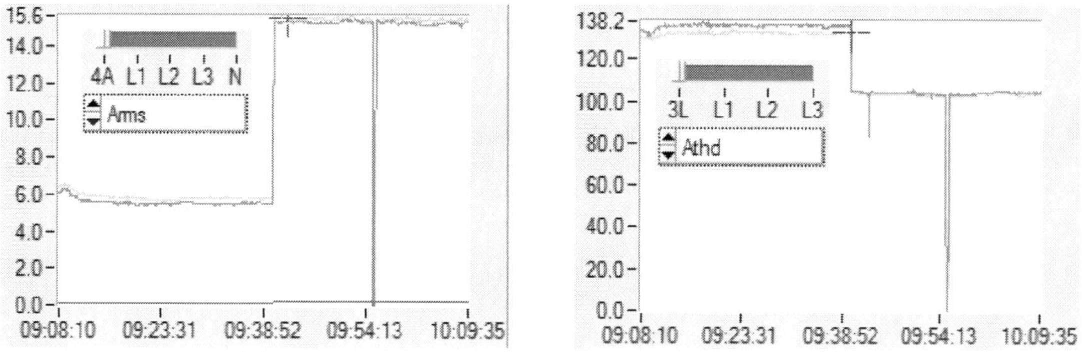

(a) (b)

Figure 2. TRMS [A] (a) and THD [%] (b) of line currents during the crucible preheating and heating stage of aluminum charge.

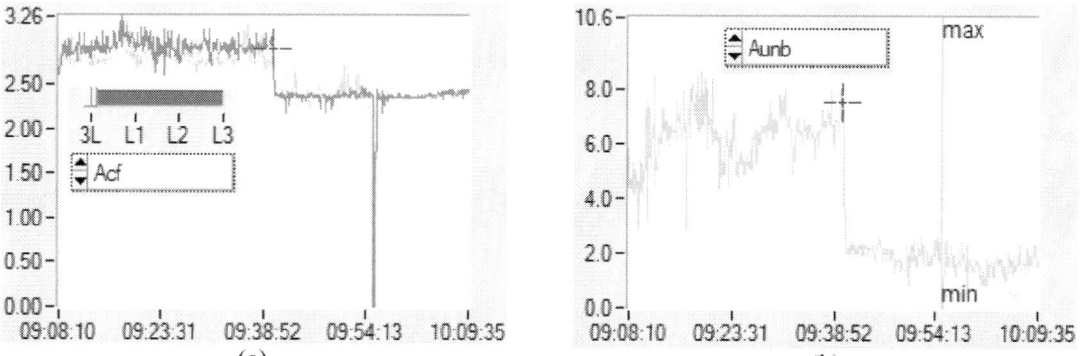

(a) (b)

Figure 3. Crest factor CF [-] (a) and unbalance [%] (b) of line currents during the crucible preheating and heating stage of aluminum charge.

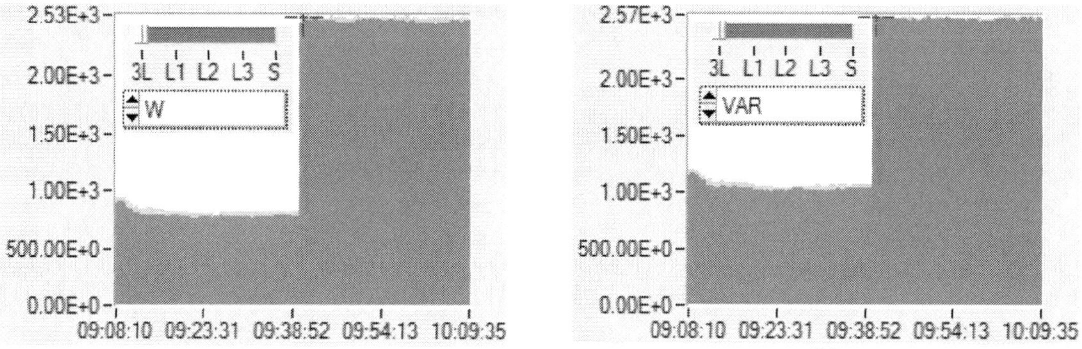

(a) (b)

Figure 4. Active power P [W] (a) and reactive power Q [VAR] (b) during the crucible preheating and heating stage of aluminum charge.

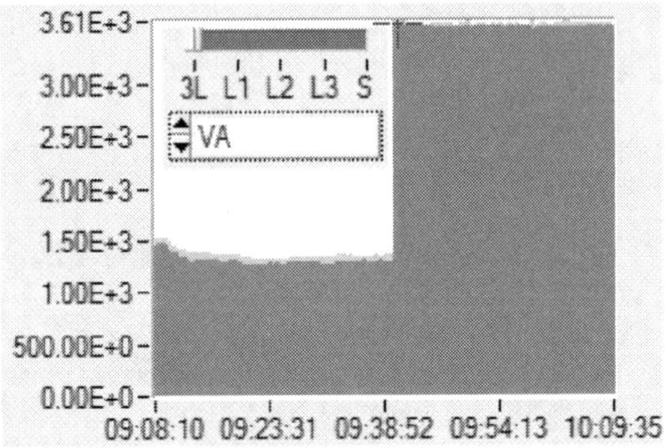

Figure 5. Apparent power S [VA] during the crucible preheating and heating stage of aluminum charge.

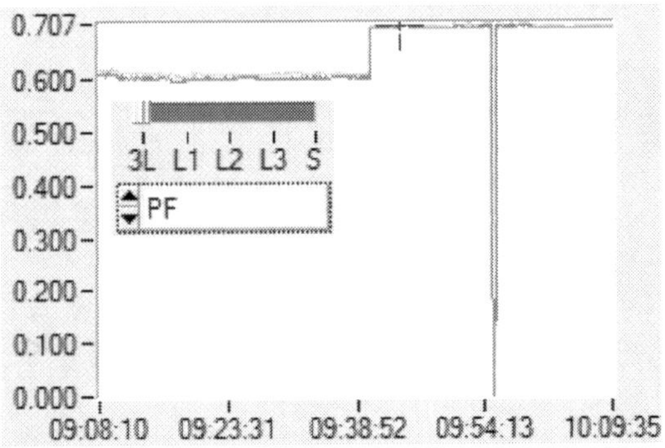

(a)

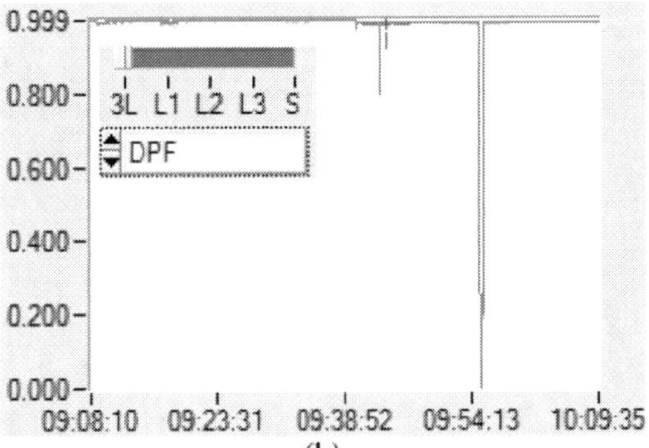

(b)

Figure 6. Power factor PF [-] (a) and displacement power factor DPF [-] (b) recorded in crucible preheating and heating stage of aluminum charge.

Figure 3.b shows that current unbalance is pronounced (4-8%) in the first heating stage, but decreases as the charge is melted (in the second stage of heating) to (0.7-2.6%). At the end of heating, current unbalance is within the limits of EMC standards (1.7%).

During the time period (9:40-10:10) has greatly increased active power consumption (Figure 4.a), because full active power of static converter was used to heat the aluminum charge (100% P_N). Figure 4.b shows that consumption of reactive power per phase was greater than the active power.

Real power factor PF (per phase) is very low in the first heating stage (9:08-9:39), about 0.6, because of the reactive power consumption and harmonic pollution of currents (Figure 6.a). After 9:39 PF increases slightly, but remains much lower (0.7) than neutral value (0.92). ,

Displacement power factor (DPF) is slightly smaller than 1 (0.999, Figure 6.b). The difference between PF and DPF indicates a large deviation from sinusoidal waveform of currents.

3.2. Melting stage of aluminum charge

THD of line currents in the melting stage of aluminum charge is slightly lower than in the previous stage (84-105%, Figure 7.b). Crest factor of the line currents in this stage is very high (2.2), indicating a very high harmonic pollution.

During the period (10:13-10:27) unbalance of line currents varies between 0-2%, and in the last five minutes of recording has a maximum value of 3.2% (Figure 8.b).

The reactive power per phase is less than active power, but still very high, indicating a insufficient power factor compensation in the installation.

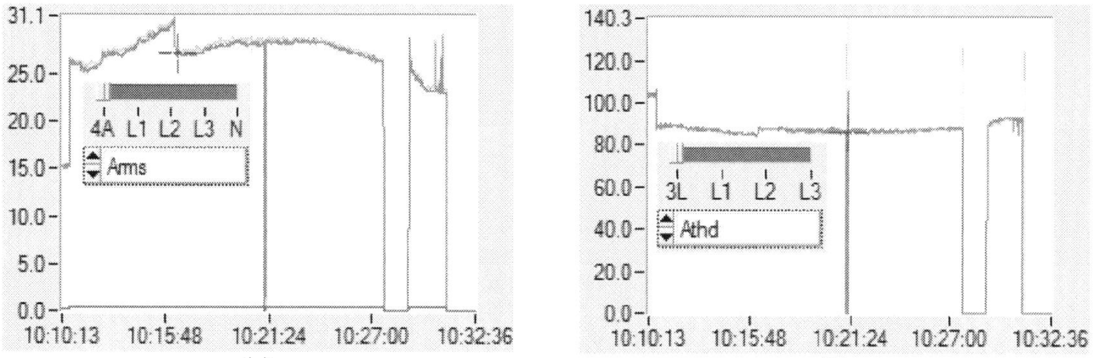

(a) (b)

Figure 7. TRMS [A] (a) and THD [%] (b) of line currents during the melting stage of aluminum charge.

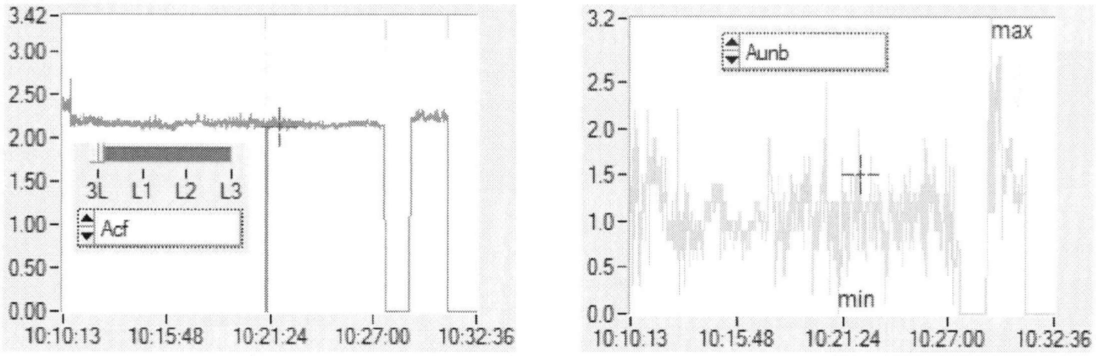

(a) (b)

Figure 8. Crest factor CF [-] (a) and unbalance [%] (b) of line currents during the melting stage of aluminum charge.

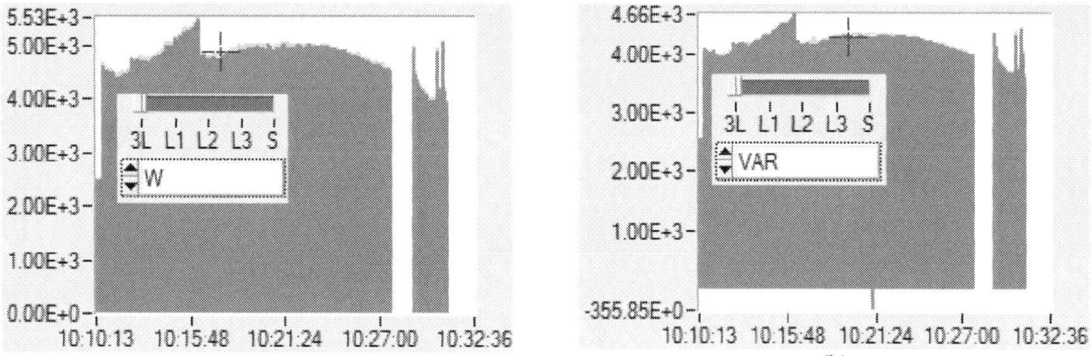

(a) (b)

Figure 9. Active power P [W] (a) and reactive power Q [VAR] (b) during the melting stage of aluminum charge.

In the melting stage, PF increase from the heating stage and reaches the value of 0.76, but still less than neutral value (Figure 11.a); DPF (Figure 11.b) is very good (0.994), higher than neutral value.

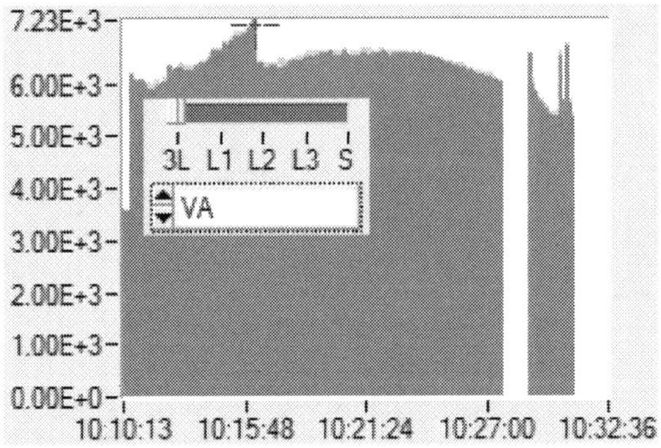

Figure 10. Apparent power S [VA] during the melting stage of aluminum charge.

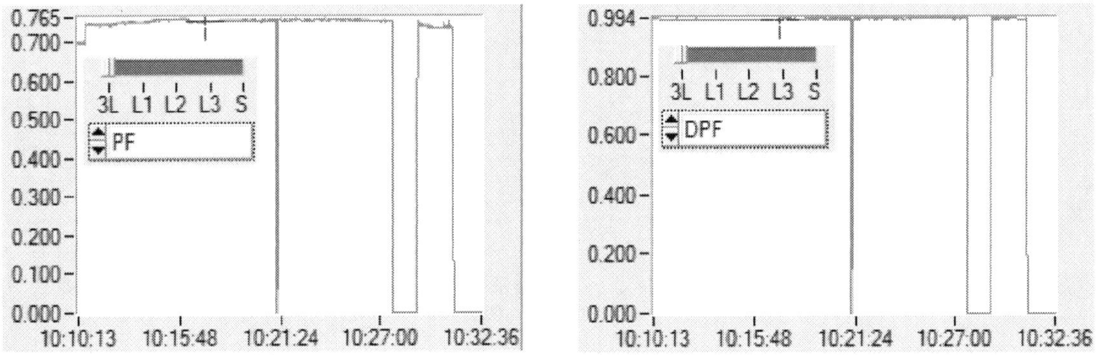

(a) (b)

Figure 11. Power factor PF [-] (a) and displacement power factor DPF [-] (b) recorded in the melting stage of aluminum charge.

Figures 12-19 show the time variation of 5th, 7th, 11th, 13th, 17th, 19th, 23th, 25th harmonics from the line currents during the melting process of aluminum charge. Levels of these harmonics exceed very much the compatibility limits [14].

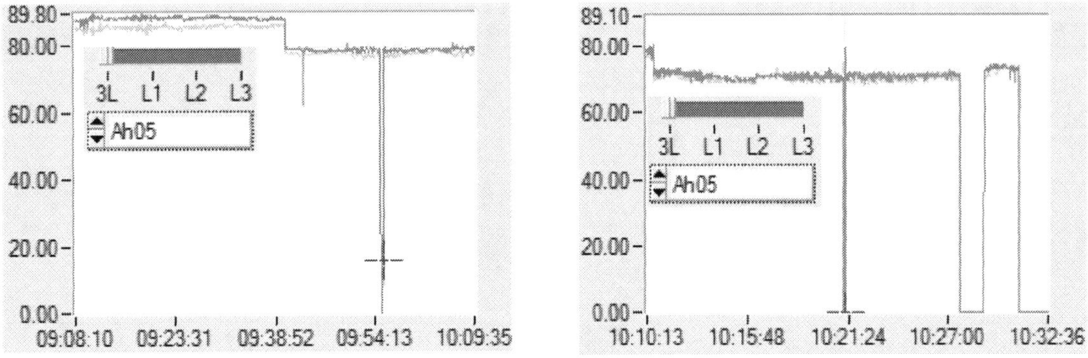

Figure 12. Time variation of 5th harmonic from line currents during the melting process.

In the first stages of aluminum charge heating: 5th harmonic level exceeds 80% (maximum 88%), 7th harmonic level exceeds 60% (maximum 77%), 11th harmonic level exceeds 25% (maximum 53%), 13th harmonic level exceeds 13% (maximum 38%), 17th harmonic level exceeds 6% (maximum 18%), 19th harmonic level exceeds 6% (maximum 10%), 23th harmonic level exceeds 2% (maximum 7%), and 25th harmonic level exceeds 1.8% (maximum 7%). In the last stage of melting: 5th harmonic level was 70%, 7th harmonic level was 50%, 11th harmonic level was 12%, 13th harmonic level was 4%, 17th harmonic level was 4%, 19th harmonic level was 1.8%, 23th harmonic was 2%, and 25th harmonic level was 1.7%.

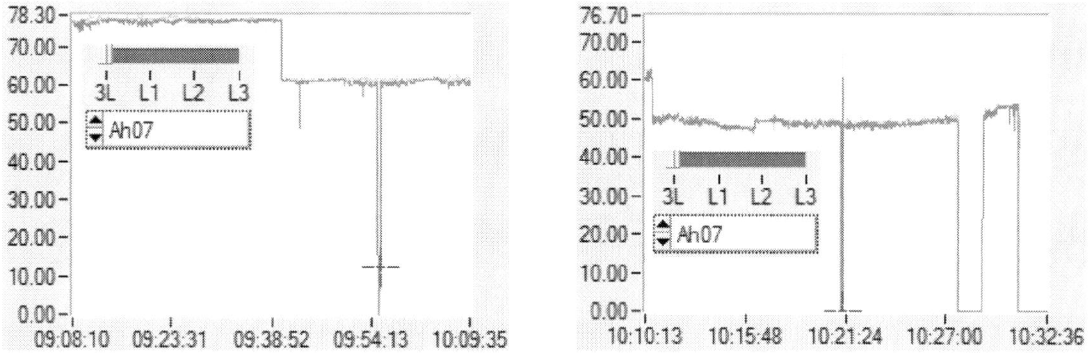

Figure 13. Time variation of 7th harmonic from line currents during the melting process.

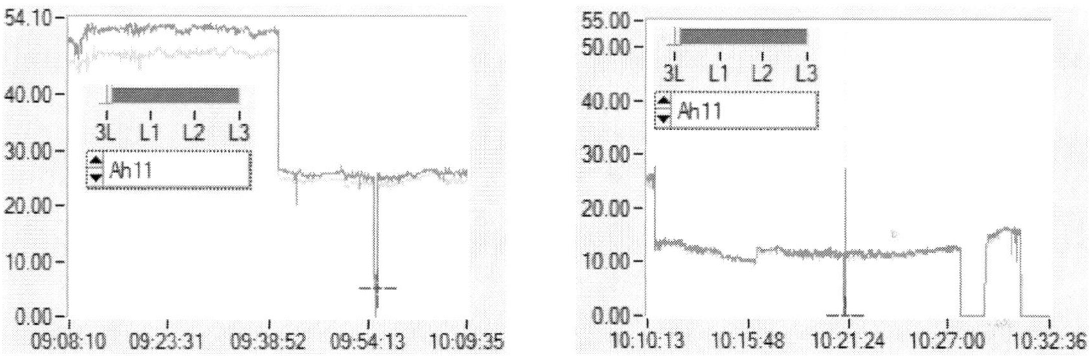

Figure 14. Time variation of 11th harmonic from line currents during the melting process.

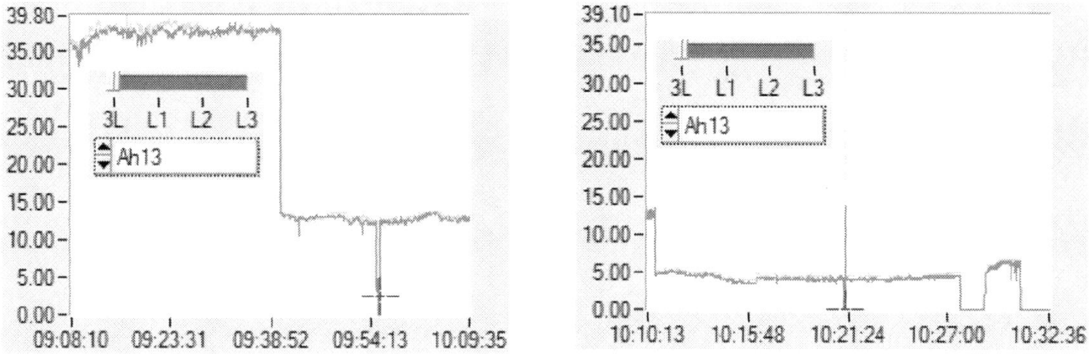

Figure 15. Time variation of 13th harmonic from line currents during the melting process.

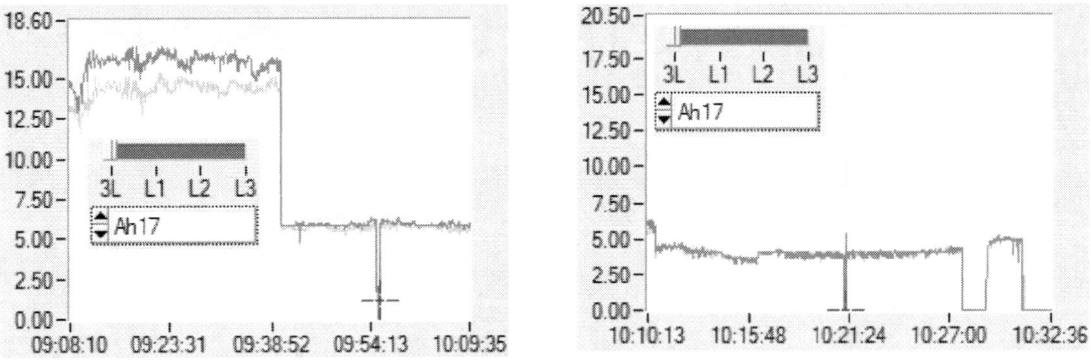

Figure 16. Time variation of 17th harmonic from line currents during the melting process.

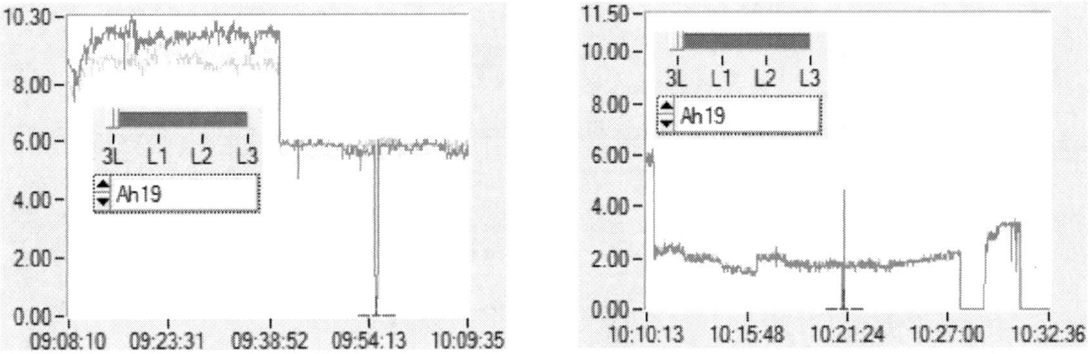

Figure 17. Time variation of 19th harmonic from line currents during the melting process.

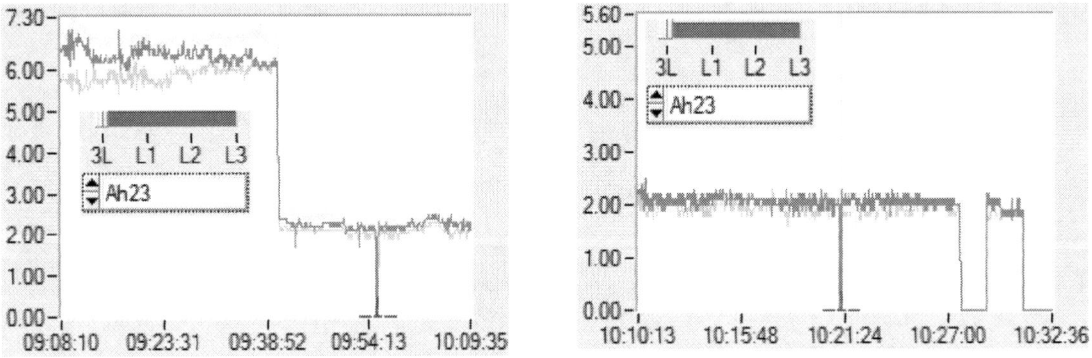

Figure 18. Time variation of 23th harmonic from line currents during the melting process.

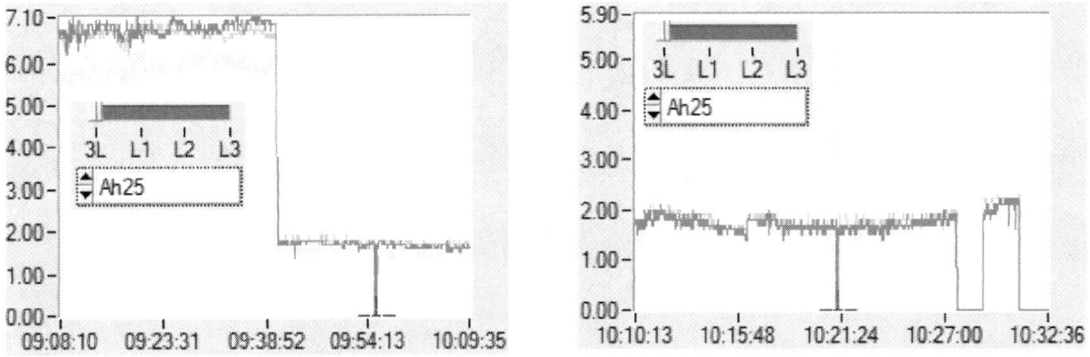

Figure 19. Time variation of 25th harmonic from line currents during the melting process.

4. Conclusions

Induction heating equipment does not introduce dust and noise emissions in operation, but cause power quality problems in the electric power system.

The measurement results showed that induction furnace operation causes unbalance and harmonics in three-phase currents absorbed from the distribution network. Harmonics in the line currents are caused mainly by static converter, and to a lesser extent by furnace load and interaction of eddy currents induced in the charge and the magnetic field of the inductor. To reduce the negative impact of the current harmonics on the distribution network is necessary the design and achievement of electrical filters for odd-order harmonics (5th, 7th, 11th, 13th, 17th, 19th, 23th, 25th).

References

[1] Rudnev V, Loveless D, Cook R and Black M 2002 *Handbook of Induction Heating*, CRC Press, Taylor&Francis Group, New York, USA

[2] Rudnev V, Loveless D, Cook R and Black M 1999 *Power quality for induction melting in metal production*, TechCommentary Electric Power Research Institute, USA

[3] Yilmaz I, Ermis M and Cadirci I 2012 Medium Frequency Induction Melting Furnace as a Load on the Power System, *IEEE Transactions on Industry Applications* **48**(4) 1203–1214

[4] Jabbar R A, Akbal M, Masood M A, Junaid M and Akram M F 2008 *Voltage waveform distortion measurement caused by current drawn by modern induction furnaces*, 13th International conference on harmonics and quality of power, Wollongong, Australia, September 28 – October 1, pp 1–7

[5] Tan A and Bayindir K C 2014 Modeling and analysis of power quality problems caused by coreless induction melting furnace connected to distribution network, *Electrical Engineering* **96**(3) 239-253

[6] Gbadamosi S L and Melodi A O 2013 Harmonic distortion from induction furnace loads in a steel production plant, *Network and Complex Systems* **3**(10) 8-16

[7] Nuns J and Yang X 2004 Les perturbations basses frequences generees sur les reseaux alimentant des installations de chauffage par induction, *Revue de l' electricite et de l' electronique* **5** 59-83

[8] Dugan R C and Conrad L E 1999 *Impact of induction furnace interharmonics on distribution systems*, Transmission and distribution conference, Los Angeles, USA, April 11-16, pp 791–796

[9] Arrillaga J, Watson N R and Chen S 2000 *Power System Quality Assessment*, John Wiley and Sons, New York

[10] ***SC AAGES Ltd. Târgu Mureş, Converter M20 ITS2 12K20. User guide

[11] ***Rectifier Diode Module DD B6U 85 N 16, Technical Information

[12] ***IGBT-modules FF300R12KS4, Technical Information

[13] ***CA8334, Three Phase Power Quality Analyser, technical handbook, Chauvin Arnoux, France, 2007

[14] ***IEC 61000-3-4, EMC, Part 3-4: Limits – Limitation of Emission of Harmonic Currents in Low-Voltage Power Supply Systems for Equipment with Rated Current Greater than 16A, 1998

IOP Conf. Series: Materials Science and Engineering **85** (2015) 012014 doi:10.1088/1757-899X/85/1/012014

Constructive and functional modernization of EAF

A Ioana[1], N Constantin[1] and P Moldovan[1]

[1] Politehnica University of Bucharest, Science and Engineering Materials Faculty, 313 Spl. Independentei, Bucharest, Romania

E-mail: adyioana@gmail.com

Abstract. This article presents the main constructive and functional modernization of Electric Arc Furnace (EAF). Schwabe is the creator of UHP (Ultra High Power) EAF. We present in this article the optimal use of EAF power and time aspects. These two main factors determine the increase in the productivity of the Electric Arc Furnace. Electric arc furnace operation can be based on either the short arc (with high intensity and low voltage) or long arc version (with low intensity and high voltage). Introduction of vault walls and water cooled has resulted in major reductions in the consumption of refractory bricks from electric arc furnace. Another important technological evolution inserted in the process of elaboration of steel in electric arc furnace is lowering the temperature of the exhaust. Using a system of intensive oxygen insufflation during melting can get many technological advantages. Another constructive and functional modernization is EAF inflatable sealing device for retention of slug.

1. Introduction

The increasing delivery capacity of steel furnaces with converters as the twin goals of productivity increase and the regulatory process required reassessment of strategy (both in terms of quantity but especially in terms of quality) of all steel products [1], [2], [6], [7].

Electric arc furnace became able to take full advantage of higher aggregate scrap resulting from the processing of the metal or from the recovery, which has seen a series of greenhouse modernization. One of the most important breakthroughs in this field has been the shift to accelerate melting through the use of high power transformers.

W.E. Schwabe, creator of UHP (Ultra High Power) EAF, has the merit of being demonstrated the capability of electric arc furnace to produce larger amount of steel through the use of short bows, of great power and by applying a high index of power and use of time [3], [4].

2. About EAF Evolution

The optimal use of electrical power and elaboration time are the two main factors that determine the increase in the productivity of the electric arc furnace. Use of power (P_u) is reflected by the ratio between the average level of power integrated in various stages of thermal cycle ($\sum_{i=1}^{n} P_i$) and the maximum possible level of power ($P_{\max}$) [5], according to the relationship (1)

IOP Conf. Series: Materials Science and Engineering **85** (2015) 012014 doi:10.1088/1757-899X/85/1/012014

$$P_u = \frac{\sum_{i=1}^{n} P_i}{P_{max}} \cdot 100\,[\%] \quad (1)$$

Optimal use of electrical power is given by:

$$P_u = 80...90\% \quad (2)$$

Electric arc furnace operation can be based on either the short arc (with high intensity and low voltage) or long arc version (with low intensity and high voltage).

Functional improvements (technology) of electric arc furnace are conditioned greenhouse modernization construction of this aggregate.

Thus, the operation of the furnace with longer arcs with high power (high-voltage, medium intensity) was possible after the development and introduction of walls and vaults have been cooled with water.

Moreover, this constructive results in upgrading another technological improvement of the policy-making process in the electric arc furnace and reducing the consumption of the electrodes.

Also, the same type of modernization (introduction of vault walls and water cooled) has resulted in major reductions in the consumption of refractory bricks from electric arc furnace (figure 1, [2]).

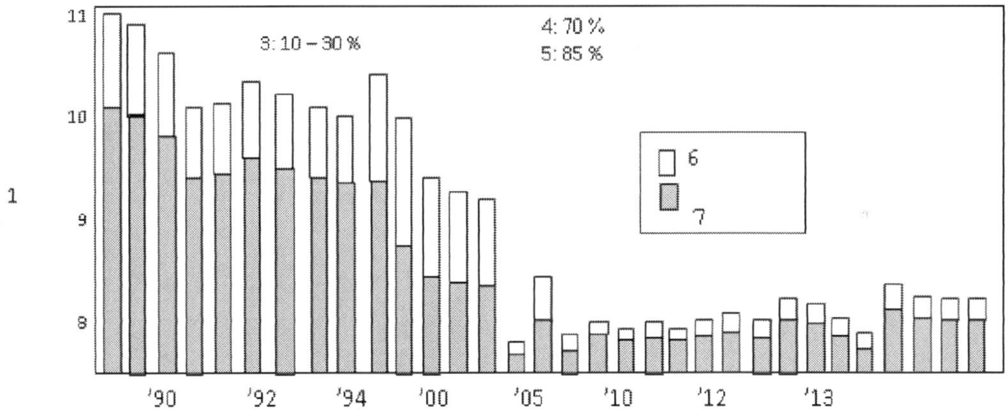

Figure 1. Reducing the consumption of refractory bricks after entering the vault walls and water cooled (ISCOR)

1 - use of bricks, kg/t; 2 - quarterly media; 3 - scrape walls with water-blocks of cast; 4 - scrape walls with water (KORF - FUCHS); 5 - arches are cooled with water (KORF - FUCHS); 6 - arches, kg/t; 7 - scrape walls, kg/t.

Specific power installed and the default operation of electric arc furnace at higher values of currents was conditioned by the increase, coupled with electrodes of siderurgy. In this area were carried out research both in terms of determining the driving technology of continuous baking furnace Riedhammer, including type of employment automation of baking in the oven, as well as for modernization of electrodes of siderurgy in baking ovens staple.

The main factors which is both siderurgy and electrode capacity efficiency economic and technical process of ripening, are:

- the thermal regime and the dynamic gas-oven baking;
- the correct and complete evacuation of volatiles of products subject to ripening;
- stripping and restoration without the baked electrodes in metal boxes of baking;
- preparing appropriate packing material;

➢ advanced mechanization difficult operations of technological flow of installation.

Another important technological evolution inserted in the process of elaboration of steel in electric arc furnace is lowering the temperature of the exhaust. This has direct implications for upgrading both the quality of steel produced and the efficiency of the regulatory process. Thus, steel produced has a lower content of gases ([O],[H],[N]), because the high temperature favors dissolving gases in steel. It also increases the productivity of the process of elaboration with 5...10% by shortening it, along with lower specific energy consumption (with 10°C reduction in temperature leads to exhaust a subtract with approx. 5 kWh/t of specific energy consumption).

The lowering the temperature of the exhaust requires preheat the container outlet (casting, VOD, cooker etc.) at higher temperatures. To this end, a team of scientists from the Institute for Scientific Research-ICEM Metallurgical S.A. Bucharest, laboratory of Fuel Economy and energy-L.E.C.E., conceived, designed, executed and experienced pilot and a wide range of different dimensions of recuperative burners for preheating and drying of casting ladles of iron and steel, with the following nominal thermal power: 63kW, 100 kW; 160kW; 250kW and 400kW.

There are also, nationally, a series of research in the field of regenerative burners. Taking into account energy efficiency regenerative combustion system, it is useful to study opportunities for the implementation of this type of burner in thermal intake facilities serving electric arc furnace.

Intensification of heat melting regime is another technological modernization of electric arc furnace. Research carried out in this area are mainly the following aspects highlighted:

➢ using a system of intensive oxygen insufflation during melting (approx. 15 m^3_N/h) you can get the following advantages:
  • increased productivity by about. 15t/h;
  • increasing the removal. 2%;
  • reducing the consumption of electric energy 25kWh/t.

➢ using oxy-fuel burners in the melting period is supported by the following test results:
  • maximum power burners 3MW;
  • natural gas flow rate: max. 300 m^3_N/h;
  • oxygen flow rate: max. 700 m^3_N/h;
  • technology: Bcc. 50% of the duration of melting, simultaneously with the electric arc;
  • economic efficiency: reduce specific energy consumption. 30kWh/t.

Excentric evacuation (by hearth furnace) and retention processes of slug constitute other greenhouse modernization construction and technological beneficial both in terms of quantity (increased productivity), and quality (high-purity of steel). Are successfully used in practice for closers for retention of slug, operated via electromagnetic detectors.

In Figure 2 we present schematic an inflatable sealing device for retention of slug.

The using of computer for management (optimization) process of steel is another level of modernization of the electric arc furnace. In this area, the world has been at the helm of the "on-line". Nationally, there were Council and develops research in this sector aimed mainly:

- electrical parameters measurement of the process;
- the study and determination of P-Q diagram;
- elaboration of the algorithm of electric system;
- determination of optimization of management system of melting.

Thus, in figure 3 we presents the evolution of specific electric power installed furnaces ("Danieli DANARC furnace") and comparative performance of different types of electric arc furnaces, and in Figure 4 is a new scheme presented electric arc furnace "Danieli" DC.

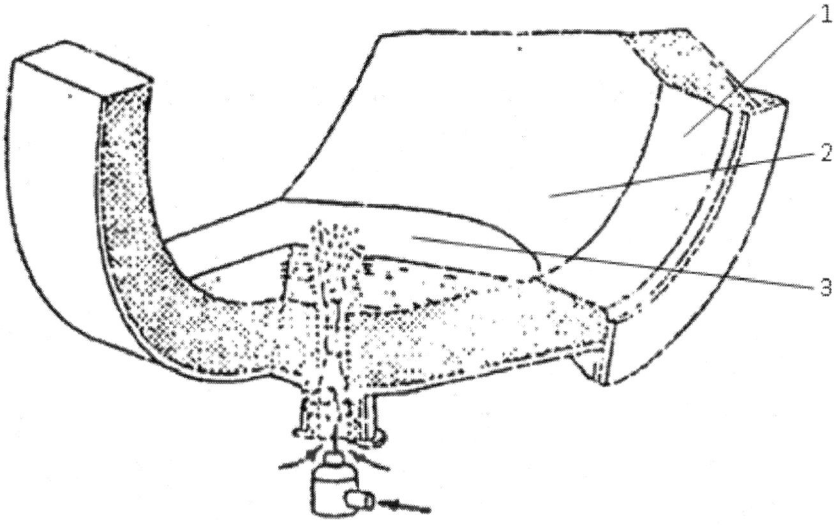

Figure 2. EAF inflatable sealing device for retention of slug
1 - Refractory lining; 2 - Metal bath; 3 - Device for retention of slug.

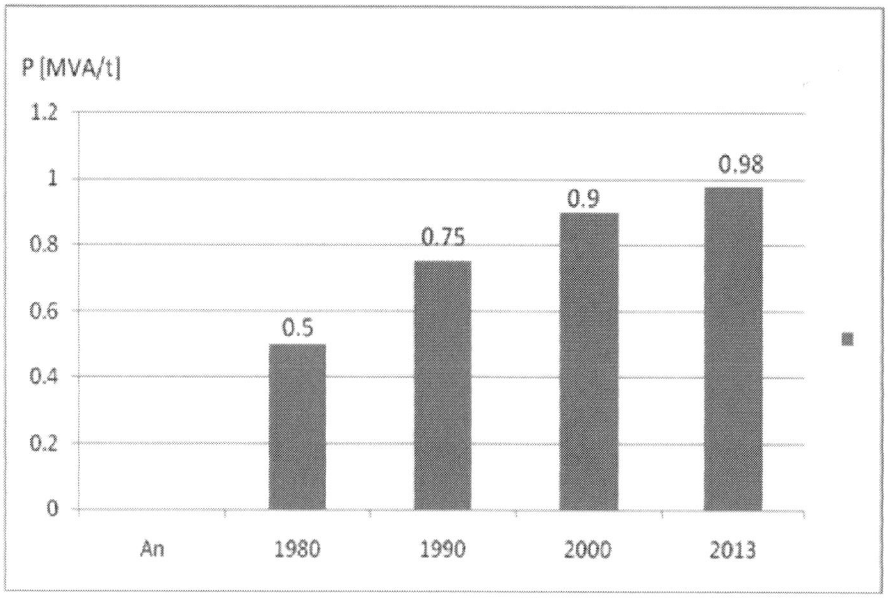

Figure 3. Evolution of specific electric power installed furnaces ("Danieli DANARC furnace").

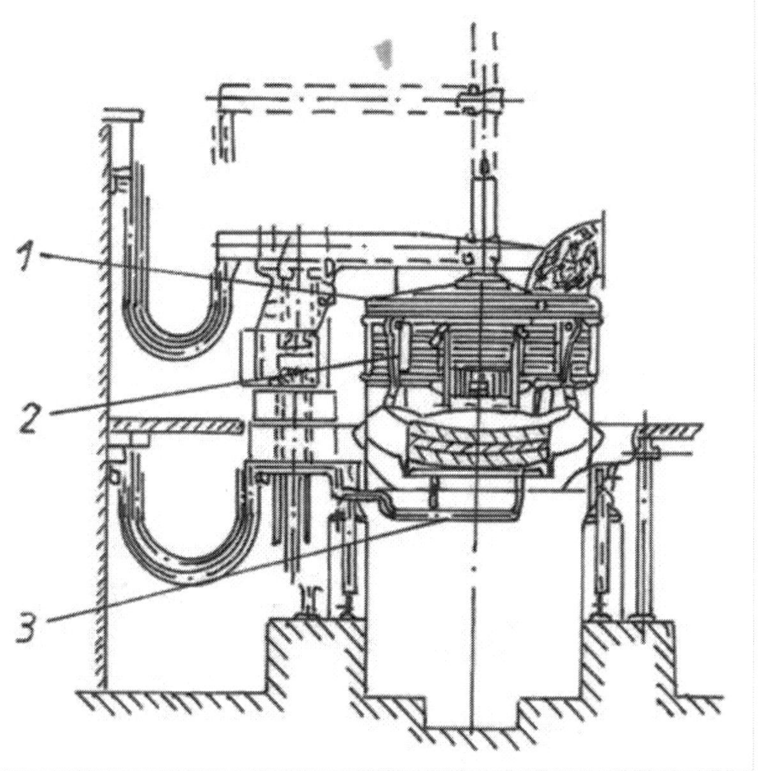

Figure 4. Scheme of EAF "Danieli" DC
1 - Chilled vault; 2 - Wall (panels) caught to cold; 3 - Installation of bulk materials supply.

3. Conclusions

Functional improvements (technology) of electric arc furnace are conditioned greenhouse modernization construction of this complex aggregate.

The most important constructive and functional modernizations of EAF are:

- Optimal use of electrical power and time of elaboration.
- Introduction of vault walls and water cooled.
- Lowering the temperature of the exhaust.
- Using of recuperative burners for preheating and drying of casting ladles of iron and steel.
- Using a system of intensive oxygen insufflation during melting.
- Using oxy-fuel burners in the melting period.
- EAF inflatable sealing device for retention of slug.

The using of computer for management (optimization) process of steel is another level of modernization of the electric arc furnace.

References

[1] Ioana A 2013 Metallurgy's Impact on Public Health, *Review of Research and Social Intervention* **43** 169-179

[2] Ioana A 2007 *Production Management of Metallic Materials Industry. Theory and Applications*, Printech Publisher, Bucharest, Romania

[3] Ioana A, Nicolae A, Predescu C, Sandu I F, Sohaciu M and Calea G G 2002 *Optimal Managing of Electric Arc Furnaces*, Fair Partners Publishing, Bucharest, Romania

[4] Ioana A 1998 *The electric arc furnaces (EAF) functional and technological performances with the preheating of the load and powder blowing optimisation for the high quality steel processing,* University Politehnica of Bucharest, Romania

[5] Ioana A and Semenescu A 2013 Technological, Economic, and Environmental Optimization of Aluminum Recycling, *Journal of the Minerals, Metals & Materials Society* **65**(8) 951-957

[6] Březinová M 2013 *Representative indicators in process management monitored by small and medium-sized business in south bohemian region*, Proceedings of "Finance and the performance of firms in science, education, and practice", Zlin, Czech Republic, April 25-26, pp 148-160

[7] Březinová M and Vrchota J 2012 *Current situation and future trends in the marketing process from the perspective of small and medium-sized enterprises in the south bohemian region*, Proceedings of "15th International Colloquium on Regional Sciences", Brno, Czech Republic, June 20-22, pp 374-378

IOP Conf. Series: Materials Science and Engineering **85** (2015) 012015 doi:10.1088/1757-899X/85/1/012015

About EAF and environment

A Ioana[1], N Constantin[1] and P Moldovan[1]

[1] Politehnica University of Bucharest, Science and Engineering Materials Faculty, 313 Spl. Independentei, Bucharest, Romania

E-mail: adyioana@gmail.com

Abstract. In this paper we present the results of industrial experiments consisted in determination, by specialized measurements, of the dust percentage in the evacuated gases from the EAF and of chemical composition of polluting burnt gases. The electric arc furnaces (EAFs), as powerful energy consumers, are also polluting emissions generators with an important environmental impact. The steel refining in the EAF is based on a complex mechanism of oxidizing chemical processes, due to the introduced oxygen and air. A first step is the diffusion and adsorption on the slag surface of molecular oxygen. The most significant polluting emissions of the EAF are metallic and oxides powders driven by emergent gases.

1. Introduction

From the point of view of reducing the impact over environment, the most important step is the optimization of powder retaining from the process gases both from the furnace and from the work-area. This optimization is necessary for improvement of work conditions and for obeying the limits imposed by the work and environment protection legislation.

A disadvantage of specific power installed growth of electric arc furnace transformers was formed by raising the level of noise in the operation of the furnace and especially from melting. Thus, in the context of the modernization works of the UHP oven steel furnances Campi-Nueva Italsider, was measured over the noise level between 63dB (80 m) and 120 dB (liturgy), [1]. This disadvantage has been removed by an enclosure of insulating panels around the oven electrically sound.

The electric arc furnaces (EAFs), as powerful energy consumers, are also polluting emissions generators with an important environmental impact.

The most significant polluting emissions of the EAF are metallic and oxides powders driven by emergent gases [1-5]. The powders are produced during the following technological operations: raw materials loading, steel melting, refining, alloying and evacuation. Generally, the driven powders contain heavy metals (Cr, Ni, Zn, Cd, Pb, Cu etc) and some metal oxides (iron, manganese, aluminum and silicon oxides) and they can reach values of more than 15 kg/t steel.

The steel refining in the EAF is based on a complex mechanism of oxidizing chemical processes, due to the introduced oxygen and air. A first step is the diffusion and adsorption on the slag surface of molecular oxygen.

Then, the presence of the adsorbed oxygen molecules (written as {O2}ads) leads to the redox chemical reaction with ferrous ions from the slag (symbolized as (Fe2+)); this is a reaction with exchange of two electrons resulting oxygen ions and ferric ions, both ionic species dissolved in the slag:

$$\frac{1}{2}\{O2\}ads + 2(Fe2^{+}) = (O2^{-}) + 2(Fe3^{+}) \qquad (1)$$

Next step is the diffusion of O2- and Fe3+ ions (within the slag and penetrating the slag–metallic melt interface).

2. EAF and Environment

From the total EAF polluting emissions, over 90% are generated during the technological operations of melting and refining [6], [7]. The chemical composition of these emissions is extremely variable and directly dependent on multiple factors, as followings:

- the melting managing way;
- type of refining process that is used (with gaseous oxygen or ore);
- desired quality degree of the elaborated steel;
- composition of the raw materials that make up the loading;
- time duration of the melting and refining steps.

Figure 1 presents the main scheme of the environment polluting system through the EAF.

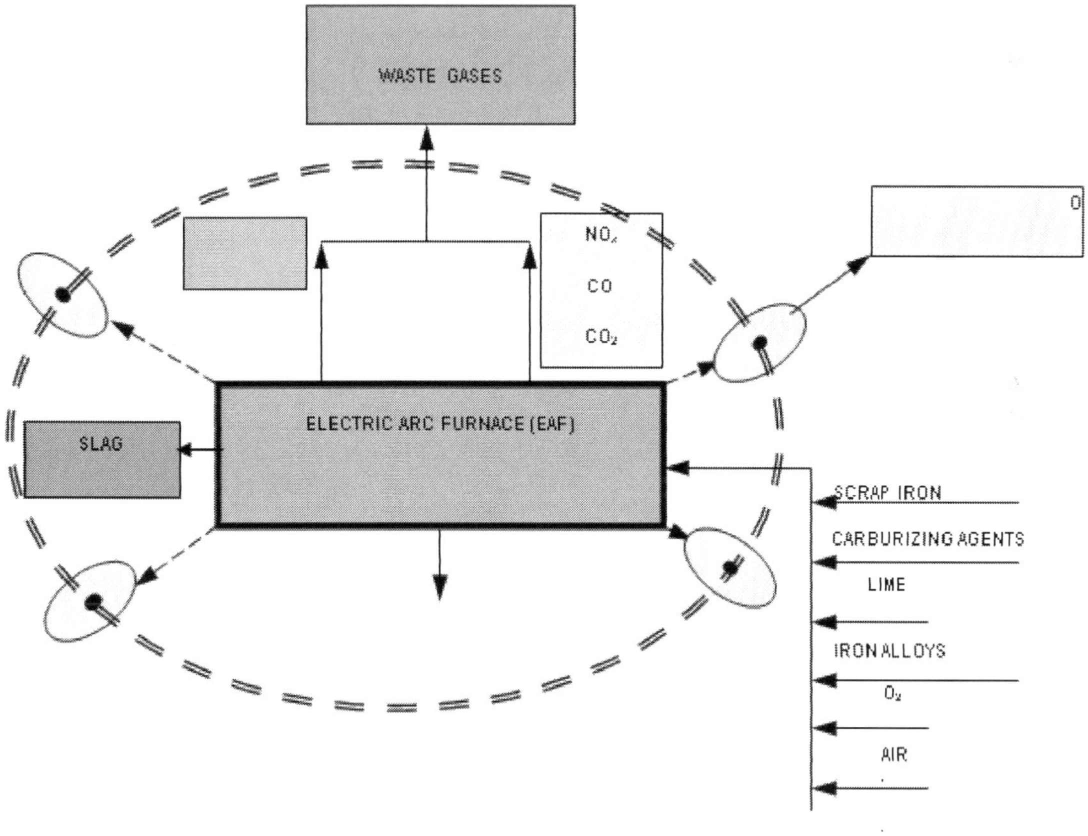

Figure 1. The scheme of the environment polluting system through the EAF.

Table 1 presents the variation limits of chemical composition for the powders generated during the steel elaboration in electric arc furnaces in the USA and Germany, from loading that consists of scrap iron, only [1-3].

Table 1. Chemical composition of the powder generated by the EAF

No.	Com-ponent	Variation limits, %		
			GERMANY	
		SUA	Plain Basic Steel	Alloy Steel
1	Fe_{total}	16.4 – 38.6	21.6 – 43.6	35.3
2	Si	0.9 – 4.2	0.9 – 1.7	17.0
3	Al	0.5 – 6.9	0.1 – 1.5	x)
4	Ca	2.6 – 15.7	6.6 – 14.5	0.4
5	Mg	1.2 – 9.0	1.0 – 4.5	1.2
6	Mn	2.3 – 9.3	0.9 – 4.8	2.0
7	P	0 – 1.0	0.1 – 0.5	x)
8	S	0– 1.0	0.3 – 1.1	0.1
9	Zn	0 – 35.3	5.8 – 26.2	1.4
10	Cr	0 – 8.2	0 – 0.1	13.4
11	Ni	0 – 2.4	x)	0.1
12	Pb	0 – 3.7	1.3 – 5.0	0.4

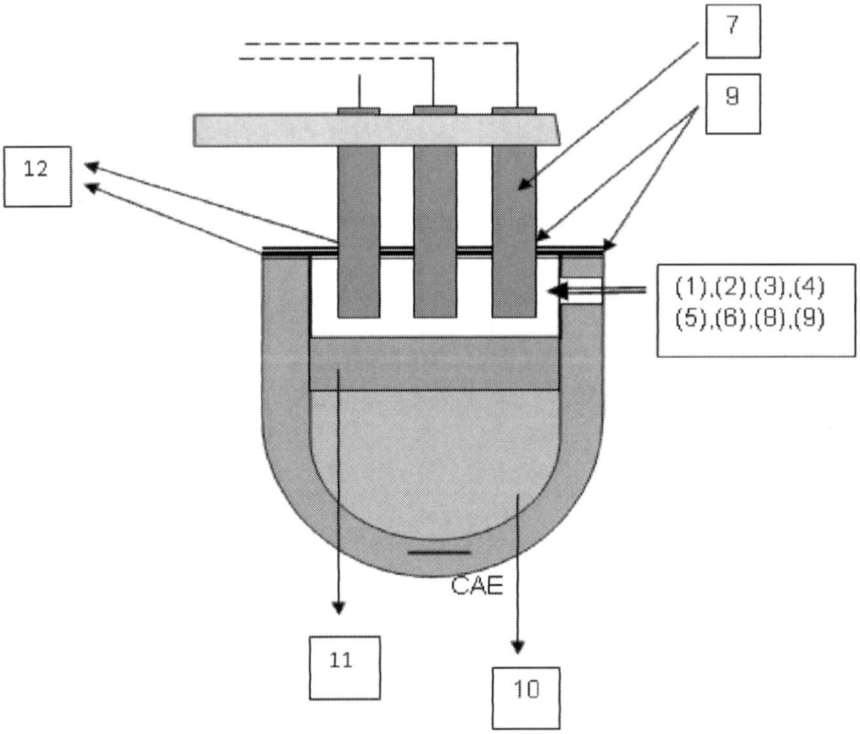

Figure 2. Scheme representation of the mathematical model concept for load calculus in the EAF
1 – Metal load (scrap iron); 2 – Iron ore; 3 - Lime; 4 – Calcium fluoride; 5 - Dolomite; 6 - Coke; 7 - Electrodes; 8 – Blown oxygen; 9 – Exhausted atmospheric air; 10 – Liquid steel; 11 – Slag (liquid); 12 – Burnt gases.

Input materials (MI,i) mainly consist of (Figure 2):
- Metal load (scrap iron)
- Iron ore
- Lime
- Calcium fluoride
- Dolomite
- Coke
- Electrodes
- Blown Oxygen
- Exhausted atmospheric air

Output materials (ME,j) are:
- Liquid steel
- Slag (liquid)
- Burnt gases

3. Conclusions

Optimizing steel making in the EAF has as a starting point the analysis of the chemical mechanisms attached to the specific transformations.

EAF polluting emissions, over 90% are generated during the technological operations of melting and refining.

The blown oxygen quantity was established by means of thermal-technical measurements, and the quantity of exhausted oxygen as a difference between the necessary oxygen for the process and the inspired oxygen. The frequently determined excess oxygen, by means of complex thermal-technical measurements, has emphasized an almost permanent supplementary exhaust of the atmospheric air inside the EAF. This aspect is directly correlated to the lack of isolation of the elaboration machinery.

References

[1] Ioana A and Semenescu A 2013 Technological, Economic, and Environmental Optimization of Aluminum Recycling, *Journal of the Minerals, Metals & Materials Society* **65**(8) 951-957

[2] Ioana A 2013 Metallurgy's Impact on Public Health, *Review of Research and Social Intervention* **43** 169-179

[3] Ioana A, Nicolae A, Predescu C, Sandu I F, Sohaciu M and Calea G G 2002 *Optimal Managing of Electric Arc Furnaces*, Fair Partners Publishing, Bucharest, Romania

[4] Ioana A 1998 *The electric arc furnaces (EAF) functional and technological performances with the preheating of the load and powder blowing optimisation for the high quality steel processing*, University Politehnica of Bucharest, Romania

[5] Ioana A 2007 *Production Management of Metallic Materials Industry. Theory and Applications*, Printech Publisher, Bucharest, Romania

[6] Březinová M 2013 *Representative indicators in process management monitored by small and medium-sized business in south bohemian region*, Proceedings of "Finance and the performance of firms in science, education, and practice", Zlin, Czech Republic, April 25-26, pp 148-160

[7] Březinová M and Vrchota J 2012 *Current situation and future trends in the marketing process from the perspective of small and medium-sized enterprises in the south bohemian region*, Proceedings of "15th International Colloquium on Regional Sciences", Brno, Czech Republic, June 20-22, pp 374-378

IOP Conf. Series: Materials Science and Engineering **85** (2015) 012016 doi:10.1088/1757-899X/85/1/012016

Using lateral cores to casting of carbon steel parts, of drive wheel type, in a metallurgical enterprise

A Josan[1], C Pinca Bretotean[1], E Ardelean[1] and M Ardelean[1]

[1] Politehnica University of Timisoara, Department of Engineering and Management, 5 Revolution Street, 331128 Hunedoara, Romania

E-mail: ana.josan@fih.upt.ro

Abstract. This paper presents the possibility of obtaining castings using the lateral cores. Steps are presented for obtaining the lateral core of the piece type drive wheel. This piece is cast carbon steel 230-450W, according to ISO 3755-95, and is part of a drive train. Using in industrial practice of such types of cores leads to significant reduction of the processing workmanship for castings, with a direct effect on the price of production.

1. Introduction

The correct realization of a castings involves strictly complying of moulding-casting technology, starting from correct dimensioning of all components. Thus, for casting of parts is required to manufacturing of set of pattern, which comprises the pattern, the core box (or boxex core, if necessary), the pouring gate, the venting, the rising head etc. With de help of the pattern is obtained a cavity in the upper in lower mould parts, and with de help the core box is obtained the core, witch ensure obtaining cavity inside of the casting. The core fixing into the mould with two extensions, called core marks [1], [2], [3].

The core marks are introduced into the space of the mark, so that it is not possible to move the core during the assembly of the mould or during the casting of the liquid alloy. With pattern mark is realized the cavity mark in moulding batch. The core mark is makes in core box, with core [1], [2], [3].

All these elements must be dimensioned corectly so that the set of pattern be made to dimensions the finished part that are to obtained by casting.

2. Making the pattern for castings, *drive wheel* type

An example of a technology that uses lateral and internal cores is used to castings of wheel type. The design of set of pattern for casting drive wheel is making based on the finished piece design (Figure 1).

Figure 1 presents a *drive wheel* cast of carbon steel in according to ISO 3755-95 [4]. This wheel is part of *the drivetrain – waggon funicular*. The wagon is suspended on a cable and drivetrain (besides welded constructions that connect the wagon bucket and raceway) has assembled 2 or 4 such rolls (depending on model and capacity). In the hub are mounted the bearings and spindle and the outside (where it is processed the concave profile) runs on cable.

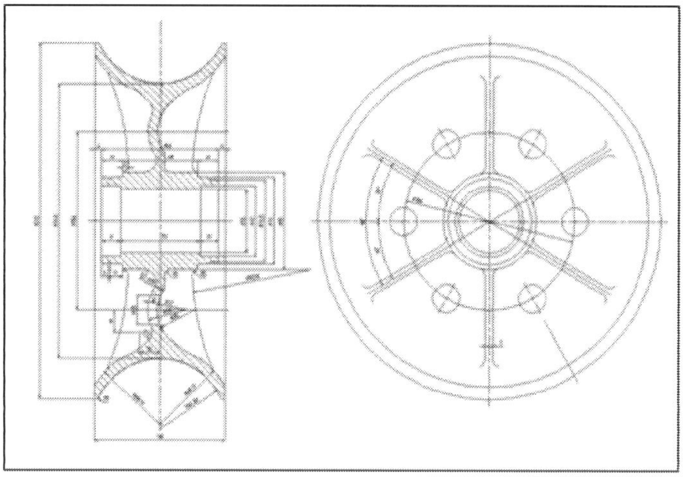

Figure 1. The finished piece design (*Drive wheel*).

To establish the optimal casting position is necessary the analysis of the emplacement the separation plan. Thus, the separation plane for such a piece is shown in Figure 2. According to this sketch the separation plane of mould is a curved line. This line follows the pattern configuration and casting position is horizontal (Figure 2) [1], [2], [3].

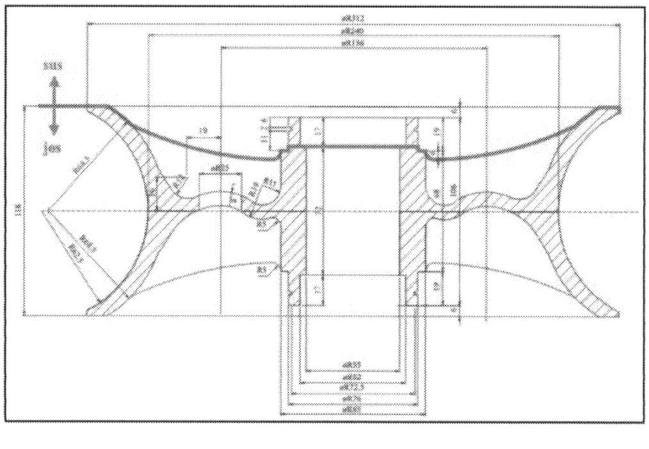

Figure 2. The separation plan for studied piece.

To establish the pattern dimensions taking into account the dimensions obtained after establishing the added shrinkage and application the added processing. After establishing the size of added shrinkage, constructive inclinations and marks a core dimension is determined the configuration and dimensions of the pattern. The pattern is made of wood and is presented in Figure 3 [1], [2], [3].

Figure 3. The pattern of wood intended for the casting.

For dimensioning of centre riser takes into account the height of the piece wall on which it is applied (i.e. 120mm). Thus, for studied casting is used a riser centrally disposed (Figure 4), whose height is 108 mm [1], [2], [3].

Figure 4. The pattern of piece and placed the centre riser.

3. Making the lateral and central cores for casting. Making the cores marks

Of the technological point of view, for piece analyzed is chosen the achievement of two cores:

- ✓ a central core, to obtain the internal hollow of piece;
- ✓ a lateral core, to determining the lateral configuration of piece.

The marks represent the technological extensions of patterns and cores, created in order to achieve internal hollow castings. *The pattern mark* is an extension, with which is achieved a hollow in the mould, where is fixed to *the core mark*. The core mark is dimensioned to resist at the stresses core, during filling the mould hollow with liquid metal. In the case too small marks is possible mould deformation, which lead to the appearance of the dimensional aberrations of the casting and the excessive size of the marks lead to an excessive consumption of core mixture, large areas for drying, amplifying dimensions of mould jackets, etc. [1], [2], [3].

For judicious of moulds dimensions and a cores conicity are used the standard provisions for horizontal and vertical marks. Generally, the cylindrical and vertical core mark, used for manual moulding, are presented in Figure 5 [1], [2], [3].

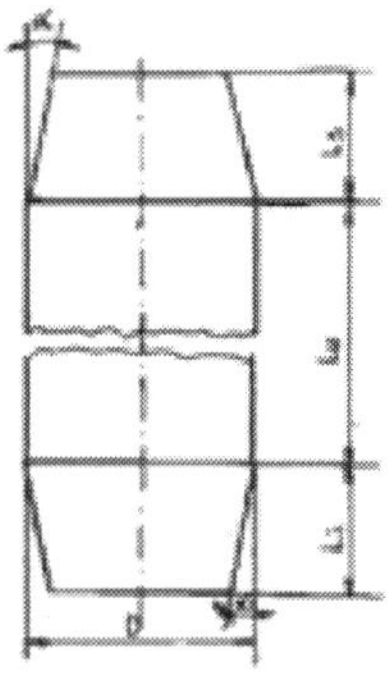

Figure 5. The cylindrical-vertical core marks.

Depending on the Lu/D to be determined dimensions of upper mark Ls and lower mark Li [1], [2], [3]. On the finished piece design were added as follows: shrinkage added, technological added, processing added and constructive inclinations. Therefore is obtained the technological design. On the technological design are determined the usable length of the central core, Lu, and the core diameter D:

$$L_u = 102\text{mm};\ D \cong 60\text{mm};\ \Rightarrow \frac{L_u}{D} = \frac{102}{60} = 1{,}7\,. \quad (1)$$

According to [1,2,3], result following dimensions:

$$\frac{L_s}{L_i} = \frac{25}{45}\text{mm}; \alpha_{sup} = 10^\circ; \alpha_{inf} = 7^\circ\,. \quad (2)$$

Thus, minimum diameter of the upper mark is 25 mm and minimum diameter of the lower mark is 45 mm. Lateral core is made separately, in box core and has the following dimensions: height 120 mm and width 35 mm. Making the lateral core, step by step, is presented in Figures 6, 7, 8.

Figure 6. Making the lateral core – stage 1 – preparation of the plate moulding.

Figure 7. Making the lateral core – stage 2 – mounting the outside of the core box.

Figure 8. Making the lateral core – stage 3 – core box assembly.

6. Conclusions

To obtain a casting, the design of set of pattern has an essential role. Also, the design the moulding-casting technology involves establishing types of cores used. Thus, for a drive wheel casting, the experiments of industrial practice show that the best technological solution is to place the separation plane, the additions and cores just as in Figure 9. Figure 10 presents the lateral cores, making according to the instructions.

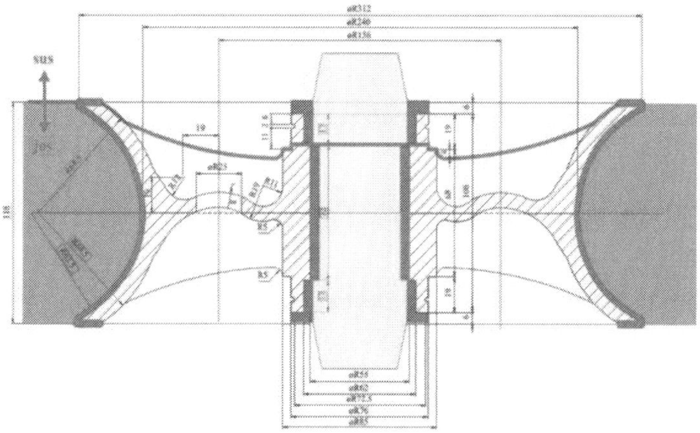

Figure 9. To position the cores and the separation plane.

Figure 10. The core of the studied piece.

The castings (drive wheel) in accordance with this technology is presented in Figure 11.

Figure 11. The casting – drive wheel.

According to this technology, which involves the use of two types of cores for establishing the internal and external of the casting, it reduces the quantity of material removed by machining, decreases the workmanship of the mechanical processing by cutting, which leads to decreased prices for fabrication of casting, with direct effects on enterprise costs.

References

[1] Josan A 2008 *Technology moulding and casting alloys*, Publishing Polytechnic, Timişoara, Romania

[2] Josan A 2012 *Casting and moulding technologies of metal parts – the project guide-book*, Publishing Polytechnic, Timişoara, Romania

[3] Socalici A, Ardelean E, Ardelean M, Heput T and Josan A 2007 *Casting and solidification of steel* Publishing Cermi, Iaşi, Romania

[4] ***SR ISO 3755:1995. – Non alloy cast steels for mechanical constructions of general use.

IOP Conf. Series: Materials Science and Engineering **85** (2015) 012017 doi:10.1088/1757-899X/85/1/012017

Using special additions to preparation of the moulding mixture for casting steel parts of drive wheel type

A Josan[1] and C Pinca Bretotean[1]

[1] Politehnica University of Timisoara, Department of Engineering and Management, 5 Revolution Street, 331128 Hunedoara, Romania

E-mail: ana.josan@fih.upt.ro

Abstract. The paper presents the possibility of using special additions to the execution of moulding mixtures for steel castings, drive wheel type. Critical analysis of moulding technology leads to the idea that most defects appear due to using improper moulding mixture. Using a improper moulding mixture leads to penetration of steel in moulding mixture, resulting in the formation of adherences, due to inadequate refractarity of the mould and core mixtures. Using only the unique mixture to the moulding leads to increasing consumption of new sand, respectively to the increase of price of piece. Acording to the dates registered in the industrial practice is necessary to use the special additions to obtain the moulding mixtures, carbonaceous materials respectively.

1. Introduction

In foundries, for achievement the moulds and the cores are used natural and synthetic moulding mixtures. In the first case are used the sands in their natural state, of quarry, with or without small special additions. In the second case are used the quartz sands (naturals or washed), to which are added necessary binders [1], [2], [3], [4].

Because in nature can be found few materials that correspond to related thermophysical properties, productivity and economicity imposed by foundry, becomes necessary to make some recipes for moulding mixtures/core mixtures cheap and good quality, corresponding to final destination (the realization of moulds and cores).

Temporary moulds contain a granular material, with high refractory, as the base material, to which is added some binders and addition materials. The product obtained by dosing and homogenization of these materials in desired proportions, is called the moulding mixture [1], [2], [3], [4].

2. The recipes of moulding and core mixtures

For piece *Drive wheel* (Figure 1) cast of carbon steel 230-450W [5], [6] is used both pattern mixture and filling mixtures, as well core mixtures.

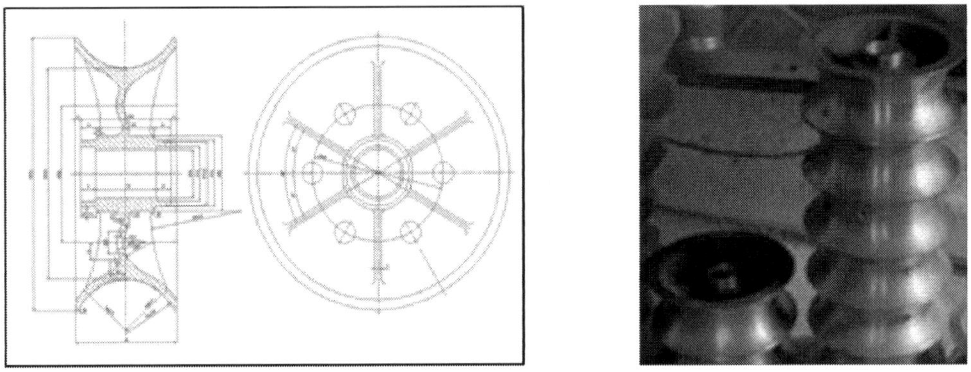

Figure 1. The casting – *Drive wheel.*

Thus, for execution the moulds, are used *green sands (as a pattern mixture)*, 90…92% new sand and 6…8% bentonite. Mixtures characteristics are [1], [2], [3], [4]: humidity: 3,5…4,0%; permeability: 100UP; compressions strength: 0,6…0,8daN/cm^2.

The filling mixture are the following recipe: reused mixture: 95%; new sand: 3,5%; bentonite: 1,5%; water: 3%; coal dust: 0,1%.

To making the core is used the core mixture, with the following composition: 98% new sand; 1,6% bentonite; 1…2,5% oil flaxen; 0,5…1,4% dextrin. The mixture characteristics are: permeability: 100UP; levigable component: 2%; humidity: 2…3%; compressions strength: 0,15…0,3daN/cm^2.

Figure 2. The lateral core.

3. Using the special additions to preparation of the moulding and core mixtures

Critical analysis of the mouding technology leads to the idea that most defects appear due to use the improper moulding mixture. Thus, using a inadequate moulding mixture, leads to penetration of steel in moulding mixture, resulting in the formation of *adherences*, due to inadequate refractarity of the pattern and core mixtures [7], [8], [9], [10].

Using just one sort of mixture (unique mixture) to moulding leads to *increased consumption of new sand*, respectively to *increase in the price of the piece*.

To prevening the chemical penetrations to steel casting and correct execution of moulds, is recommended changing the recipe used for moulding mixtures. According to data registered in the industrial practice, is ideally suited to use special additions in composition of the moulding mixtures, respectively *coal materials*.

About these aspects, among additions of moulding mixtures an important role has the "*additions against of adherences*". Their role is to produce a protectiv film (gas or solid) between the metal and the wall moulding, during casting and solidification of the alloy [7], [8], [9].

Mineral coal (pitcoal, brown coal or lignit) are used in grinding state to casting in green-sand mould. The protective action of coal is carried at 1000^{0}C, to contact surface metal-mould [7], [8].

The coal used as an addition in moulding mixtures is characterized by the following:

- Low ash content (max. 11%);
- To sift in proportion of 80% through sievc of 0,063 mm and 95% through sievc of 0,16 mm;
- Content of volatile substance, min. 35%;
- Humidity, max. 10%.

Recipe the preparation of moulding misture contains [1], [2], [3], [4], [10]:

a) *The composition of the starting mixture:*

Quartz sand Vălenii de Munte	84%
Bentonite from Valea Chioarului	12%
Grinding coal	4%
Ethyleneglycol	0,5%
Water, max.	3,2%

The mixing time – 6 min.

b) *The recipe of refresh* – the calculation of refresh recipe will be based on levigable component (LC) and mixture degradation, which will be determide by ratio mixture/metal.

For a content of 11% LC, refresh recipe is:

New sand	2%
Bentonite	10,5%
Grinding coal	0,3%
Glycol	0,1%
Water	depending on residual moisturein the misture.

In the preparation of the unique mixture, based on refresh recipe, the order to introduce the materials in the mixer as follows:

- Recirculated mixture (after reconditioning);
- New sand;
- Bentonite;
- Grinding coal,

these are then mixed for one minute, then added:

- Water;
- Glycol,

and mixture was further mixed 1,5 minute; follows:

- Passing through disintegrator (the aerator/mousseur);
- The control of moulding mixture.

4. Using refractory paints / powders to moulds and cores

To prevent the appearance of adherences and obtain smooth surfaces, with commercial aspect, is realized by used of *refractory paints,* which prevents penetration of molten alloy into the mould pores and does not react with these.

The refractory paints are applied on the surface of moulds and cores in order to produce a refractory layer to contact surface of mould with molten alloy; the refractory paint must be fluid and dense, to form a continuous layer, thin and uniform, on the surface of mould and core.

The refractory paint recommended is composed of:

- dusty component (zircon dust – is also called zircon flour $ZrSiO_4$) – has high refractivity (melting temperature - 2500°C) and a low coefficient of thermal expansion;
- binder (synthetic resins with fast solidification);
- dispersion medium (organic solvent, isopropyl alcohol).

The refractory paint used to reduce defects is of type **FOUNDRYLAC ZIRCONIU**. This type of paint is characterized by high strength and optimum dyeing, with leveling effect, which conferring a very good degree the final finishing of the surface (figure 3).

The powders protection

Prevention of adherences is realized using anti adhesive powders, which are used only for casting into the green-sand moulds, this forming a layer with high refractory on surface of the green-sand mould. This layer prevents direct contact between molten alloy and walls mould. Refractory powders most frequently used are: graphite powder, quartz powder; milled talc, Portland cement.

Figure 3. The mould parts (upper and lower) performed.

4. Conclusions

In the case of piece analyzed (*drive wheel*), the defects registered most frequently are *adherences*, due to the processes occurring in the moulds, during casting of molten steel. These types of defects are categorized as the surface defects of type D [7].

Adherences are defects characterized by a rough surface and is a defect is found to visual inspection, being the defect that causes the more rejects. This defect is the effect of metal penetration into the moulding mixture, and the causes of the appearance of adherences can be mentioned as follows:

- inadequate moulding mixture (high grain-size, low refractivity);
- inadequate mould stamping, especially at intersections of surfaces;
- using paints and powders of low refractory.

In order to improve the moulding-casting technology for analyzed casting, has making a series changes and following these changes the consumption of material will vary as follows:

- ➢ decreases the consumption of sand quartz (Valeni) and bentonite
- ➢ increase the consumption of refractory paint;
- ➢ is added the consumption of coal and ethylene glycol.

After the economic calculations, the total value of materials used in the moulding-casting, production price and price of the analyzed piece in the two technologies are presented in Figure 4.

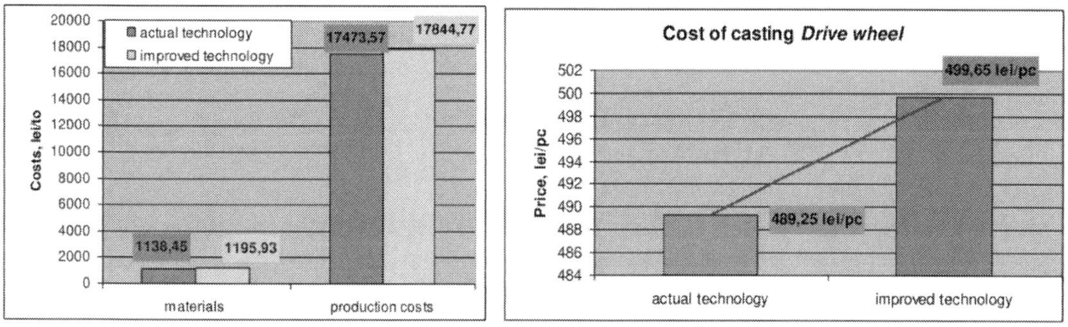

Figure 4. Comparative analysis of the two moulding technologies.

After performing changes in moulding-casting technology for the casting study results:

- for the studied casting (*Drive wheel*) the price is 499,65 lei; in the case of changed moulding-casting technology result a price increase with 10,4 lei/pc, respectively 2,12%
- the same price increase in not significant, because the percentage of rejects registered due to defects decreases by approximately 3,2%.

References

[1] Josan A 2002 *Technology moulding and casting alloys*, Publishing Polytechnic, Timişoara, Romania

[2] Josan A 2012 *Casting and moulding technologies of metal parts – the project guide-book*, Publishing Polytechnic, Timişoara, Romania

[3] Socalici A, Ardelean E, Ardelean M, Heput T and Josan A 2007 *Casting and solidification of steel,* Publishing Cermi, Iaşi, Romania

[4] Stefanescu C and Cosneanu C 1989 *Mixtures moulding systems for foundry,* Technical Publishing, Bucuresti, Romania

[5] ***SR ISO 3755:1995. – *Non alloy cast steels for mechanical constructions of general use*

[6] Chesa I 1984 *Choosing and using of steels* Technical Publishing, Bucuresti, Romania

[7] * * * 1977 *International Atlas of casting defects* Technical Publishing, Bucureşti, Romania

[8] Stefanescu D M, Piwonka T S, Giese S R, Barlow K O and Haynes K 2000 The penetration of the ferrous alloys in the moulding mixtures, *The Journal of Foundry* **3** 4-10

[9] Josan A, Pinca Bretotean C, Ratiu S and Pauca A 2011 *Research on changing technology of molding casting of a part cast steel for reducing rejects registered in industrial practice,* Fourth International Conference on Advanced Materials and Structures AMS2011, Timişoara, Romania, October 27-28, pp 67

[10] Josan A, Pinca Bretotean C and Pauca A 2012 *The Use of Regenerated Mold Mixtures in Manganese Steel Piece Casting* Proceedings of the 3-rd International Conference on Energy, Environment, Devices, Systems, Communications, Computers (INEEE '12): Mathematical Modelling and Simulation in Applied Sciences, Rovaniemi, Finland, April 18-20, pp 162-164

IOP Conf. Series: Materials Science and Engineering **85** (2015) 012018 doi:10.1088/1757-899X/85/1/012018

Develop virtual joint laboratory for education like distance engineering system for robotic applications

T S Latinovic[1], S I Deaconu[2], M T Latinović[3], N Malešević[4] and C Barz[5]

[1] University of Banja Luka, Faculty of Mechanical engineering, Mechatronic Department, 78000 Banja Luka, Republic of Srpska, Bosnia and Herzegovina
[2] Politehnica University of Timisoara, Department of Electrical Engineering and Industrial Informatics, 5 Revolutiei Street, 331128 Hunedoara, Romania
[3] Agency for Information Society of the Republic of Srpska, Bosnia and Herzegovina
[4] University of Banja Luka, Faculty of Mechanical engineering, 78000 Banja Luka, Republic of Srpska, Bosnia and Herzegovina
[5] Technical University of Cluj-Napoca, North University Centre from Baia Mare, Department of Electrical, Electronic and Computer Engineering, 62A Victor Babes Street, 430083, Baia Mare, Romania

E-mail: tihomir.latinovic1@gmail.com

Abstract. This paper work with a new system that provides distance learning and online training engineers. The purpose of this paper is to develop and provide web-based system for the handling and control of remote devices via the Internet. Remote devices are currently the industry or mobile robots [13]. For future product development machine in the factory will be included in the system. This article also discusses the current use of virtual reality tools in the fields of science and engineering education. One programming tool in particular, virtual reality modeling language (VRML) is presented in the light of its applications and capabilities in the development of computer visualization tool for education. One contribution of this paper is to present the software tools and examples that can encourage educators to develop a virtual reality model to improve teaching in their discipline. [12] This paper aims to introduce a software platform, called VALIP where users can build, share, and manipulate 3D content in cooperation with the interaction processes in a 3D context, while participating hardware and software devices can be physical and / or logical distributed and connected together via the Internet. VALIP the integration of virtual laboratories to appropriate partners; therefore, allowing access to all laboratories in any of the partners in the project. VALIP provides advanced laboratory for training and research within robotics and production engineering, and thus, provides a great laboratory facilities with only having to invest a limited amount of resources at the local level to the partner site.

1. Introduction

A large number of applications that are based on a platform for distance learning developed in recent years [1], [2]. They made major changes and realize the potential for improving training and robot engineering.

Presented platform in operation, the goal of this application is the design and programming of the remote control. The application of industrial robots that are accessed over the Internet for this purpose, is the program creates witch emulates a practical laboratory in which engineering students can confirm the theoretical results by comparing them to the right paths affected by the robot. Students also have the same conditions as in a real lab, because the task and manage real laboratory.

The wide expansion of the World Wide Web and the Internet has created all necessary conditions for the adoption of this powerful means of learning in the common collaboration arena. E-learning can be a way to overcome the geographical isolation and physical distance from the source or learning centers. In these situations, the use of the Internet and connect via Ethernet realistic laboratory receives its full meaning. Robotics is a suitable problem domain, because it essentially requires multidisciplinary knowledge. Robotics covers topics such as mechanical engineering, electronics, control, communications, computer vision, real-time parallel computing and systems design.

- Robotics is very interesting for both children and adults
- Robotics events and contests to motivate the study of technical subjects
- Robotics provides hands-on examples for teaching science, technology and engineering

An increasing number of academic institutions are searching for new methodologies and techniques, teaching that allow students to face real professional situations [3]. This is especially important for interdisciplinary fields such as robotics that require the integration of knowledge from different fields, and a few skills for effective learning. Specifically, the paper combining project-oriented learning, virtual labs and remote labs realistic manageable realized innovative the courses in which students learn the basics of programming industrial robots robot for the competition.

Distance laboratories, houses laboratories and every kind of remote is not exactly new technology. It has previously been done, and some successful results have been received [4], [5]. Much remote laboratory projects after the initialization and a display time, and the process is not included in the paper. In this case, I will try to provide an integrated approach to the study of engineering fields (mechatronics, robotics) remotely without losing the practical hands-on experience at the same time.

In the next section we will introduce VALIP system that provides the basis for all development and implementation of the procedures described in this paper. Subsections 2.1 describes the concept of software components (their behavior and communications) and VIRCA (one of the main software tools that are used to represent virtual environment to the user), respectively. A practical example that describes a common scenario where VALIP system used is shown in Section 3 Finally, Section 4 contains the conclusions of the review and the work presented in this paper.

1.1. VALIP (Virtual Joint Laboratory for Advanced ICT)

The interactive visualization tools operate by accepting input values from users through a graphical user interface developed using a visual language such as Visual Basic, Visual C++, or any other object and graphics oriented language. The main program then creates a VRML file description of the visualization tool and inserts the user supplied values where necessary to create the desired virtual reality display and interactive animation which is then passed to the internet browser for viewing.

VALIP is an acronym which stands for Virtual Joint Laboratory for Advanced ICT (Information and Communication Technology) in Production. In more readable terms, it represents a common place where all collaborators can communicate and exchange information. It represents a virtual copy of the real world and real environment, e.g. laboratory or production environment with machines, robots and other industrial equipment, allowing the collaborators to remotely access the resources that would otherwise be unavailable to them.

Joint virtual laboratory – creating infrastructure for administration, maintenance and functioning of the program revolving around joint laboratories – including acquiring equipment for two laboratories

IOP Conf. Series: Materials Science and Engineering **85** (2015) 012018 doi:10.1088/1757-899X/85/1/012018

(remote sensing and remote controlling), joint software purchasing and development; integrating laboratories into teaching and research activities at UBL

VALIP uses a range of tools to accomplish this goal and it consists of software and hardware devices that all communicate together in order to display required information to the participants and allow them to interact with the environment, directly affecting the real world state through virtual workspace. 3D visualization can be supplemented by video-camera streaming over the Internet, providing in-depth information and a complete picture about the state of the environment.

VALIP provides an advanced laboratory for education and research within robotics and production engineering, and thus, offers access to huge laboratory facilities with only needing to invest a limited amount of resources locally at the partner's location.

The important part of this virtual reality is the safety of the robots, equipment and personnel. The primary idea is to have a program that will communicate with safety PLC on one side, and virtual reality generator on other side. This program will represent one of the main components in VALIP system.

When the safety light beams are blocked, PLC will stop the robot and send a signal to designated component in VALIP system which will further inform the whole system about the new state of the cell. Virtual reality generator will reflect this information to visual presentation of the cell. The reverse communication will also be possible. Person who is remotely operating the cell will be able to send commands to the PLC to power the robots or the conveyors, through user interface rendered in virtual reality.[9], [10], [11] The visualization of this program is represented in the Figure 1.

Figure 1. Control workstation with multiple screens, displaying VirCA application window, virtual reality windows, and real laboratory Control workstation with multiple screens, displaying VirCA application window, virtual reality windows, and real laboratory

2. Object components

Inside of VALIP environment, every physical or logical object is represented by software component which describes the object and their two most important aspects:

- The behavior of the object, taking into account its internal behavior and the response to actions from other objects connected to it, and
- The ports or connectors through which the object communicates with other objects in the system.

Since all objects (machines, robots, sensors, logical processes, etc.) in the real world exchange the information between them, it is crucial that their virtual representation has the same ability. This is accomplished by implementing custom ports to software components, based on how they interact with the environment in the real world, through OpenRTM-aist [6], a software platform for development of component-based robotic systems. The components can be developed in several programming languages, most notably C++ and Python.

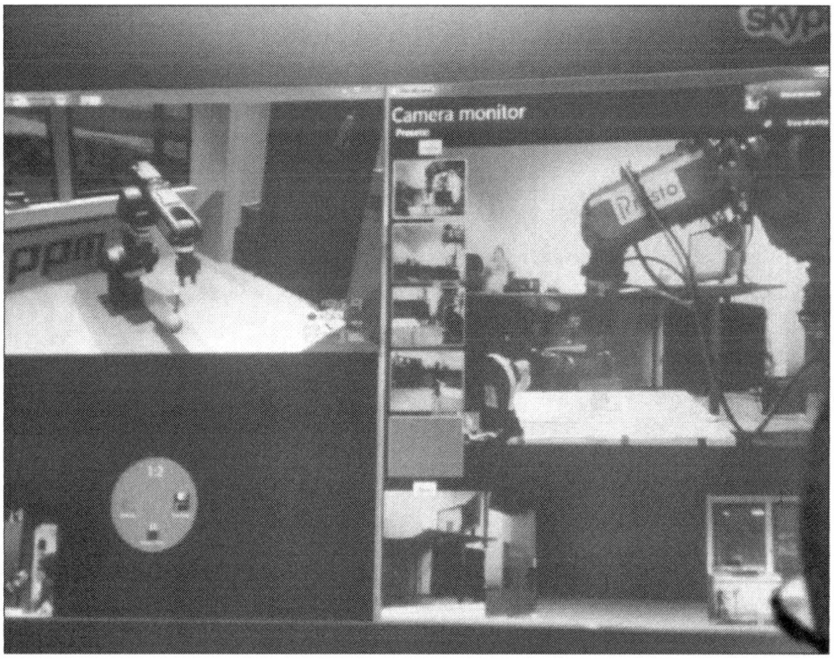

Figure 2. Control workstation with multiple screens, displaying VirCA application window, virtual reality windows, and real laboratory Control workstation with multiple screens, displaying VirCA application window, virtual reality windows, and real laboratory

The communication between the components is established through one of the following port types, which can be based on either CORBA or ICE implementation [1]:

- Data Port – used for continuous exchange of information between the components (CORBA based),
- Service Port – used for interface-based communication defined by IDL (Interface Description File) (CORBA based), and
- Ice Port – used for interface-based communication defined by Slice (Specification Language for Ice) port (Ice based).

In addition, each port can either act exclusively as a provider or a consumer. This means that in order for components to have a two-way communication between them, each of them has to have a set of provider/consumer ports. The example for component layout and their connections will be presented in section 3.

If you would like to register a picture into VirCA you have to place a new node in the "scene" file of the room and have to create a new "material" file in textures folder.

Add this node to the scene file of the room:

```
<node name="Pic1" >
  <position x="349" y="150" z="0" />
  <scale x="2" y="2" z="2" />
  <rotation w="1" x="0" y="0" z="0" />
  <plane name="Pic1"
        castShadows="false"
        receiveShadows="true"
        distance="0"
        width="40"
        height="65"
        xSegments="1"
        ySegments="1"
        numTexCoordSets="1"
        uTile="1"
        vTile="1"
        material="04pic01"
        normals="true"
        movablePlane="true">
        <normal x="-1" y="0" z="0" />
        <upVector x="0" y="0" z="-1" />
  </plane>
</node>
```

In every example, the configuration of specified displays is based on the real circumstances of the environment, where VirCA should be visualized. The origin of the coordinate system, in which the displays should be determined, is the user's actual point of view. That is the point where the user will watch the display(s) from. [22] Axes of the coordinate system are defined by the conventions of the Ogre3D engine. Namely the X dimension means the width (or horizontal axis) on the display, the Y refers to the height (or vertical axis) on the display, and the Z dimension means the depth on the display, in the way, that the negative direction of the axis points towards the display.

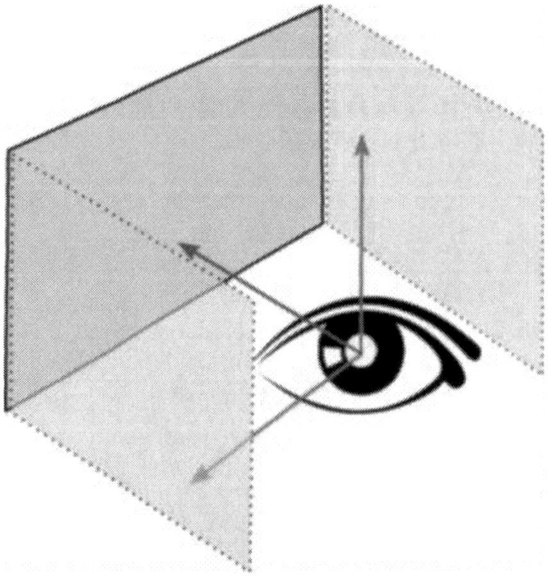

Figure 3. COORDSYS.

2.1. VIRCA - Virtual environment

VALIP uses software package VirCA[23] as its primary tool for presenting the information to the user [2] [7]. VirCA is an acronym and stands for Virtual Collaboration Arena. It is a software application which renders a 3D environment inside of which virtual reality can be presented to end users. Collaborators can interact with the objects displayed in application window and inspect the current state of the virtual world (which, as already mentioned, represents the copy of the real world) in real-time [3].

Figure 4. VIRCA room.

The content inside of virtual workspace is defined by components that are connected to VirCA component. These components constantly send information about their virtual representation allowing VirCA to render their 3D models or other graphical details inside of virtual world.

The user is able to interact with the components through VirCA application window or by using dedicated control components. Sample tasks include turning on the conveyer belt or instructing the robot to pick up an object.

The Virtual Collaboration Arena is a modular built and component based, distributed, interactive virtual reality manager system for connecting virtual and real rooms, displaying and manipulating the devices in them, realizing collaboration that way.

[20] The VirCA system - as virtual reality manager system - handles on the one hand physical, on the other hand communication rooms in order to realize distributed, interactive collaboration. To the virtual room different real rooms can be connected, providing the relation to the reality that way. In the virtual reality, there are representations of virtual and real devices as well as static and dynamic virtual objects, to which last physics simulation can be applied. The user can display the virtual room with any kind of display devices and manipulate it with any kind of input devices, e.g. drag'n'drop dynamic virtual objects. The virtual and real devices can refresh their representations through so-called trackers and can retrieve information from the virtual room through virtual sensors.

[21] In the communication room, the devices can share their own knowledge, which can be some kind of information, service or ability. Hereby the devices in the room can use the knowledge of each other, realizing the knowledge plug n play that way between intelligent entities. So the knowledge implemented once need not be implemented again for the given application, but as a component, it is available on the internet, can be reused. The user can access the abilities of the single devices through

a standard interface, the so-called commander, while the devices can be controlled from the communication room through a standard interface as well, the so-called controller.

[18] The novelty of the VirCA system is on the other hand the connection of virtual and real physical rooms. The user can see - with real 3D as well - and manipulate the different, from each other physically far devices together, on a standard surface, so VirCA provides information integration. With the help of the virtual objects, the cooperation of such real devices far from each other can be tested without the development of a simulator, but with greater precision, the transport or purchase of which would be expensive otherwise, so VirCA is an efficient test environment. [19] The sensors created in the virtual room can be virtual representations of existing, but very expensive, or even not existing (future sensors) sensors, which operate transparent for the devices, identical to the real sensors, so VirCA enables fast development.

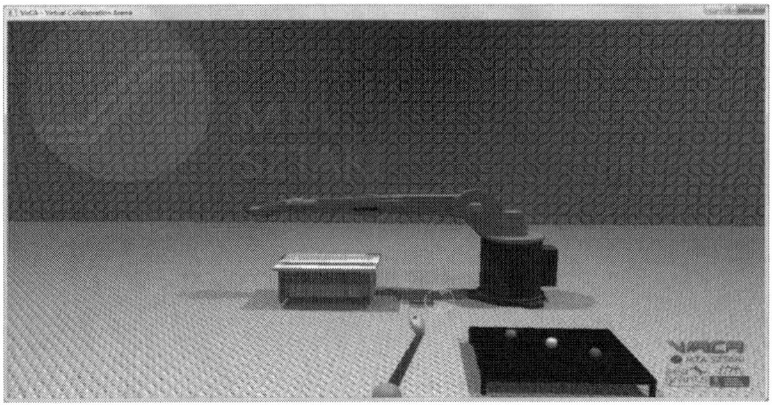

Figure 5. VIRCA room with Robot.

One of the two VirCA has to be the slave and the other has to be the master. You can set it in the rtc.conf file - just change the value (true or false) of this row:

```
VirCA.IsMaster: true
```

If you set true then your VirCA is a master, if you set false your VirCA is a slave.

Note: If you use VirCA and start the slave VirCA then you have to choose a slave number of it between 0 – 3.

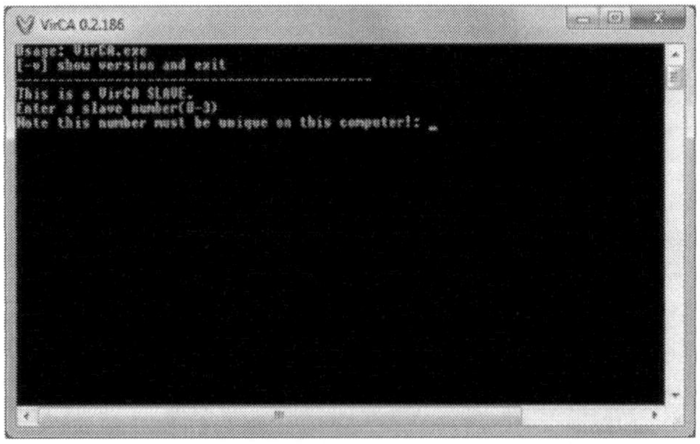

Figure 6. VIRCA basic screen.

3. Practical example

Practical example of VALIP-based system is demonstrated by performing a remote operation on robotic cell located at Narvik University College (NUC). The task described represents a basic functionality of the robotic cell and contains procedures which are implemented in any other task robots commonly perform in industrial environments. Allmost all other types of robotic assignments are constructed upon simple instructions of robot movement, gripper actuation (analogously welding control) and sensor reading, all of which are included in the steps of this task.[8]

For the sake of simplicity, the cell consists of a single robot with the gripper attached to it, one conveyer belt equipped with sensors and a simple table. The robot task consists of the following steps:

- picking up the object from specific position on the table,
- placing the object on the conveyer belt,
- waiting for the object to trigger one of the detection sensors located on conveyer belt,
- positioning the robot at object pickup location,
- waiting for the object to approach the position,
- picking up the object,
- Placing the object back at the original pickup position on the table.

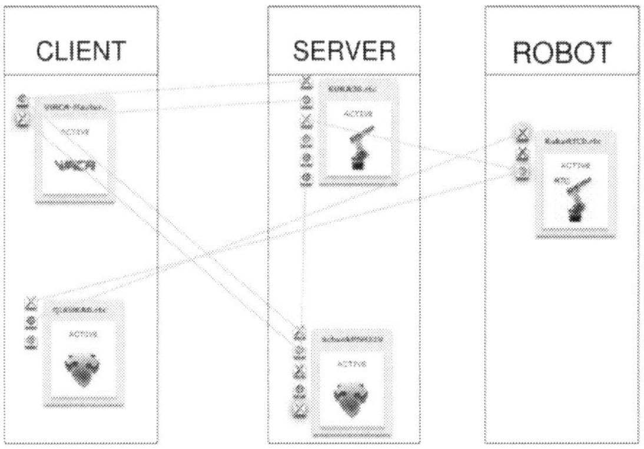

Figure 7. Component layout and connections between them.

The system consists of multiple software components which are run on three computers. Server computer is located at NUC and provides a name server to which all of the components connect and communicate through. [17] Robot computer is also located at NUC and is directly connected to KUKA robot. Finally, client computer is located at UBL and is used to display virtual reality to remote users. Each of the components is described below.

- Karat. Used for robot control, this component is run on a dedicated robot computer. This computer is directly connected to the robot and also connected to the server computer via separate LAN network or Internet connection. The components allow VALIP to communicate to robot and exchange data with it.
- Kuka3. Used for providing VirCA with the information about robot position and orientation, allowing it to render robot model. This component is run on server computer.
- SchunkPSH32Vis. Used for providing VirCA with the information about gripper position and orientation, allowing it to render gripper model. This component is run on server computer. It is designed as a separate component to allow gripper to be easily replaced with different model (with its own visualization component).
- QtKUKA. Used to provide additional GUI to allow further control by the user. While it is also possible to control the task directly through VirCA, this GUI presents more advanced options to the user, making the whole user experience more intuitive. This component is run at client computer, allowing for remote operation.

- VirCA. Software tool used to render the environment and component models as a virtual reality. This component is run at client computer, allowing for remote operation.

All components are connected to each other, exchanging the information between them. The connections are established using System Editor, yet another software tool, located at server computer. Ice ports are marked in blue color and are mainly used for visual information exchange between visualization components and VirCA. CORBA Data ports are marked in green color and are used for continuous feed of information between components. Red colored CORBA data ports are used for IDL-based control interfaces.

Figure 8.1 Physical robotic cell.

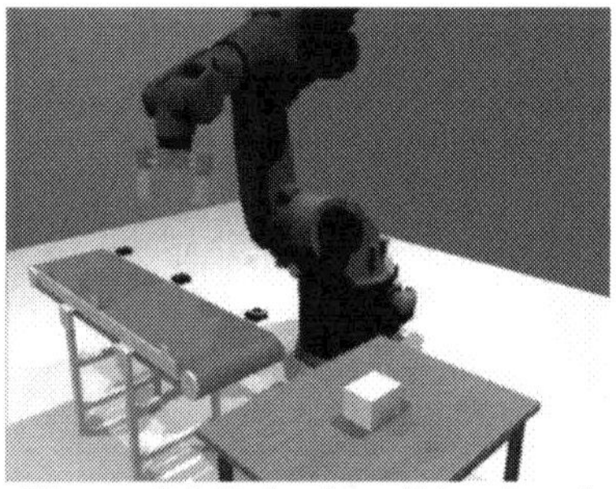

Figure 8.2 Robotic cell in virtual reality as rendered by VirCA.

Since the robot is already programmed for this task, the process is being monitored through VirCA from Laboratory for Intelligent Systems at University of Banja Luka (UBL) which is equipped with required hardware and software. More importantly, the whole task can be manually started, positions and other parameters modified and process controlled from this location, more than 2.600 km away from the physical robotic cell (or any other location world-wide, for that matter).

4. Conclusion

In this article, has been presented real-time controlled Laboratory through the Internet. The main use-value of the system is that it helps the student to perform practice experiments remotely. With the virtual lab developed, students can learn robotic concepts such as direct/inverse cinematic, path planning, dynamics and programming. The user interface is very user-friendly, and the graphical simulation very realistic.

[14] The remote capabilities of the application allow users to experiment with real equipment through the Internet. Finally, the system presented collects a lot of interesting virtual and remote features (complete robot simulation, robot dynamics, remote power and robot control, augmented reality, etc.); the use of this new technology facilitates a reduction in the impact of the growing number of students on the familiar constraints of laboratory time and cost.

[15] This kind of virtual laboratory allows the teacher and the students of an on-line course through the Internet to use shared simulations in order to experiment practical concepts in a coordinated way [8]. Also, if we built Virtual Joint Laboratory, then we are the part of ISPACE. The iSpace Laboratory Network is a worldwide scientific non-profit network of professors, researchers and laboratories engaging in active research in the intelligent space-related

This paper was focused on describing VALIP as a feasible solution that fills in the gaps that are present in conventional methods of control and monitoring. [16] VALIP represents highly modular and customizable product, and offers tools that allow end users to easily integrate components of robotic cells, production lines and other similar environments, allowing personnel to remotely interact with them.

References

[1] Foss B A, Malving K E and Eikaas T I 2004 *Remote Experimentation–New Content in Distance Learning,* International Conference on Engineering Education, Oslo, Norway, August 6 – 10, 2001, pp. 8D1-1–8D1-6, Session 8D1

[2] Sanchez J, Dormido S, Pastor R and Morilla F 2004 A Java/Matlab–Based Environment for Remote Control System Laboratories: Illustrated with an Inverted Pendulum, *IEEE Transactions on Education* **47**(3) 321-329

[3] Tzafestas C S, Alifragis M, Palaiologou N, Thomopoulos S C A, Brahman M and Exarchou A E 2004 *Development and Experimental Evaluation of a Remote Laboratory Platform for Teaching Robot Manipulator Programming,*International Conference on Engineering Education ICEE'2004, Florida, USA, October 16-21, pp 322-324

[4] Martín M 2002 *El modelo educativo del Tecnológico de Monterrey* Ed. Tecnológic de Monterrey, Nuevo León, México, April 20-22, pp 17-29

[5] Noguez J, Sucar L E and Ramos F 2003 *A probabilistic relational model for virtual laboratories* International Conference on Artificial Intelligence in Education, Sydney, Australia, July 20-24, pp 533-534

[6] Lopez D, Cdazo R, Sinchez F and Sebastian M 2007 *CICLOPE ROBOT: A Remote Laboratory for Teaching Embedded Real Time Systems*, Industrial Electronics ISIE - IEEE International Symposium, Madrid, Espania, June 4-7, pp 1200-1205

[7] Murtra M, Jansà G, Martínez H, Domingo J, Gámiz J and Grau A 2007 *A Proposal of Remote Laboratory for Distance Training in Robotic Applications*, Emerging Technologies & Factory Automation ETFA - IEEE Conference, Patras, Greece, September 25-28, pp 124-128

[8] Malešević N, Sziebig G, Solvang B and Latinovic T 2013 Simulation of robotic tasks with VALIP system – Practical application, *Annals of Faculty Engineering Hunedoara – International Journal of Engineering* **XI**(4) 305-308

[9] Kerezović T, Sziebig G, Solvang B and Latinovic T, Human safety in robot applications–review of safety trends, Acta Technica Corviniensis **7** 113-118

[10] Latinovic T, Deaconu S I, Labudski R and Topor M 2013 *Intelligent approach for mobile robot simulator with robosim software,* DEMI 2013 Conference Proceedings, Banja Luka, Bosnia and Herzegovina, May 30- June 1, pp 857-862

[11] Latinovic T, Deaconu S I, Djurdjevic M and Dobrnjac M 2011 *The basics of designing controllers for industrial robots, eg. Robots ABB IRB 2000,* DEMI 2013 Conference Proceedings, Banja Luka, Bosnia and Herzegovina, May 30- June 1, pp 101–104

[12] Latinovic T, Deaconu S, Popov G and Topor M 2011 Intelligent workcell robot system in manufacturing with ABB IRB 2000, *MTM JOURNAL ONLINE* **12** 43-45

[13] Sell R and Otto T 2008 *Remotely controlled multi robot environment*, Proceedings of 19th EAEEIE Annual Conference, Tallinn, Estonia, May 30-31, pp 20-25

[14] Galambos P and Baranyi P 2011 *VirCA as Virtual Intelligent Space for RT-Middleware*, IEEE/ASME International Conference on Advanced Intelligent Mechatronics, Budapest, Hungary, July 3-7, pp 140-145

[15] Sziebig G and Øritsland T A 2012 *Navigating in 3D Immersive Environments: a VirCA usability study*, 10th IFAC Symposium on Robot Control International Federation of Automatic Control, Dubrovnik, Croatia, September 5-7, pp 380-384

[16] Solvang B G 2012 *On Industrial Robots and Cognitive Info-Communication,* 3rd IEEE International Conference on Cognitive Info-communications, Kosice, Slovakia, December 2-5, pp 459-464

[17] Aryania A, Thomessen T and Sziebig G 2012 *New trends in industrial robot controller user interfaces,* 3rd IEEE International Conference on Cognitive Info-communications, Kosice, Slovakia, December 2-5, pp 365-368

[18] Familia R 2005 *A Virtual Laboratory for Cooperative Learning of Robotics and Mechatronics,*

ITHET 6th Annual International Conference, Juan Dolio, Dominican Republic, July 07-09, pp 126-128

[19] Haslina A, Jaslinda J and Shahnorbanun S 2010 *Teaching Robot Kinematic in a Virtual Environment*, WSEAS Proceedings of the World Congress on Engineering and Computer Science, San Francisco, USA, October 20-22, p 1001

[20] Reza N J 2010 *Theory of Applied Robotics: Kinematics, Dynamics, and Control*, 2nd Edition, Springer, Melbourne, Australia

[21] ***National Institute of Advanced Industrial Science and Technology OpenRTM-aist, from http://www.openrtm.org/openrtm/en/node/629

[22] ***Joint Laboratory for Emerging Information Technology in Manufacturing (ITM). VirCA from http://www.virca.hu

[23] ***http://virca.hu/vircapedia

Influence of volumic heat treatments upon cavitation erosion resistance of duplex X2CrNiMoN 22-5-3 stainless steels

L M Micu[1], I Bordeasu[1], M O Popoviciu[2], M Popescu[1], D Bordeasu[3] and L C Salcianu[1]

[1] Politehnica University of Timisoara, 2 Piata Victoriei, 300006 Timisoara, Romania
[2] Academy of Romanian Scientists, 1 Bvd. Mihai Viteazul, Timisoara, Romania
[3] Via Universisty College, Horsens, Denmark

E-mail: lavimicu@yahoo.com

Abstract. The stainless steels Duplex 2205 with austenite and ferrite structure have mechanical characteristics close to those of martensite stainless steels but a better corrosion resistance; these steels are very sensitive on the heat treatments. Present work studies the cavitation erosion for those steels for three different heat treatments: simply quenched, annealed at 475°C post quenching and annealed at 875°C. The researches were undertaken at Timisoara "Politehnica" University in the Laboratory of Material Science and the Laboratory of Cavitation, using the T2 facility which integrally respects the recommendation of ASTM G32-10 Standard. The best results were obtained with the specimens annealed at 875°C. In comparison with the stainless steel 41Cr4, with very good cavitation erosion qualities, all tested steels presented also good erosion resistance. So, Duplex 2205 steels can be used for details subjected to cavitation. The best results are obtained by increasing both the hardness and the quantity of the structure constituent with better cavitation erosion resistance, in our case the alloyed austenite.

1. Introduction

For manufacturing marine structures carrying chemical aggressive substances frequently are used the stainless steels Duplex 2205/ X2CrNiMoN 22-5-3 because it has very good mechanical properties (ultimate resistance 736 N/mm^2, elongation 34%, yield point 570 N/mm^2) but also because it has excellent welding qualities. The welded joints have both good tenacity and corrosion resistance [1], [5], [6], [7], [13], [15], [18]. Important is also the fact that after the detail wear out, the material can be melted and reused. This steel is very sensitive to heat treatments and the effect of these treatments must be thoroughly known [9]. By quenching the Duplex 2205 with heating at a temperature of (1049…1149) °C the structure obtain a small quantity of ferrite, regardless of the cooling velocity and by tempering at temperatures between (449…599)°C a good structural hardening is obtained [9]. For the same content in Cr and Ni, the behavior of the Duplex stainless steel to pit corrosion is superior to those of austenitic steels [5]. In present, there are undertaken numerous studies for increasing also the mechanical properties of Duplex steels, through volume heat treatments. Taking into consideration these tendencies the present paper reveals the importance of adequate heat treatments in order to obtain an increased resistance to cavitation erosion of the X2CrNiMoN22-5-3 Duplex stainless steel.

The cavitation researches were performed in Timisoara Cavitation Laboratory using a vibratory facility, strictly respecting the recommendation of ASTM G32-10 Standard. To ascertain the level of cavitation erosion resistance, the obtained values were compared with those of the stainless steel 41Cr4, a material with good cavitation behavior both in laboratory and processing plants.

2. Researched material and applied treatments

The researched material is Duplex 2205 stainless steel, symbolized X2CrNiMoN22-5-3 in conformity with the European Standard EN 10088 [12]. For comparisons, was chosen the stainless steel 41Cr4, considered with good cavitation erosion behavior [1]. The chemical composition is presented in table 1 and the mechanical characteristics in table 2.

Table 1. Chemical composition (%) of the researched steels [1], [9], [13]

Steel	C	Mn	P	S	Si	Ni	Cr	Mo	N
X2CrNiMoN22-5-3	0.017	1.837	0.024	0.0002	0.313	5.019	22.083	2.585	0.1502
41Cr4	0.445	0.561	-	0.09	0.28	-	0.94	-	-

Table 2. Mechanical characteristics before heat treatment [1], [14]

Steel	Hardness HB	Yield point $Rp_{0,2}$ (N/mm^2)	Ultimate strength R_m (N/mm^2)	Elongation at fracture A_5 (%)
X2CrNiMoN22-5-3	240	450	Min.650	25
41Cr4	238	790	808	12

From Table 1 it can be seen that in comparison with 41Cr4, the stainless steel X2CrNiMoN22-5-3 has a smaller content of carbon and consequently a better welding ability. In the same time 41Cr4 has better mechanical characteristics ($Rp_{0,2}$ and Rm) which improve the cavitation erosion resistance [1], [2], [17], [18]. The heat treatments were done in the Laboratory for Material Science and the cavitation erosion tests in the Cavitation Erosion Laboratory, both belonging to Timisoara "Politehnica" University. The cavitation erosion tests respect the Standard ASTM G32-2010 [10], [11].

The tested material X2CrNiMoN22-5-3 was subjected to three different heat treatments:

A. quenching for putting into solution by heating at 1060°C during 30 minutes and after that cooled in water;
B. quenching in the same way but subjected to annealing by heating at 475°C during 4 hours, followed by air cooling;
C. quenching in the same way but with annealing by heating at 850°C during 2 hours, followed by air cooling.

After the heat treatments the samples hardness was measured. The mean results are:

- 275 HV1 for sample A
- 330 HV1 for sample B
- 362 HV1 for sample C

For each type of heat treatment were subjected to cavitation erosion three specimens, following the recommendations of the Standard ASTM G32-2010 [4], [10]. The mean value of these three results was considered representative for cavitation erosion. Figure 1 present the surface aspect of the heat treated specimens before the cavitation tests.

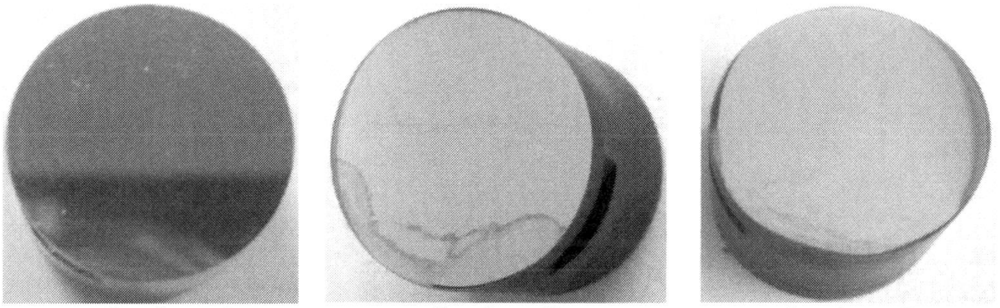

Specimen A Specimen B Specimen C

Figure 1. Surface aspect before cavitation erosion exposure (minute 0)

3. Experimental results

From each sample were extracted five specimens, three of them were tested respecting the ASTM G32 Standard. The total duration of the cavitation test was 165 minutes divided in 12 periods (the first two were shorter 5 and 10 minutes the remaining ten were of 15 minutes). The test facility of the Timisoara Polytechnic University Cavitation Laboratory is presented in Figure 2 and respect integrally the ASTM Standard. All the parameters of the testing facility were permanently controlled and maintained at same level [2], [3]. As testing liquid was used drinking water from the urban net. The testing temperature was maintained at 22-23°C.

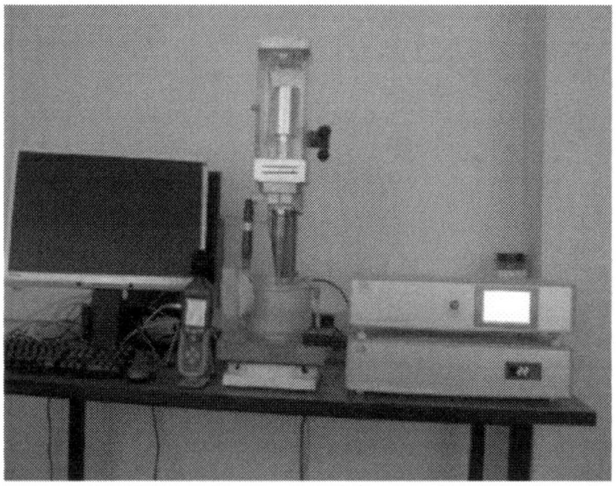

a)

b)

Figure 2 Test devices T2 with piezoelectric crystals
a) – general view, b) – details: beaker, specimen and horn

After each of the 12 testing periods the specimens were carefully washed, dried and weighted. The mean value of these three tested specimens was considered the mean depth of erosion and was used for plotting the characteristic curves MDE(t) (mean depth erosion against time). A similar procedure was used for MDER(t) (mean depth erosion rate against time) [16], [17], [8]. One specimen from each group was longitudinally cut-off (see Figure 3) in order to analyze the shape of the eroded area, the maximum depth of the erosion and possible cracks propagation in the structure (Figure 6). For structural microscopic analyze the cut specimens were embedded in resin (Figure 3). To evaluate the cavitation erosion resistance were used the mean depth erosion MDE, Figure 4, and mean depth erosion rate MDER, Figure 5, as recommended by ASTM G32-10 [16]. In Figure 4 are given photographic images after three exposure times (60, 120 and 165 minutes). Those images, in correlation with figure 6 show clearly the effect of the volume heat treatment upon the behavior at cavitation erosion.

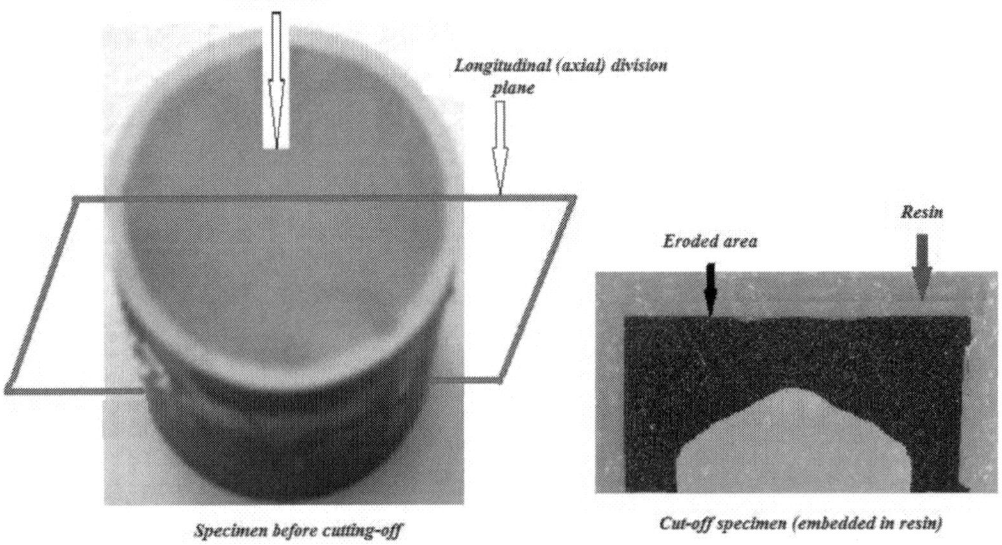

Figure 3. Specimen preparation for microscopic analyse.

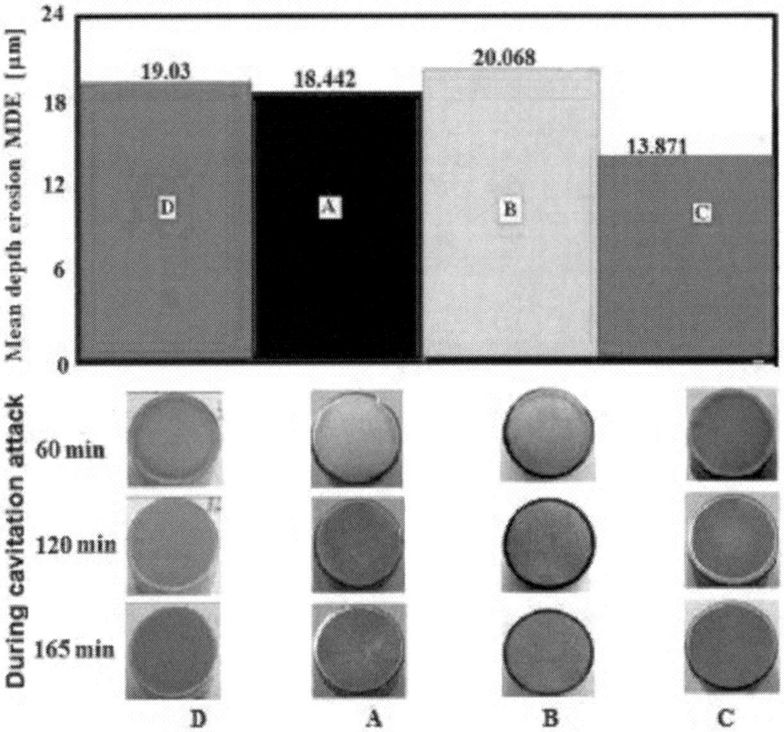

-Eroded areas for three different exposures-

Figure 4. Mean depth erosion obtained after 165 minutes of cavitation exposure. There were used the following symbols:

A) Duplex 2205 Quenched and cooled in water, **B**) Duplex 2205 Quenched/ annealed at 475°C/air cooled, **C**) Quenched specimen/ annealed at 850°C/ air cooled and **D**) 41Cr4 Stainless steel

IOP Conf. Series: Materials Science and Engineering 85 (2015) 012019 doi:10.1088/1757-899X/85/1/012019

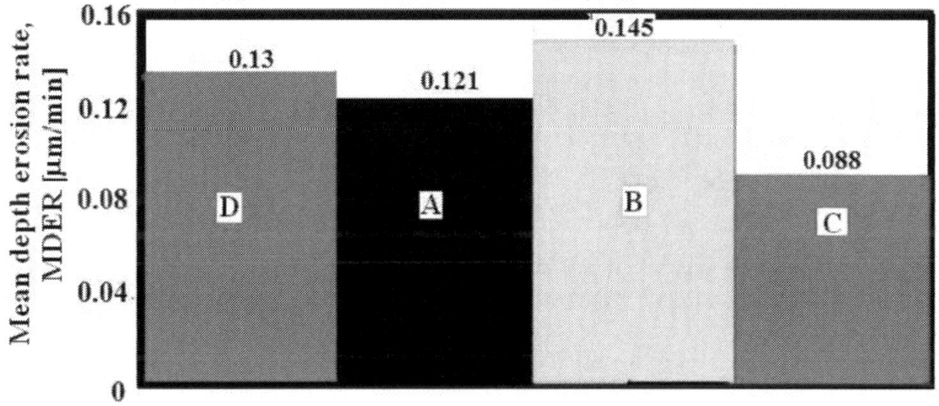

Figure 5. Mean depth erosion rate at final stage (same symbols). Symbols used: **A**) Duplex 2205 Quenched and cooled in water, **B**) Duplex 2205 Quenched/ annealed at 475°C/air cooled, **C**) Quenched specimen/ annealed at 850°C/ air cooled and **D**) 41Cr4 Stainless steel

By comparison with the control specimen 41Cr4 results that all heat treated Duplex steels have good cavitation erosion behavior. The best results were obtained for the heat treatment C (annealed by heating at 850°C and maintained 2 hours at this temperature). Those specimens present also the greatest hardness 362 HV1. The worst results were obtained for the heat treatment B (annealed by heating at 475°C and maintained 4 hours at this temperature) even if the hardness is 330 HV1, greater that of the specimens simply quenched (hardness 275 HV1). This result can be explained by the structure of the specimens, which has increased quantities of ferrite. The specimen C has cavitation erosion enhanced with approximate 10% in comparison to the control specimen (even if the hardness is better with 27%) because of the content of the structural component ferrite.

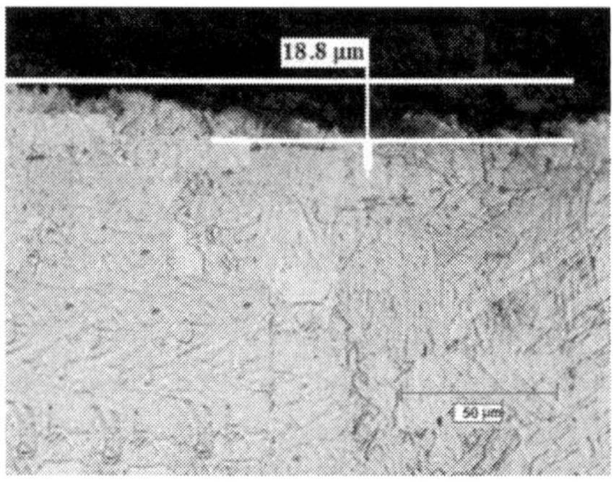

Heat treatment A

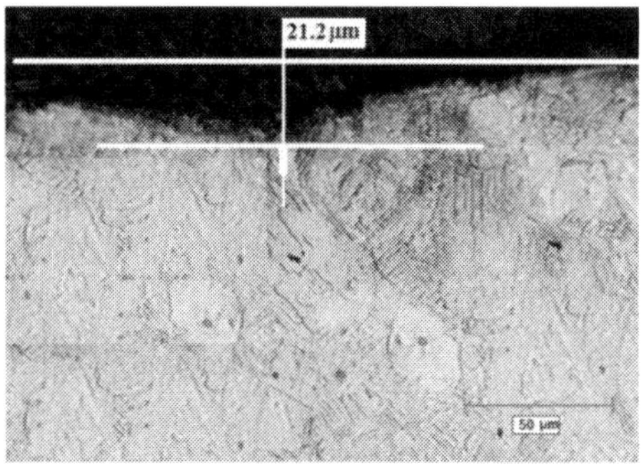

Heat treatment B

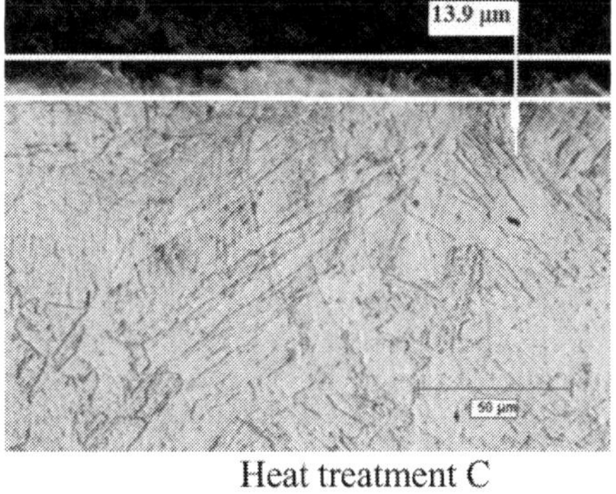

Heat treatment C

Figure 6. Sections through specimens subjected to final cavitation exposure (8x).

The better resistance of the specimen C is given not only by the greatest hardness but also by the structure in which the austenite is dominant. In Figure 6 are given the microscopic aspects of the eroded area, in the cross sections, near the center of the specimen (zone in which the erosion has its maximum depth). Comparing the maximum erosion depth (from Figure 6) with the mean depth erosion (computed from the lost mass divided with the material density and the specimen area), Figure 4, resulted incredible small differences. When the material is a weak one, the differences increase (in our case till approximate 5.6%). This result can be explained by the fact that in same cases an entire ferrite grain can be expelled. For the resistant materials, the erosion occur by expelling only small parts of the austenite cracked grain, so the difference between MDE and h_{max} is in the range of 1%. The classification of cavitation resistance is the same regardless the procedure adopted (using MDE or h_{max}).

The images in figure 6 present a settlement of the material layer in close vicinity with the layer suffering cavitation erosion. This settlement characterize the materials with excellent cavitation behavior because lead to the reduction of the mean depth erosion rate [1].

4. Conclusions

1. Adequate heat treatments improve cavitation erosion resistance of Duplex stainless steels.
2. The best results are obtained by simultaneously increasing the hardness and improving the structure of the steels. In our case the increase of the alloyed austenite enhanced the behavior of steels annealed at 850°C. The increase of hardness without obtaining the adequate structure can worsen the results.
3. For Duplex steels with reduced cavitation erosion, the differences between the computed MDE and the measured h_{max} are insignificant.

References

[1] Bordeasu I 2006 *Cavitation Erosion of Materials*, Timisoara Polytechnic Publishing House, Timisoara, Romania
[2] Bordeasu I and Popoviciu M O 2012 Improving Cavitation Erosion Resistance through Surface and Structural Hardening, *Machine Designe* **4**(3) 171-176
[3] Bordeasu I and Mitelea I 2012 Cavitation Erosion Behavior for some Stainless Steels with Constant Nickel and Variable Chromium Content, *MP Material Testing* **1** 53-58
[4] Frank J P and Michel J M 2004 *Fundamentals of Cavitation*, Kluwer Academic Publishers, Boston, USA
[5] Micu L M, Bordeaşu I, Mitelea I, Ghera C and Salcianu L 2014 Researches upon the Cavitation Erosion of the Stainless Steel X2CrNiMn22-5-3 heat treated, *Science and Engineering* **26** 425-430
[6] Mitelea I 2009 *Material Science*, Vol.1, Timisoara Polytechnic Publishing House, Timisoara, Romania
[7] Mitelea I, Bordeasu I and Hadar A 2005 The Effect of Nickel Content Upon Cavitation Erosion for Stainless Steels with 13% Chromium and less than 0,1% Carbon, *Chemical Review* **56**(11) 1169-1174
[8] Oanca O V, Bordeasu I, Mitelea I and Craciunescu C 2013 *Phenomenology of Degradation by Cavitation for Heat Treated CuNiAlFe Bronzes*, 22-th International Conference on Metallurgy and Materials, Brno, Czech Republic, May 15-17, pp 1561-1566
[9] Trusculescu M and Ieremia A 1983 Stainless and Refractory Steels, Facla Publishing House, Timisoara, Romanian
[10] ***2010 Standard method of vibratory cavitation erosion test, ASTM, Standard G32
[11] ***2009 ASTM Standard G 32. Cavitation Erosion Using Vibratory Apparatus. ASTM International,West Conshohocken, PA
[12] ***http://www.inoxservice.hu/index.php/ro/rozsdamentesacel
[13] ***http://www.creeaza.com/tehnologie/tehnica-mecanica/Sudarea-otelul-inoxidabil-DUPL261.php

[14] ***http://www.outokumpu.com/SiteCollectionDocuments/Outokumpu-Duplex-Stainless-Steel-Data-Sheet.pdf

[15] ***http://tehnologie/tehnica-mecanica/PROIECT-Studiu-tehnico-economi843.php

[16] Bordeasu I, Popoviciu M O, Micu L M , Salcianu L C and Bordeasu C 2014 *Cavitation Erosion Researches upon two AMPCO Bronzes*, KOD 2014, Balatonfüred, Hungary, June 12-15, pp 249-254

[17] Bordeasu I, Popoviciu M O, Mitelea I, Ghiban B, Ghiban N, Sava M, Duma S T and Badarau R 2013 Correlations between mechanical properties and cavitation erosion resistance for stainless steels with 12% Chromium and variable contents of Nickel, *IOP Conference Series: Materials Science and Engineering* **57** 012006

[18] Bordeasu I, Popoviciu M O, Mitelea I, Ghiban B and Ghiban N 2013 Cavitation erosion resistance of two steels with the same percentage of Chromium and Nickel but different Carbon content, *IOP Conference Series: Materials Science and Engineering* **57** 012007

Simulation of the bimetal cast in the case of milling rolls

G Mihut[1] and E Popa[1]

[1] Politehnica University of Timisoara, Department of Engineering and Management, 5 Revolution Street, 331128 Hunedoara, Romania

E-mail: gabriela.mihut@fih.upt.ro

Abstract. In the paper it is proposed, in main, to obtain of a model of numerical simulation, valid general and applicable the whole peculiars cases of bimetal casting, model with which help can be studied through the computer, the optimization possibility of flowing working condition of liquid alloy of the distribution of temperatures field, of the liquid phase and contraction during the solidification, with the minimum price (necessary reimbursement of the software and calculus equipment) in very short time etc.

1. Introduction

Bimetals cylinders rolling destination are make, in Romania, by casting of a two alloys types into a shape from making (for inferiors axels and maselote) combined with metallic one (for pane), this casting could be made static or dynamic (S.C. Fortus S.A. Iaşi).

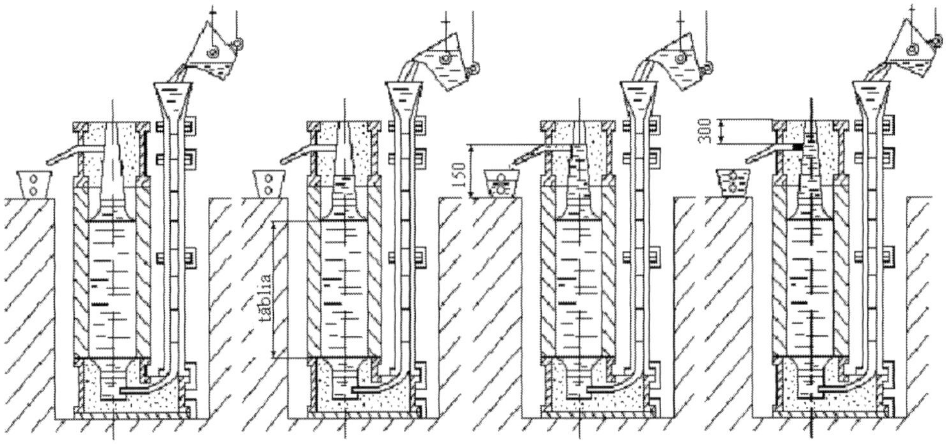

Figure 1. Casting duplex proceeding scheme of a rolling cylinders.

Static casting presume the casting of an alloy named “primary”, alloy whit who it is wanted to obtain of an hard crust, and an alloy named “secondary”, who is whicker alloyed and leads to the obtain of a middle and a inferiors axels with mechanical characteristics of an lower resistance, but tenacious. This variant of casting, like in Figure 1, presume more stage: primary alloy casting trough the casting net, until the filling of a certain part of casting shape (over the chill mould level); after this operation, it is stay a time part to realization the hard crust solidification, from the chill mould thru the shape center, next moment a refer to beginning of gradually washing of the middle with secondary alloy, and then the finale stage – continuous washing, until the axels and middle obtaining [1].

In this paper has as purpose the technology optimization of bimetal casting, in static variant, as a result of economic technical analysis of the technology from S.C. Cilindrul S.A. Călan, on a 78 cylinders ϕ 750x1740 mm and ϕ 928x3300 mm lot, and also, by originals contributions bringed casting ansamble, by numerical modeled and simulation of the shape filling processes and alloys solidification. In sequel, under surname "communication vessels system", it is will be present the purposed bimetal casting technology [2].

2. Experiments

2.1. The establish of the bimetal casting technology (of a communication vessels), in porous metal consumption reduction and taking out index increasing

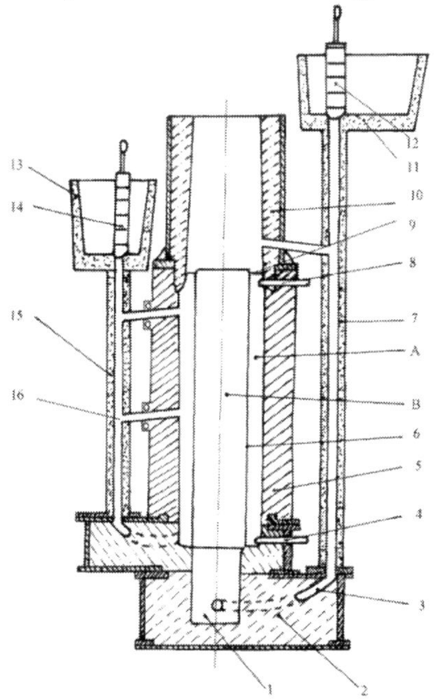

Figure 2. Proceeding and plant for casting of bimetals rolling cylinders, with the possibility of primary alloy casting on ranged net:
1 - inferior axle shape; 2 - inferior frame; 3 - alimentary channel; 4 and 8 - level signaling device; 5 - metallic shape (chill mould); 6 - dissolvable membrane; 7 and 15 the foot of casting net; 9 - membrane supports; 10 - superior frame (for making of superior axle); 11 and 13 - casting reservoir; 12 and 14 - stopper; 16 - the alimentary of ranged net.

The presented proceeding in Figure 2 remove the classical method short comings, anterior remembered, by using of un exact quantity of an secondary and primary alloys, realization an hard crust and passing stratum dimension well checked, removing of cracks appearance to warm or/and to could and the cylinder relating, on base internal tension decreasing.

The installation for the proceeding realization it is composes from an inferior frame 2, provided with an alimentary channel 3 whitch is in relation with an alimentary net 7 (in cascade, with two alimentary). This from behind distribute secondary liquid alloy from casting reservoir 11, assured with an stopper 12, which role is to begin and interrupt the shape supply and, also, retain the slag in reservoir 11.

Over inferior axle shape 2 it is set the metallic shape 5, having vertical separation surface to allowed the making out and extract casted raw cylinder. In her interior is dispose cooling membrane dissolvable of cylinder shape 6, which symmetrical axe correspond to with the child mould one, her role being the one of a separate this space in two compartments: exterior compartment **A** destinated primary alloy casting, for making hard crust of the pane, and interior compartment **B** is destinated secondary alloy casting, which will make the cylinder middle. Separating membrane is centered at superior part of the chill mould by three supports 9 (manufactured from the same material as the membrane) fixed in shape wall, which have also the role to stop membrane raising during shape alimentation with the two alloy. Separating membrane and fixing supports it is realized from alloys with thermo-physical characteristics of well-determinated material, will alled the dissolved during the solidification of two types of pig iron (with which comes in contact), having measured thickness from the condition of thermic balance.

To the inferior part of the chill mould it is dispose a second casting net (in cascade, with three supply, vertically set) mint to the primary alloy casting, adequate dimensioned, from the shape filling condition, in compartments **A** and **B** of the chill mould 5, with the same speed:

$$s_{alloyI} \cong s_{alloyII}, \tag{1}$$

in which s_{alloyI} and $s_{alloyII}$ are increasing speeds in the shape of the two alloys, primary and secondary. Casting reservoir 13, with who are connected the net 15, is assured with stopper 14, having the same role as the one in the position 12.

Presented proceeding allowed concomitant casting of the liquid alloys, by different ways, for the liquid alloy, which will make the axles and the cylinder middle, to rise in shape at the same level, with the alloy which will make hard crust or active part of the cylinder [3].

After the both casting reservoirs are supplied with adequate alloys, at the optimum temperatures of casting, it is rise de stopper 12, beginning filling sequence of the supply net 7 with secondary alloy, followed by the inferior shape filling 1. When liquid alloy touched the inferior level of the cylinder pane, begin also the casting sequence of the primary alloy, by the second net of casting 15. This sequence it is initiated by rising de stopper 14, in the moment of touching realization between the secondary liquid alloy and level signaling device 4. This signaling device may by a immersion thermocouple tide to an indicator device of temperature (SPEDOMAX), placed on the inferior shape at the adequate quote of the inferior level of the pane.

Departing membrane it is dimensiong in such way the not dissolve herself, on her height, before the chill mould filling with the both liquid alloys. Because of the alloys themperature, still in liquid phase, with which comes in contact, the membrane it is dissolve, taking place a miscibilitate a tho two alloys, realizing the passing zone between the hard crust and the tenacious middle of the cylinder, on a thickness not bigger than 20 … 40 mm, because of the solidification, in species by meeting of the two crystallization fronts (from the chill mould 5 to the cylinder axe and from the cylinder axe to the cavity periphery **A**).

Next sequence of the proceeding it is refer to the stopping of the primary alloy casting, when this level is in the right of the second signaling device 8, identical from building point of view with the signaling device 4, established in the superior frame, at the superior quote of the cylinder pane. The casting stopping it is realizes by obturating supplies net 15, by descending of the stopper 14. In this time, the secondary alloy casting continues, until the filling of the superior shape, when it is descending also the stopper 12, and the supply net 7 are so obdurate.

By comparation with the classical proceeding of the bimetals cylinders casting, the major advantage of this method consist in the fact that it is realizing the alimentation with liquid alloy at the constant themperature on shape height, eliminating so the nondisolving risk of the separation membrane at the superior part of the pane, where it is usually registrated lost themperature of secondary alloy to approximately 100 °C, face to the themperature which enter in ahape.

If the alimentation net with primary alloy would have the alimentary disposed ranged, differenced, no in the same vertical plane, the rotation move imparted to the liquid alloy will be accelerated, then eventual impurity attracted in the time of casting processes will be decanted in the exterior stratum of the pane and in the superior part of the cylinder maselot, parts which, anyway, are removed by mechanical processing (of the ulterior unthichen).

The method disadvantage it is refers, in first place, to manufacture necessity of the casting equipment (chill mould), with more orifices mint to the alimentation, in this case appearing locals overheats in the alimentary zones (considerated cushion for cracks to heat). Although, exist the possibility of lowering the number of the casting cylinders be described aparature and frequetility replacement of the chill mould 5 [3].

2.2. Bidimensional numerical model (with finished differences) of bimetal cylinder solidification

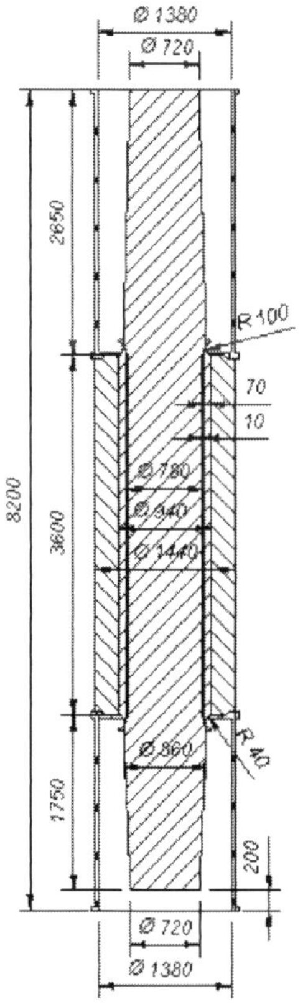

Figure 3. The dimensions casted raw cylinder and the casting ansamble.

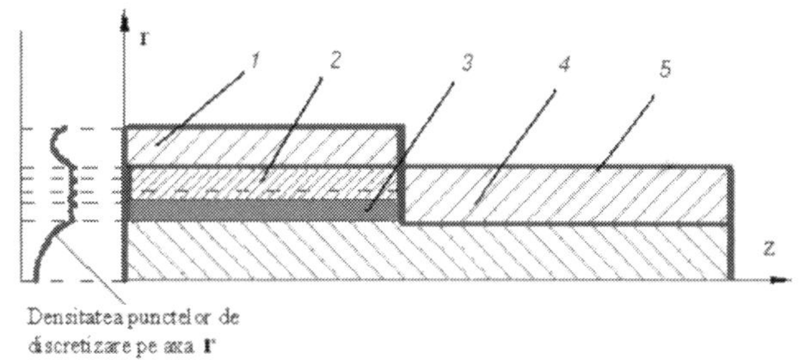

Figure 4. Quality schem of thermic system used in simulation:
1 – chill mould; 2 – hard crust; 3- cooling membrane;
4 – middle and cylinder axle; 5 – axle shape.

It is purpose a model of numerical simulation using the method of finished differences, of the solidification phenomenon of the bimetal cylinders rolling [4]. In this purpose, the analysis field, represented by a continuous medium (Figure 3) is transformed in a discontinuous medium (discreet), formed by a points net (discretionary net), of which density is given by the choosen steps for each coordinates axe (Figure 4). The themperature in each knot represents the medium themperature of the knot adjacent surface. In those knots it is written the differential equations of the transforming heat transmition in equations with finished differences.

Differential equation of the heat transmition after the three axes has the form [4]:

$$\frac{\partial t}{\partial \tau} = a \cdot \nabla^2 t \tag{2}$$

where: t - themperature, [°C]; τ - time, [s]; a - thermic difuzivity, [m^2/s]; ∇^2 - Laplace operator.

The realization of numerical model imposes the establishment of the simplifications ipotesis:

1. It is considering that the system has perfect axial symmetry (after z axe). As a result would not be transmitted heat only radial and vertical.

IOP Conf. Series: Materials Science and Engineering **85** (2015) 012020 doi:10.1088/1757-899X/85/1/012020

2. Because the cylinder has a complex shape, this will be simplified, removing the conics and connection rays.

3. It is neglecting the density variation, so of volume of the materials which form the system.

4. The heat transmition to the chill mould surface and the frames to the ambient medium take place by convection and radium.

5. It is neglecting the heat release by the superior surface of the maselot and the inferior one of the inferior axle frame of the cylinder.

6. The hit transmitions by removing surfaces (middle – membrane – hard crust – chill mould) take place by conduction.

7. The latent heat release of melting. It is make during liquidus – solidus, proportional with the themperature.

3. Results and discussions

The simulation was realized for the casting ansamble from Figure 3 (the casting shape and casting raw cylinder), which geometrical dimensions are ϕ1440x8200 mm. As the number of discretisation knots are bigger (for the casting shapes, as for the bimetal cylinder), respective enthalpy variation maximum in a smaller iteration, so as the real time of simulation is bigger. In the case of presented data, de simulation time was of 20 minutes for a real time of flowing process, solidification and cooling of 500 minutes [3].

The program could be enterupted (time step), with the mention that can't be turn on from the same time step, but must be rolled the program from the beginning. For a better illustration of the program functions, we effectuated captures of the computer screen, at the differentious time steps, and the figures such as resulted are presented next, from them obtaining regarding information to themperatures from diverses knots of the discretionary net and the real time (of calculation) covered of the simulated process. At the stopping of simulation process, the program presents the bill sticking option of simulated parameters variation in time.

Table 1. The coordinates of sections planes and points in discretionary net.

Quote on **z** axe	Mesuring points from cylinder inside	Mesuring plan from discretonary net inside of cylinder
z_1=1750	1 (r_1 =0; z_2/2 =2675), placed in cylinder axe	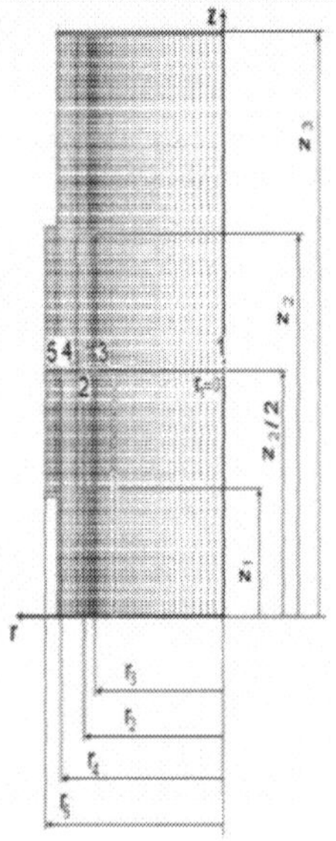
z_2=5350	2 (r_2 =455; z_2/2=2675), placed in cylinder crust	
z_3=8000	3 (r_3 = 415; z_2/2=2675), placed in membrane	
	4 (r_4 = 500; z_2/2=2675), placed to the interface of chill mould - crust	
	5 (r_5 = 700; z_2/2=2675), placed to the interface ambient field – chill mould	

The realized program allows the quality seeing of the thermic field, for differentious time moments, in planes section of the cylinders ϕ928x3300 mm, perpendiculars on z axe of the choosen reference system (in cylindrical coordinates). z axe is identical with symmetry axe of the bimetal cylinder, and the section are situated at the differentious heights face to the base plan, with whom those are parallels (for example z = 0, the plan is situated at the inferior axle base of the rolling cylinder).

In Table 1 are presented the planes quotes of the discretionary net section from numerical model field with finished differentious, those having correspondent with de planes from real field (casting ansamble ϕ1440x8200 mm, for bimetal rolling cylinder).

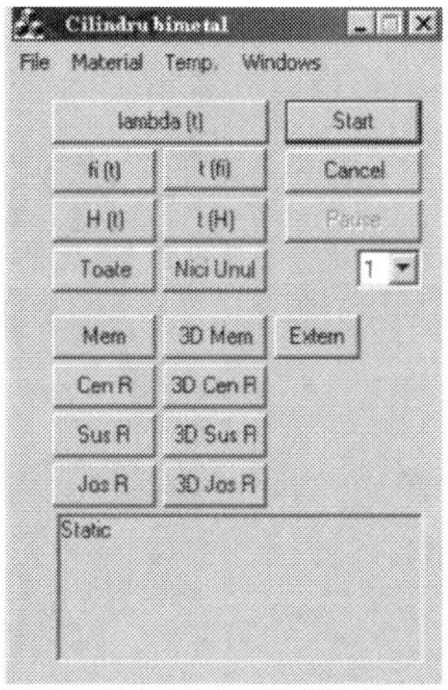

Figure 5. The principal dialog window.

In Figure 5 is presented the dialog window which is open with the initialization moment of the program "Bimetal cylinder" it is realizes pushing the button "Start/Stop". In the very next moment, the obtained image allowd the seeing of the discretionary net (Figure 6), of the simulation time and of the themperatures from diferentious zones of bimetal cylinder, as also the level to which are arrived the alloys in shape. The themperatures are showed trough color gradient, with the values: red for casting themperature, blue for ambient thermic themperature and green for their mediate. Any intermediate themperature is a combination of those.

In Figure 6, the 5 color gradient allowds an quality analysis of the solidification in time. So, the coloured zones in red are liquidus, tore in blue are solids, and the zone in greed shades show the distribution of biphasic zones (liquid and solid), with other words liquidus – solidus interval.

From Figure 6 it is could be observe that at the 4848 iteration (mining 156,458 seconds simulated time) secondary alloy, arrived to the table level. Next step, to 158,08 seconds from the simulation beginning, mentions the beginning of 2 crust alimentation with primary alloy, to the themperatur of t_4 = 1370°C (in figure 6, indication $t_4 = 0$ show that still it is not begin the primary alloy casting).

The membrane themperature is introduced in simulation with initial value t_3 = 250°C, so as the one of the chill mould t_5 = 250°C. The themperature t_1 is measured to the superior part of the cylinder superior axle, and the indication $t_1 = 0$, from Figure 6, is refers to the fact that, at this moment of simulation, the liguid alloy have not arrived to the respective level.

Notes j = 34 respective lev = 1738,42, from Figure 6, show the level (z and r coordinates) to which has arrived the fusion, meaning in 34 point of the discretionary net at the 1738,42 quote.

For a correct analysis, as a quality point of view but also quantitatively, in 7. a, b, c, d, e and f figures are presented the liquids alloys evolutions (primary and secondary) during the shape filling and in solidification time.

Is observe from Figures 7.a and b the filling of two cavity of chill mould (2 crust and 4 middle) with the same speed and, also, moving forward of the solidification from to the cylinder axe (Figure 7, c, d, e and f). The program cannot establish, so, the place of contraction defaults and either their measure.

From the tridimensional representations of the thermic field it is could be made the conexions between the level (z quote) to which had increase the alloys in shape, the themperatures and the simulated time (fig. 8). This imagines tipes are obtained by activating the dialog window **Metal3DGraph**, or of the butons from main window (3D Cem R etc) [3].

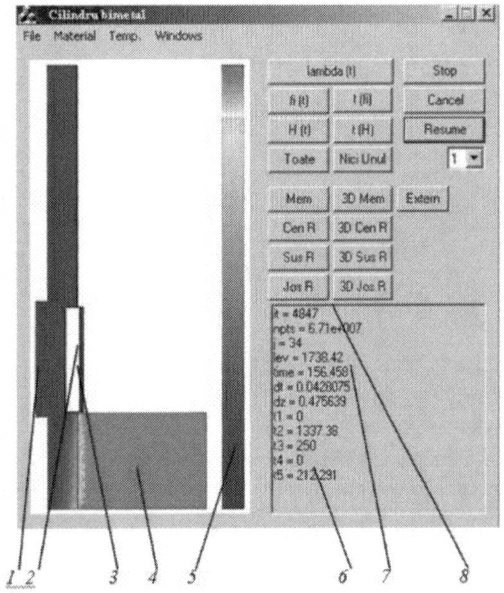

Figure 6. Seeing the simulation steps of the solidification and filling processes of bimetal cylinder: 1–chill mould; 2- hard crust; 3–disolvable cooling membrane; 4– inferior axle; 6–touche temperatures in differentious zones of the casting ansamble; 7–simulated time; 8–iteration.

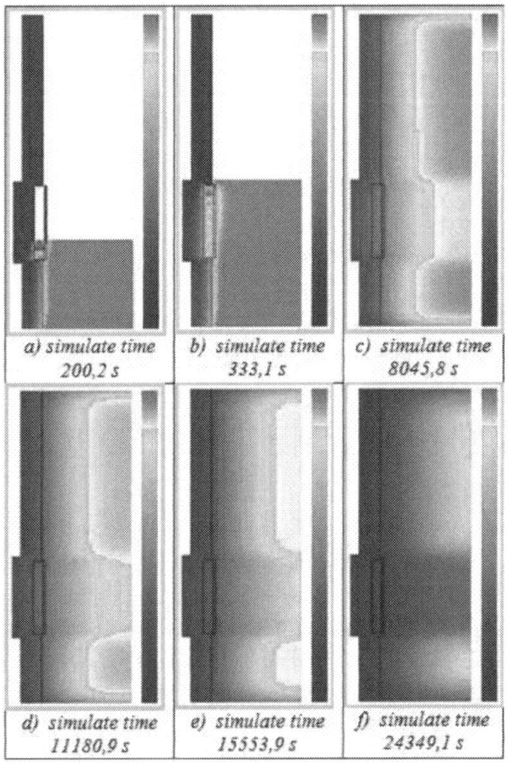

Figure 7. The liquid phase distribution on filling up way of the shape and during the solidification of bimetal casted cylinders.

Observation: The variation graphics of the time, temperature, at the differentious planes of the section in long axes r and z, use the procentual expression of the coordinates.

The evolution in time of the 1, 2, 3, 4 and 5 points themperature (Table 1) may be fallowed by activating the window **MetalGraph**, or directly, pushing the main window buttons (Cen R, Sus R etc), presented in Figure 9.

The obtained data can be exported in purpose of their comparation with the obtained values by casuals experimental measures, realized with the help of immersion thermocouples (plased in points of real field corresponding to the points of definite measure in discretionary net, simulated with the help of finished differentious method [3].

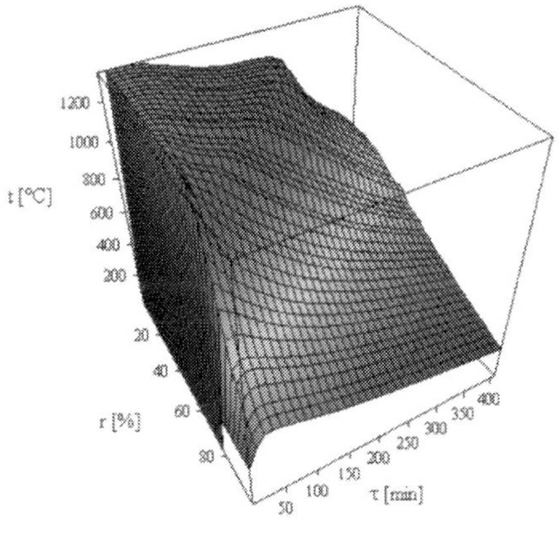

Figure 8. The temperature variation in time, determinated at the quotation $z_2/2 = 2675$, after 400 minutes from the simulation beginning

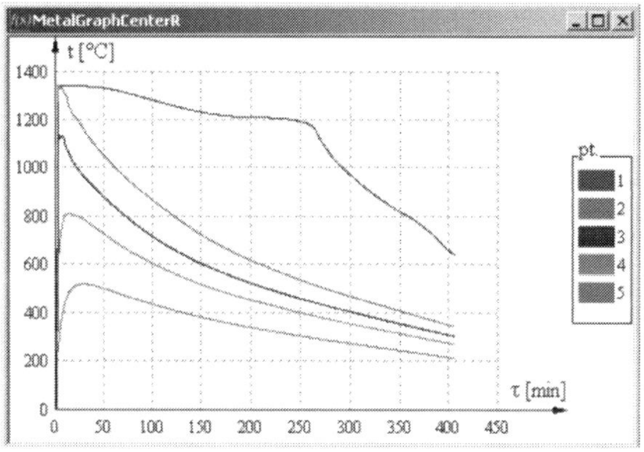

Figure 9. The graphic presentation of temperature evolution at the chill mould middle ($z_2/2 = 2675$), after 400 minutes from the simulation beginning

4. Conclusions

As a result of critical analysis of actual technology of bimetal casting, in static regime, of the big diameters cylinders, destinated pane rolling, it were proposed three variants (author own conception) of technology improving, final, establishing as optimum the vessels communication, which use a dissolvable micro-cooler membrane and of ranged casting nets.

Beginning from filling up principle of the two cavity of chill mould (A – destinated to the tenacious middle obtaining of the cylinder pane and B – destinated the hard crust realization, of equal thickness on pane generatrix) at the same level, on filling up whole way, it was dimensioned casting system, compose from the independent alimentary two nets.

Taking account of micro-cooler effect played by the dissolvable membrane, introduced in chill mould, was effectuated a moulding study of solidification process of bimetals cylinders (casting from A and B fusions), in this hypothesis.

The numerical and analytic moulding of the flowing processes solidification and cooling constitue a base instrument, useful as in conception phase such as in the analysis one of the metallurgical processes, which, combined with computers use, allowds the establish of the optimums regims of those development.

The results and the obtained conclusions after numerical moulding with the help of finite differences method can be extrapolated with a high veracity at the industrial processes which take place at the solidification conduction of bimetal cylinders of big diameters, in porous of their quality improving in exploitation.

In the presented context, one of the main objectives (the quality improving of big diameters bimetal cylinders) was realized by solidification research and the links establishment regarding to technologic parameters various (casting temperature, specific consume of micro-coolers and those dimensions) depending on operational variables specific of solidification processes, to determinate the influence modalities of those in view of obtaining of some pieces (bimetals rolling cylinders) of superior quality, reproducibility conditions and maximum economic – technical efficiency.

References

[1] Mihuţ G, Prejban I and Maksay S 2003 *The influence of the elements C, Cr and Mo upon the hardness of the bimetallic casted iron pig milling rolls*, 7th International Symposium Interdisciplinary Regional Research, Hunedoara, Romania, pp 634-639

[2] Mihuţ G 2000 Studyes and researches on the replacing of carbide rolls for wire rolling with bimettalics rolls, *Universitaria ROPET 2000* 165-170

[3] Mihuţ G 2005 *Optimization of bimetal casting technology for large diameter cylinders destined rolling*, Politehnica University of Timisoara, Romania, Doctoral Thesis

[4] Minkowycs W J 1996 *Advances in Numerical Heat Transfer*, Volume I, Taylor & Francis, Washington, USA

Studies about the attempt frequency influence on the effective relaxation time in a system of nanoparticles for magnetic hyperthermia

M Cacciola[1] and M Osaci[2]

[1] University "Mediterranea" of Reggio Calabria, DICEAM, Via Graziella Feo di Vito, I-89100, Italy
[2] Politehnica University of Timisoara, Department of Electrical Engineering and Industrial Informatics, 5 Revolutiei Street, 331128 Hunedoara, Romania

E-mail: mihaela.osaci@upt.ro

Abstract. This paper presents a series of studies about the modality in which the attempt frequency influences the Néel relaxation time and thus the effective relaxation time, in a spherical-nanoparticle-saturated nanofluid in external magnetic field. The nanoparticles have a random distribution, and their magnetic moments have a magnetic dipole-dipole interaction, along with the distributions of sizes and anisotropy constants. In such a system, the energy of the equilibrium states depends on the total magnetic field experienced by the particle, which includes the magnetic field applied and the dipolar magnetic field produced by the surrounding particles. Because of the high complexity level of the issue, it seems useful to consider the numerical experiments by computational simulation.

1. Introduction

Currently, in nanomagnetism, the study of magnetic nanoparticle systems and especially the iron oxides γ-Fe_2O_3 and Fe_3O_4 occupies a special place because of the biomedical applications, in particular for the nanofluid-based magnetic hyperthermia. In the magnetic hyperthermia with nanofluid, several heating mechanisms are possible, associated with susceptibility loss and hysteresis loss. The susceptibility loss has two associated relaxation processes: the Néel relaxation process and Brownian relaxation process. The Néel relaxation process is the solid state mechanism that takes place inside the nanoparticle, corresponding to the magnetic moment rotation between two equilibrium positions. The nanoparticle is fixed in space, the Brownian relaxation process corresponding to the nanoparticle rotation within the fluid environment. The change in the orientation of the particle causes changes in the orientation of its magnetization. The two processes are treated separately. The Néel relaxation time, τ_N, is usually described for a single barrier, by the Arrhenius law:

$$\tau_N = \frac{1}{f_0}\exp\left(\frac{E_b}{k_B T}\right) \quad (1)$$

where k_B is the Boltzmann constant, T is the temperature, and E_b is the energy barrier corresponding to the minimum energetic state. The pre-exponential factor f_0 is the so-called attempt frequency, which generally depends on the material properties. For spherical particles, the Brownian relaxation time is usually described by $\tau_B = \frac{3V_H\eta}{k_BT}$, where V_H is the hydrodynamic volume of the particle and η is the coefficient of dynamic viscosity. The relaxation mechanism which dominates the magnetic behaviour of the colloidal suspension is determined by the nanoparticle properties. Hence, the effective relaxation time can be described as

$$\frac{1}{\tau_{eff}} = \frac{1}{\tau_N} + \frac{1}{\tau_B} \tag{2}$$

In many papers, the attempt frequency is considered to be $f_0 = 10^9$ s^{-1} [1] or to fall within various ranges, such as $(10^{13}\text{-}10^9)$ s^{-1} [2], and to depend only on the material properties, although, in most cases, the actual system conditions are beyond the single particle model in the absence of the external magnetic field.

2. The Néel-Brown theoretical model for thermal instability of magnetic moments for monodomain nanoparticles with uniaxial anisotropy

The ferromagnetic or ferrimagnetic nanoparticles show an unstable behaviour of their magnetic moments due to thermal agitation. Currently, the thermal fluctuations and relaxation of the magnetic moments of the monodomain nanoparticles play a major role in the biomedical applications. A monodomain nanoparticle is in a state of uniform magnetization for any field applied. The magnetic moment of the monodomain nanoparticle is [3]:

$$\vec{\mu} = M_s v \hat{\mu} \tag{3}$$

where M_s is the spontaneous magnetization, v_i is the particle volume and $\hat{\mu}$ is the unit vector of the magnetic moments.

The Gilbert equation, for the motion of the magnetization augmented by a random field [4], is:

$$\dot{\hat{\mu}}(t) = \hat{\mu}(t) x \left[\gamma\mu_0 \vec{H}(t) - \alpha\, \dot{\hat{\mu}}(t) + \gamma\vec{b}(t) \right] \tag{4}$$

where γ is the gyromagnetic ratio and α is the dimensionless damping (dissipation) parameter.

Now, please let us consider

$$\vec{H}(t) = -\frac{1}{\mu_0}\frac{\partial U}{\partial \vec{\mu}} \tag{5}$$

as the local magnetic field acting on the nanoparticle, where U is the Gibbs free energy (characterizing the magnetic anisotropy and Zeeman energy of the nanoparticle).

If the magnetic field is applied along the easy axis, the reduced potential (U/k_BT) is

$$V(\cos\vartheta) = -\sigma\cos^2\theta - \xi\cos\theta = -\sigma\left(\cos^2\theta + 2h\cos\theta\right) \tag{6}$$

where

$$\xi = \frac{\mu_0 M_s v H}{k_B T},\ \sigma = \frac{Kv}{k_BT},\ h = \frac{\mu_0 M_s H}{2K} \tag{7}$$

In equation (4), $\vec{b}(t)$ is a random Gaussian field with white noise properties:

$$\overline{b_i(t)} = 0,\ \overline{b_i(t_1)b_j(t_2)} = \frac{2k_BT\alpha}{v\gamma M_s}\delta_{ij}\delta(t_1 - t_2) \tag{8}$$

Here, i, j = 1, 2, 3 in Kronecker's delta δ_{ij}, h_i are components of $\vec{b}(t)$ in the Cartesian coordinate system, $\delta(t)$ is the Dirac delta function, and overbar means the statistical average over an ensemble of particles which all have the same magnetization at time t. The random field accounts for the thermal fluctuations of magnetization of an individual nanoparticle without which the random orientational motion would not be sustained. Then, Brown deduces from the equation (4) the appropriate Fokker-

Plank equation for the distribution function $W(\theta,\varphi,t)$ of the orientation of magnetic moments vector on the surface of unit sphere [5]. In an axially symmetric potential V, the complicated Fokker-Planck equation [6], for the distribution function $W(\theta,t)$, is actually a one-space-variable equation:

$$2\tau_{0N}\frac{\partial W}{\partial t}=\frac{1}{\sin\theta}\frac{\partial}{\partial\theta}\left[\sin\theta\left(\frac{\partial W}{\partial\theta}+W\frac{\partial V}{\partial\vartheta}\right)\right] \quad (9)$$

$$\tau_{0N}=\frac{vM_s(1+\alpha^2)}{2\gamma\alpha k_BT} \quad (10)$$

In equation (10) τ_{0N} denotes the free diffusion time of magnetization (of order 10^{-11}-10^{-8}s) [6].

For low damping (α<<1), the precessional motion is the dominant motion. Therefore [6]:

$$\tau_{0N}=\frac{vM_s}{2\gamma\alpha k_BT} \quad (11)$$

For high damping (α>>1) alignment is dominant and [6]:

$$\tau_{0N}=\frac{vM_s\alpha}{2\gamma k_BT} \quad (12)$$

For magnetite nanoparticle with 9 nm diameter, the damping (dissipation) parameter value resulting from experimental measurements is 0.16 [7].

With the potential equation (6), in the high barrier approximation [5] $\sigma(1-h)^2>>1$, and h<1, we obtain:

$$\tau_N=\frac{\tau_{0N}\sqrt{\pi}}{\sigma^{3/2}(1-h^2)}\left[(1+h)e^{-\sigma(1+h)^2}+(1-h)e^{-\sigma(1-h)^2}\right]^{-1} \quad (13)$$

For $h=0$,

$$\tau_N=\frac{\tau_{0N}\sqrt{\pi}}{2\sigma^{3/2}}e^{\sigma}=\frac{1}{f_0}e^{\sigma} \quad (14)$$

where $f_0=\frac{2\sigma^{3/2}}{\tau_{0N}\sqrt{\pi}}$ is the attempt frequency without magnetic field.

Equation (13) can be written:

$$\frac{1}{\tau_N}=\frac{\sigma^{3/2}(1-h^2)}{\tau_{0N}\sqrt{\pi}}\left[(1+h)e^{-\sigma(1+h)^2}+(1-h)e^{-\sigma(1-h)^2}\right]=f_{01}e^{-\sigma(1+h)^2}+f_{02}e^{-\sigma(1-h)^2} \quad (15)$$

where

$$f_{01}=\frac{\sigma^{3/2}(1-h^2)}{\tau_{0N}\sqrt{\pi}}(1+h) \text{ and } f_{02}=\frac{\sigma^{3/2}(1-h^2)}{\tau_{0N}\sqrt{\pi}}(1-h) \quad (16)$$

f_{01} and f_{02} are the real attempt frequency factors in the presence of magnetic field.

Usually in the literature [8, 9], the equation (13) is approximated as:

$$\tau'_N=\frac{\tau_{0N}\sqrt{\pi}}{\sigma^{3/2}}\left[e^{-\sigma(1+h)^2}+e^{-\sigma(1-h)^2}\right]^{-1}=\frac{1}{f'_0}\left[e^{-\sigma(1+h)^2}+e^{-\sigma(1-h)^2}\right]^{-1} \quad (17)$$

where $f'_0=\frac{\sigma^{3/2}}{\tau_{0N}\sqrt{\pi}}=f_0/2$ is the approximate attempt frequency factor in the presence of magnetic field. This approximation is motivated by the fact that the pre-exponential factors change with the field - in the equation (13) -, is much smaller than the change produced in the exponential factor [9].

We intend to verify this statement by numerical simulation, to see if the approximation (17) is valid when working in external magnetic fields typical for applications in hyperthermia (0-14 kA/m), in two cases: if we do not neglect the internal dipolar field on the particle, and in the real case where the sizes of particles and effective anisotropy constants show a distribution.

In the case of magnetic dipolar interactions among the nanoparticles, if using the Dormann-Bessais-Fiorani approximation [10], only the component on the external magnetic field direction ($\vec{H}_{ext}$) of the dipolar magnetic field ($H_{dipolar}$), is taken into account in the above relations $H = H_{dipolar} + H_{ext}$.

3. Numerical simulations

Thus, we start from the original Néel-Brown model in an external magnetic field parallel to the aligned easy axes of anisotropy of the spherical nanoparticles. As we showed in the previous paragraph, this model provides a field-dependent expression for the attempt frequency. Then, we generalise the model by introducing, in a stochastic manner, the magnetic dipole–dipole interactions among the nanoparticles. The local magnetic dipolar field produced by the surrounding particles will be calculated using the Dormann-Bessais-Fiorani approximation. We will analyze the Néel relaxation time, simulated by taking into account the customary value for the attempt frequency and the attempt frequency value given by the Néel theory generalized to the local magnetic field. So, we will show how the attempt frequency factors vary depending on the external magnetic field applied.

The simulation was realised for spherical magnetic nanoparticles of uncoated iron oxide, with the saturation magnetization M_s=4.46·10^5 A/m, average diameter d_m=10 nm and uniaxial magnetic anisotropy, and the average value of the effective anisotropy constant K=10^4 J/m^3. Either the diameters or the effective anisotropy constants show a lognormal distribution with the parameters μ_d = -18.545 & σ_d = 0.198 (for diameters), and μ_K = 0.198 & σ_K =0.0099997 (for effective anisotropy constants). The external magnetic field applied vary within the range 0 - 14 kA/m, typical values for magnetic hyperthermia applications.

The analysed systems contain N=100 and N=1000 nanoparticles, at the temperature T=298K in a base liquid with dynamic viscosity 6·10^{-3} Pa·s. The nanoparticles have been located within a base-centred cubic lattice. Among all the possible positions, the particles have been centred in randomly selected suitable locations, according to the ratio between the volumes of the considered nanoparticles and the whole volume of the simulation box. The periodic boundary conditions have been applied on all sides of the simulation box. The external magnetic field intensity varied in the range (0-14 kA/m), typical values used in magnetic hyperthermia.

4. Results and discussions

This simulation aims to verify the statement made in the literature according to which, the pre-exponential factors modification in the equation (13) with the field is much smaller than the change produced in the exponential factor [9]. For this, because the attempt frequency factors depend also on the volume of nanoparticles, effective anisotropy constants and temperature, we keep unchanged the parameters of the corresponding distributions and the system temperature.

The first study refers to the modality in which the attempt frequency factors vary with the external magnetic field intensity. Thus, we are monitoring the variation of f_0, which is the attempt frequency without magnetic field, f'_0, which is the approximate attempt frequency factor in the presence of magnetic field, f_{01}, and f_{02}, which is the exact attempt frequency factor in the presence of magnetic field. The results obtained for a system made of 100 nanoparticles with the volume fraction f = 0.08 are presented in Figure 1.

It can be seen that f_{01} shows a peak at about 6500 A/m, and then decreases with increasing external magnetic field. About f_{02}, we can say that it decreases with increasing external magnetic field. If we increase the number of nanoparticles in the system, we see that the increase in the nanoparticle number leads to a decrease of the exact attempt frequency values due to the dipolar internal magnetic field variation – Figures 2, 3.

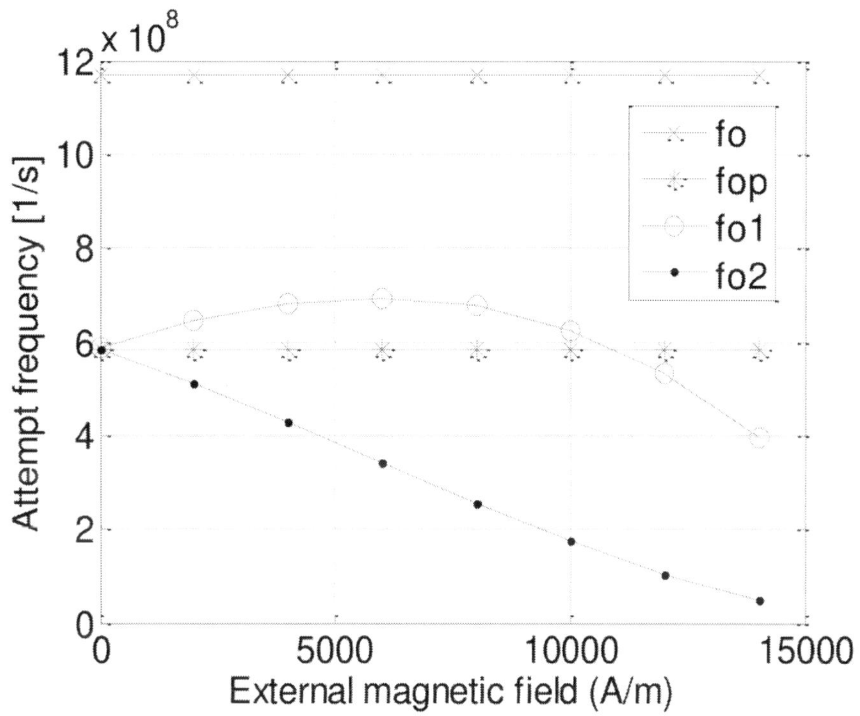

Figure 1. The attempt frequency (f_0, f'_0 (fop), f_{01} and f_{02}) versus the external magnetic field, at the nanoparticle volume fraction $f = 0.08$, for 100 particles in the system.

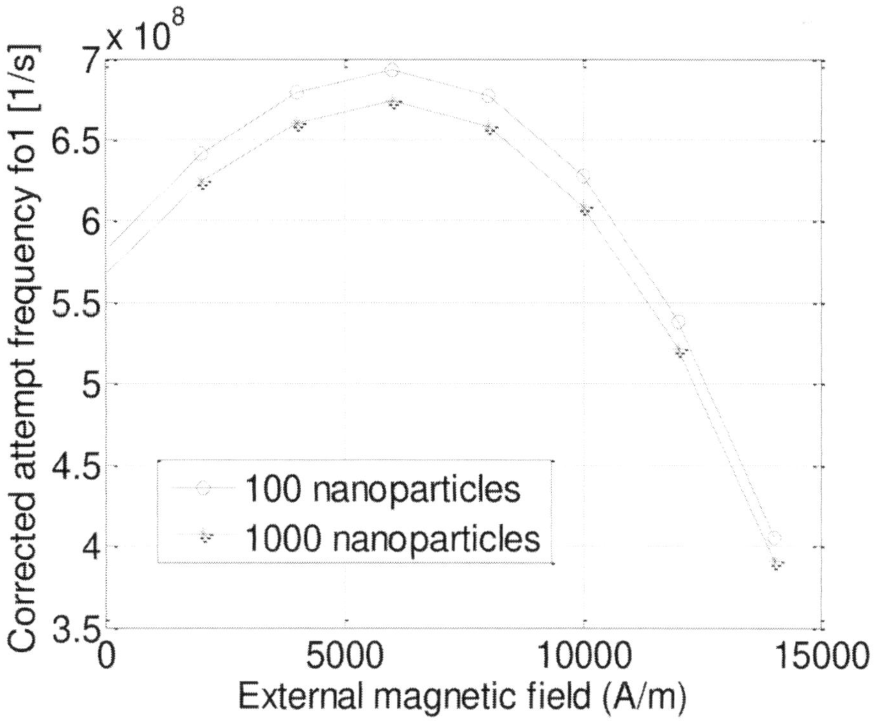

Figure 2. The exact attempt frequency f_{01} versus the external magnetic field, at the volume fraction $f = 0.1$

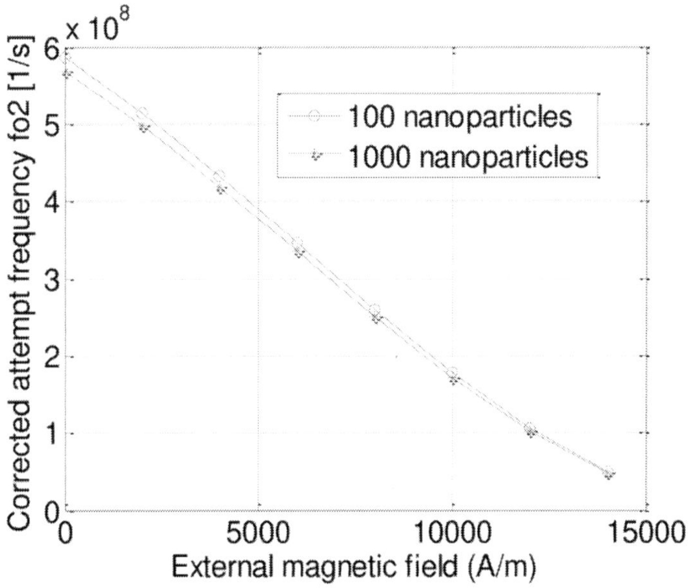

Figure 3. The exact attempt frequency f_{02} versus the external magnetic field, at the volume fraction $f = 0.1$

The second part of the study aims to find how the Néel relaxation time is affected by the attempt frequency variation with the external magnetic field intensity. The results are shown in Figure 4 for a system made of 100 nanoparticles with the volume fraction $f = 0.08$.

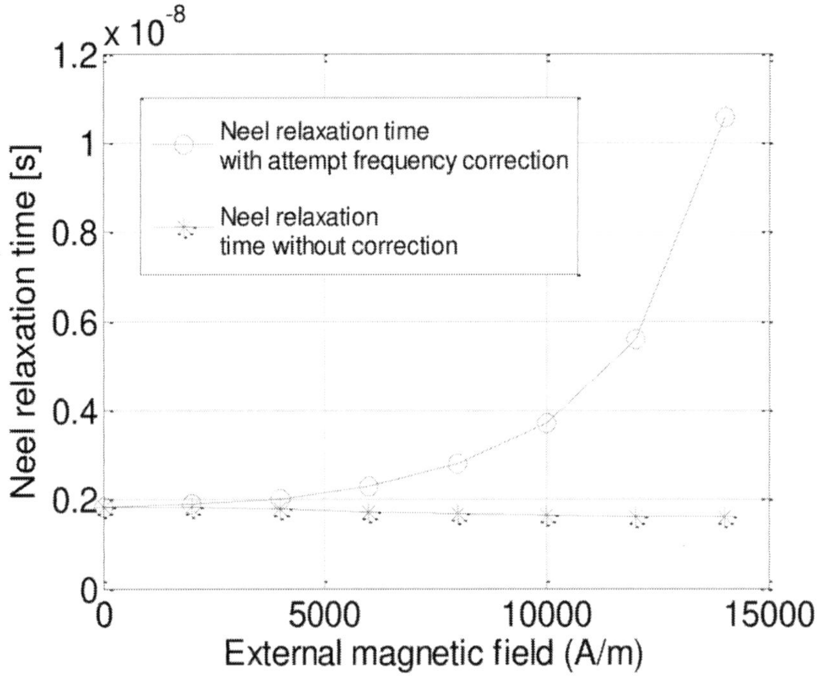

Figure 4. Néel relaxation time versus external magnetic field, at the volume fraction of nanoparticles $f = 0.08$, for 100 particles in the system.

It can be seen that, at the small external magnetic fields, there is no significant variation, which means that the constant attempt frequency approximation is valid. But, at more intense fields, there is an almost exponential increase of the relaxation time, calculated using the relation (15), compared with the Néel relaxation time calculated with approximate constant value of the attempt frequency which decreases slightly with increasing external magnetic field intensity.

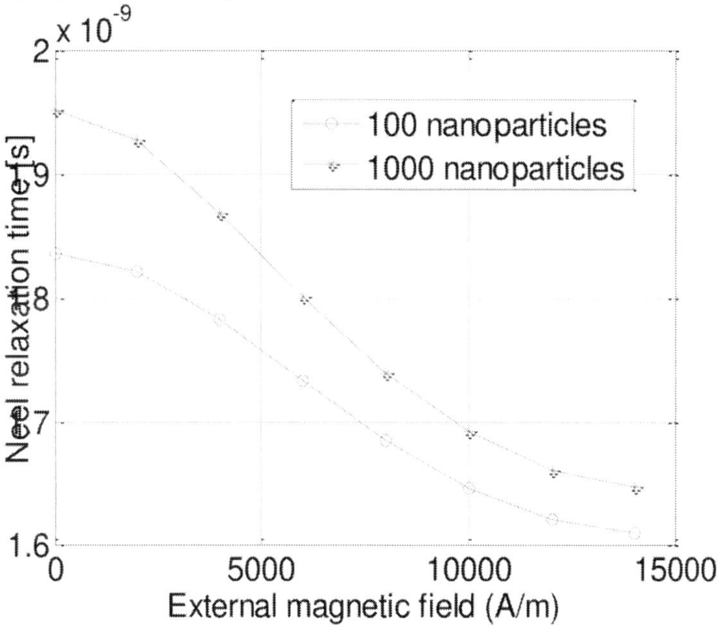

Figure 5. The approximate Néel relaxation time versus the external magnetic field, at the volume fraction $f = 0.1$, for 100 and 1000 nanoparticles.

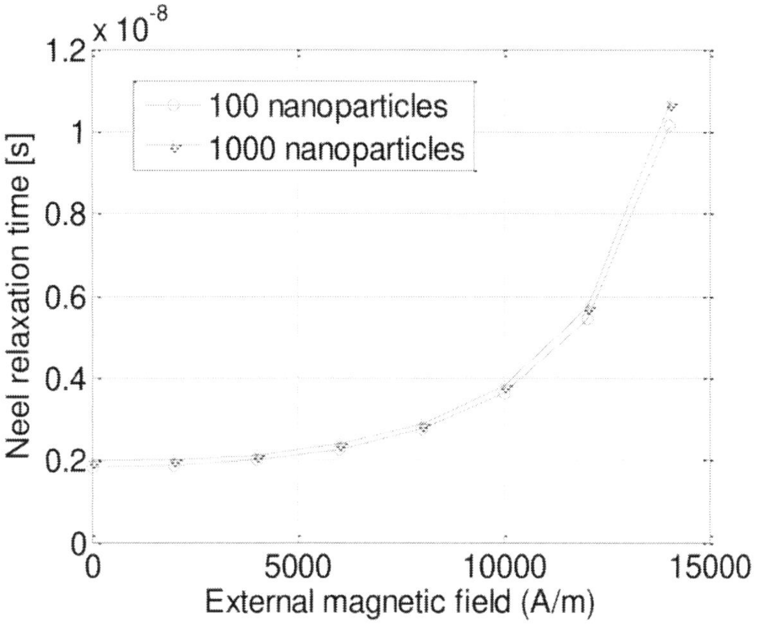

Figure 6. The exact Néel relaxation time versus the external magnetic field, at the volume fraction $f = 0.1$, for 100 and 1000 nanoparticles.

IOP Conf. Series: Materials Science and Engineering **85** (2015) 012021 doi:10.1088/1757-899X/85/1/012021

If we vary the number of particles in the system, it is found that, due to the dipolar internal magnetic field variation, the relaxation time increases with increasing number of particles in the system – Figures 5 and 6.

We then simulated the exact Néel relaxation time (calculated using the equation 15) at two different concentrations, $f = 0.02$ and $f = 0.03$. It was found a slight decrease in the exact Néel relaxation time with increasing concentration – Table 1.

Table 1. Exact Néel relaxation time at f=0.02 and f= 0.03 at different values of external magnetic field

External magnetic field (A/m)	Exact Néel relaxation time (s) at f= 0.02	Exact Néel relaxation time (s) at f= 0.03
0	1.836091589093639e-09	1.836060931144190e-09
2000	1.880709887411698e-09	1.877547492722059e-09
4000	2.020740803674264e-09	2.013649949075247e-09
6000	2.293565438087405e-09	2.280551065183938e-09
8000	2.784823253592053e-09	2.761393623573384e-09
10000	3.696013390832159e-09	3.651093288259277e-09
12000	5.582427455707154e-09	5.482641827927936e-09
14000	1.050582994121399e-08	1.020631945091540e-08

This fact was found experimentally in diluted systems. The Mössbauer spectroscopy measurements [10], [11], [12] on weakly interacting γ-Fe_2O_3 nanoparticles showed a decrease of the relaxation time with increasing interparticle interaction. The publication [12] related about the specific absorption rate (SAR) of a maghemite-based ferrofluid, measured at 315 K, 3 kA/m, and 109 kHz. The magnetic characterization of three different concentrations of this ferrofluid revealed the decrease of the initial susceptibility values and the Néel relaxation times, τ_N, with increasing concentration.

Given equation (2) and simulation parameters, in the same conditions, no differences between the mean Néel relaxation time and the mean effective relaxation time because Brownian relaxation time is of the order 10^{-6}s.

5. Conclusions

This paper presents a series of studies regarding the attempt frequency influence on the Néel relaxation time, and thus the effective relaxation time, through their dependence on the external magnetic field and dipolar magnetic field. The studies showed that the statement "the pre-exponential factors change with the field is much smaller than the change produced in the exponential factor", justifying that the choice of constant exponential prefactors (which depend on the field variation) is truthful only for relatively low external magnetic field intensity. Another obtained result is that the exact relaxation time increases when the number of particles in the system increases, because the variation of particle number in the system leads to the variation of the dipolar internal magnetic field.

It is found also a slight decrease in the exact Néel relaxation time with the concentration increase, in agreement with experimental findings published in the literature. For our simulation parameters, in the same conditions, no differences between the mean Néel relaxation time and the mean effective relaxation time.

References

[1] Beke D L 1998 Intrinsic and Domain Magnetism in Nanocrystalline Materials, *Cryst. Res. Technol.* **33**(7-8) 1039-1059

[2] Morup S, Hansen M F and Frandsen C 2010 Magnetic interactions between nanoparticles,

Beilstein Journal of Nanotechology **1** 182-190

[3] Haase C and Nowak U 2012 Role of dipole-dipole interactions for hyperthermia heating of magnetic nanoparticles ensamble, *Physical Review B* **85**(4) 045435-1-045435-5

[4] Coffey W T and Kalmykov Y P 2012 Thermal fluctuations of magnetic nanoparticles: Fifty years after Brown, *Applied Physics Reviews Section of the Journal of Applied Physics* **112** 121301-1-121301-47

[5] Brown W F Jr 1963 Thermal Fluctuations of a Single-Domain Particle, *Physical Review* **130** 1677-1686

[6] Coffey W T and Cregg P J 1993 On the Theory of Debye and Néel Relaxation of Single Domain Ferromagnetic Particles, *Advances in Chemical Physics* **83** 263-464

[7] Fannin P C, Marin C N, Raj K, Couper C and Barvinschi P 2012, An experimental study of the dynamic properties of nanoparticle colloids with identical magnetization but different particle size, *Journal of Magnetism and Magnetic Materials* **324**(21) 3443–3447

[8] Prozorov R and Yeshurun Y 1999 Magnetic irreversibility and relaxation in assembly of ferromagnetic nanoparticles, *Physical Review B* **59**(10) 6959-6965

[9] Kechrakos D 2010 *Magnetic Nanoparticle Assemblies. Handbook of Nanophysics,* vol 3: Nanoparticles and Quantum Dots, CRC Press Taylor & Francis Group, Danvers, USA

[10] Hansen M F and Morup S 1998 Models for the dynamics of interacting magnetic nanoparticles, *Magnetism and Magnetic Materials* **184** 262-274

[11] Prené P, Tronc E, Jolivet J P, Livage J, Cherkaoui R, Noguès M, Dormann J L and Fiorani D 1993 Magnetic properties of isolated γ-Fe2O3 particles, *IEEE Transactions on Magnetics* **29** 2658-2660

[12] Urtizberea A, Natividad E, Arizaga A, Castro M and Mediano A 2010 Specific Absorption Rates and Magnetic Properties of Ferrofluids with Interaction Effects at Low Concentrations, *Journal of Physical Chemistry C* **114**(11) 4916–4922

IOP Conf. Series: Materials Science and Engineering **85** (2015) 012022 doi:10.1088/1757-899X/85/1/012022

Microcontroller interrupts in frequency controlling

C Panoiu[1], R Rob[1] and M Panoiu[1]

[1] Politehnica University of Timisoara, Department of Electrical Engineering and Industrial Informatics, 5 Revolution Street, 331128 Hunedoara, Romania

E-mail: caius.panoiu@gih.upt.ro

Abstract. Present paper represents an educational software application that permits to measure the signals frequency. The application is accomplished in MicroC compiler that also permit to user to program a microcontroller PIC18F45K22 using a developing board EasyPICv7. An important issue of present application is to measure the frequency of an external signal that is acquired by an input pin of the microcontroller. The frequency value is displayed on the seven segment display provided by the developing board.

1. Introduction

PIC microcontrollers are a very useful and versatile tool for use in many electronic projects. They are very inexpensive and easy to find. They are also very powerful and many are capable of speeds up to 64 MIPS using the internal oscillator block, about 16 times faster than most comparable AVR microcontrollers. PICs are also easy to program, however getting the project set up can often be tricky.

An interrupt is a means of interrupting the main program flow in response to an event, so that the event can be handled. The event (referred to an interrupt source) can be internal to the PIC, such as a timer overflowing, or external, such as a change on an input pin.

There are 19 registers used to control interrupt operation. In general, interrupt sources have three bits to control their operation [1], [5]. They are presented in the followings:

- Flag bit to indicate that an interrupt event occurred.
- Enable bit that allows program execution to branch to the interrupt vector address when the flag bit is set.
- Priority bit to select high priority or low priority.

Interrupt Priority

The interrupt priority feature [2] is enabled by setting the IPEN bit of the RCON register. When interrupt priority is enabled the GIE/GIEH and PEIE/GIEL global interrupt enable bits of Compatibility mode are replaced by the GIEH high priority, and GIEL low priority, global interrupt enables [6]. When set, the GIEH bit of the INTCON register enables all interrupts that have their associated IPRx register or INTCONx register priority bit set (high priority). When clear, the GIEH bit disables all interrupt sources including those selected as low priority. When clear, the GIEL bit of the INTCON register disables only the interrupts that have their associated priority bit cleared (low priority). When is set, the GIEL bit enables the low priority sources when the GIEH bit is also set. When the interrupt flag enables bit and appropriate Global Interrupt Enable (GIE) bit are all set, the interrupt vector will immediately address 0008h for high priority, or 0018h for low priority, depending on level of the interrupting source's priority bit. Individual interrupts can be disabled through their corresponding interrupt enable bits.

Interrupt Response

When an interrupt is responded to, the Global Interrupt Enable bit is cleared to disable further interrupts. The GIE/GIEH bit is the global interrupt enable when the IPEN bit is cleared. When the IPEN bit is set, enabling interrupt priority levels, the GIEH bit is the high priority global interrupt enable and the GIEL bit is the low priority global interrupt enable. High priority interrupt sources can interrupt a low priority interrupt. Low priority interrupts are not processed while high priority interrupts are in progress. The return address is pushed onto the stack and the PC is loaded with the interrupt vector address (0008h or 0018h). Once in the Interrupt Service Routine, the source(s) of the interrupt can be determined by polling the interrupt flag bits in the INTCONx and PIRx registers. The interrupt flag bits must be cleared by software before re-enabling interrupts to avoid repeating the same interrupt. The "return from interrupt" instruction, RETFIE, exits the interrupt routine and sets the GIE/GIEH bit (GIEH or GIEL if priority levels are used), which re-enables interrupts. For external interrupt events, such as the INT pins or the PORTB interrupt-on-change, the interrupt latency will be three to four instruction cycles. The exact latency is the same for one-cycle or two-cycle instructions. Individual interrupt flag bits are set, regardless of the status of their corresponding enable bits or the Global Interrupt Enable bit.

2. Application description

The application described here is able to acquire a digital signal provided by an external source and to count its pulses. Signal is acquired by port B of PIC18F45K22 microcontroller. The value of the measured frequency is displayed on the seven segments display connected on the development board. The display uses the A and D ports of PIC18F45K22 like in Figure 1 [8].

The soft application is based on generating of an interrupt to TMR0 timer. Using INTCON register, TMR0 generates one transition on each second. In these conditions, the application counts the external signal transitions during one second.

Therefore, the application accomplished in MicroC compiler [7] declares two variables: one for incrementing the transition counter (*cnt*) and one for incrementing the overflow counter (*ovrflw*).

Table 1 presents the INTCON register [2].

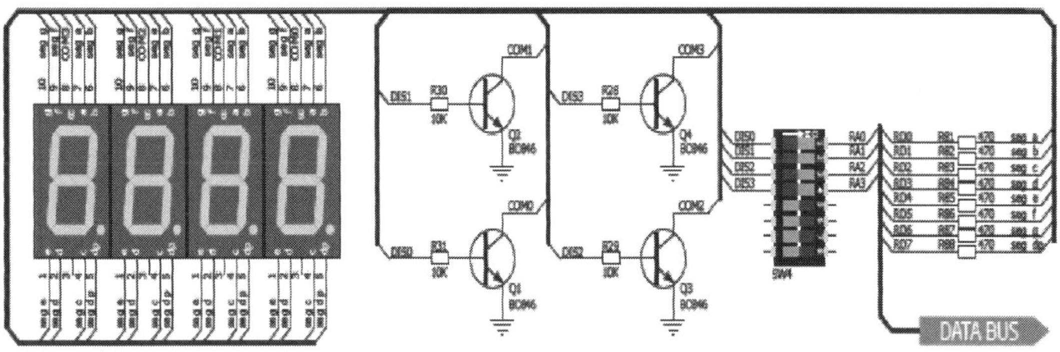

Figure 1. Connection of seven segments display.

Table 1. INTCON register configuration.

R/W-0	R/W-0	R/W-0	R/W-0	R/W-0	R/W-0	R/W-0	R/W-x
GIE/GIEH	PEIE/GIEL	TMR0IE	INT0IE	RBIE	TMR0IF	INT0IF	RBIF
bit 7							bit 0

IOP Conf. Series: Materials Science and Engineering **85** (2015) 012022 doi:10.1088/1757-899X/85/1/012022

Data from Table 2 is detailed in the followings:
bit 7 **GIE/GIEH:** Global Interrupt Enable bit
When IPEN = 0:
1 = Enables all unmasked interrupts
0 = Disables all interrupts including peripherals
When IPEN = 1:
1 = Enables all high priority interrupts
0 = Disables all interrupts including low priority
bit 6 **PEIE/GIEL:** Peripheral Interrupt Enable bit
When IPEN = 0:
1 = Enables all unmasked peripheral interrupts
0 = Disables all peripheral interrupts
When IPEN = 1:
1 = Enables all low priority interrupts
0 = Disables all low priority interrupts
bit 5 **TMR0IE:** TMR0 Overflow Interrupt Enable bit
1 = Enables the TMR0 overflow interrupt
0 = Disables the TMR0 overflow interrupt
bit 4 **INT0IE:** INT0 External Interrupt Enable bit
1 = Enables the INT0 external interrupt
0 = Disables the INT0 external interrupt
bit 3 **RBIE:** Port B Interrupt-On-Change (IOCx) Interrupt Enable bit**(2)**
1 = Enables the IOCx port change interrupt
0 = Disables the IOCx port change interrupt
bit 2 **TMR0IF:** TMR0 Overflow Interrupt Flag bit
1 = TMR0 register has overflowed (must be cleared by software)
0 = TMR0 register did not overflow
bit 1 **INT0IF:** INT0 External Interrupt Flag bit
1 = The INT0 external interrupt occurred (must be cleared by software)
0 = The INT0 external interrupt did not occur
bit 0 **RBIF:** Port B Interrupt-On-Change (IOCx) Interrupt Flag bit**(1)**
1 = At least one of the IOC<3:0> (RB<7:4>) pins changed state (must be cleared by software)
0 = None of the IOC<3:0> (RB<7:4>) pins have changed state

The source code of the application is presented in the followings:

```
#include "Display_Utils.h"
unsigned short shifter, portd_index;          //Declaring variable for 7 segments display
unsigned int   digit;                         //Declaring variable digit
unsigned short portd_array[4];                //Display array
  unsigned int cnt;                           // Transition counter variable
  unsigned int ovrflw;                        // Overflow counter variable
unsigned short mask(unsigned short num) {
  switch (num) {
    case 0 : return 0x3F;
    case 1 : return 0x06;
```

```
  case 2 : return 0x5B;
  case 3 : return 0x4F;
  case 4 : return 0x66;
  case 5 : return 0x6D;
  case 6 : return 0x7D;
  case 7 : return 0x07;
  case 8 : return 0x7F;
  case 9 : return 0x6F;
 }
}
void interrupt() {                        // Interrupt function
 if (INT0IF_bit) {
  cnt++;                                  // Transition counter increment
  INT0IF_bit = 0;                         // Clear interrupt flag to enable next call

  }
  else if (TMR0IF_bit) {
      ovrflw++;                           // Overflow counter increment
      TMR0IF_bit = 0;                     // Clear interrupt flag to enable next call on overflow
       }
}
void main() {
 ANSELA = 0;                              // Configure port A as digital
 ANSELD = 0;                              // Configure port D as digital
 TRISA = 0;                               // Configure port A as output
 LATA  = 0;                               // Disabling port A
 TRISD = 0;                               // Configure port D as output
 LATD  = 0;                               // Disabling port D
 TMR0L = 0;                               // Disabling TMR0, least significant byte
 digit = 0;
 portd_index = 0;
 shifter = 1;
 GIE_bit = 1;
 TMR0IE_bit = 1;
```

```
ANSELB = 0x00;                         // Configure RB0 pin as digital
TRISB = 0xFF;                          // Configure RB0 pin as input
T0CON = 0xD8;                          // Configure T0CON register, no prescaler
do {
LATA = 0;                              // Switch-off for 7 segment display
LATD = portd_array[portd_index];       // Assign an adequate value for port D
LATA = shifter;
   cnt = 0;
   ovrflw = 0;
GIE_bit = 1;                           // Enable general interrupt
INT0IE_bit = 1;                        // Enable external interrupt
TMR0IE_bit = 1;                        // Enable overflow interrupt
while(ovrflw <=7813);
GIE_bit=0;                             // Disable general interrupt
 cnt=cnt % 10000;
 digit = cnt / 1000u ;                 // extract thousands digit
portd_array[3] = mask(digit);          // and store it to PORTD array
digit = (cnt / 100u) % 10u;            // extract hundreds digit
portd_array[2] = mask(digit);          // and store it to PORTD array
digit = (cnt / 10u) % 10u;             // extract tens digit
portd_array[1] = mask(digit);          // and store it to PORTD array
digit = cnt % 10u;                     // extract ones digit
portd_array[0] = mask(digit);          // and store it to PORTD array
} while(1); }
```

In Figure 2 the block scheme of the application is presented.

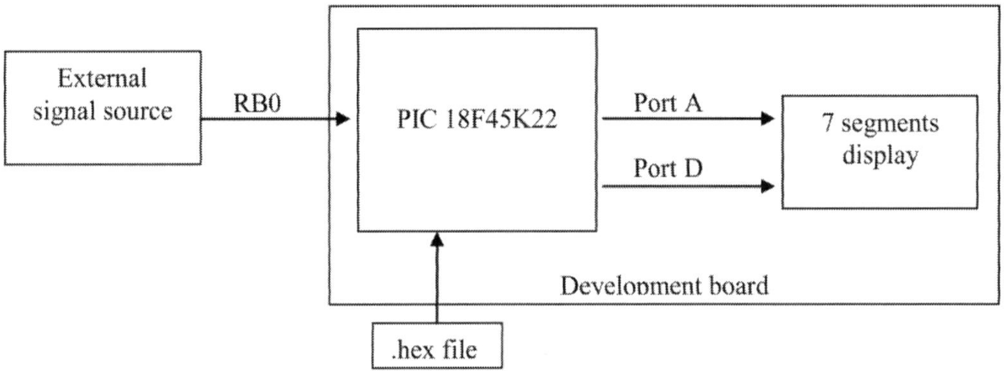

Figure 2. Block scheme of the application.

3. Conclusions

Using interrupts the following advantages can be achieved: the microcontroller can serve many devices, but not all in the same time. Each device can get the attention of the microcontroller based on the assigned priority [3], [4].

The program presented in this paper has a various applications from measuring the frequency of an input signal, to study its variation on a graphical display. The range of this frequency meter can increase with the increasing of the 7 segment digits.

References

[1] Mitescu M and Susnea I 2005 *Microcontrollers in practice,* Springer, Berlin-Heidelberg-New York
[2] ***http://ww1.microchip.com/downloads/en/DeviceDoc/41412F.pdf
[3] ***http://www.best-microcontroller-projects.com/hardware-interrupt.html
[4] ***http://www.electronics-base.com/avr-tutorials/external-interrupts
[5] Popa M 206 *Sisteme cu microcontrolere orientate pe aplicații*, Orizonturi Universitare, Timişoara, Romania
[6] Calcutt D, Cowan F and Parchizadeh H 2004 *8051 Microcontrollers: An Applications Based Introduction,* Newnes, Amsterdam, Hollande
[7] ***http://www.mikroe.com/pdf/mikroc/mikroc_manual.pdf
[8] ***http://www.mikroe.com/easypic/

Video streaming technologies using ActiveX and LabVIEW

M Panoiu[1] C L Rat[1] and C Panoiu[1]

[1] Politehnica University of Timisoara, Department of Electrical Engineering and Industrial Informatics, 5 Revolution Street, 331128 Hunedoara, Romania

E-mail: manuela.panoiu@fih.upt.ro

Abstract. The goal of this paper is to present the possibilities of remote image processing through data exchange between two programming technologies: LabVIEW and ActiveX. ActiveX refers to the process of controlling one program from another via ActiveX component; where one program acts as the client and the other as the server. LabVIEW can be either client or server. Both programs (client and server) exist independent of each other but are able to share information. The client communicates with the ActiveX objects that the server opens to allow the sharing of information [7]. In the case of video streaming [1] [2], most ActiveX controls can only display the data, being incapable of transforming it into a data type that LabVIEW can process. This becomes problematic when the system is used for remote image processing. The LabVIEW environment itself provides little if any possibilities for video streaming, and the methods it does offer are usually not high performance, but it possesses high performance toolkits and modules specialized in image processing, making it ideal for processing the captured data. Therefore, we chose to use existing software, specialized in video streaming along with LabVIEW and to capture the data provided by them, for further use, within LabVIEW. The software we studied (the ActiveX controls of a series of media players that utilize streaming technology) provide high quality data and a very small transmission delay, ensuring the reliability of the results of the image processing.

1. Introduction

Remote monitoring and processing of real-time images can represents quite a challenge, mainly because the possibilities of performing these tasks are quite limited. Transmitting images of a phenomenon in real-time and making these images available to the environment in which we want to process them is complicated, especially if we desire high quality data in order to ensure the reliability of the results of the processing. Our chosen environment for data processing is NI LabVIEW because it provides high performance toolkits and modules for image processing [5], [6], [8]. Unfortunately, the LabVIEW environment provides little if any possibilities for video streaming, and the methods it does offer are usually not high performance. Hence, we chose for avoiding this inconvenience by using existing software, specialized in video streaming along with LabVIEW and to capture the data provided by it, for further use, within LabVIEW. The aforementioned software is formed of ActiveX components that can be embedded in and controlled from LabVIEW.

But these components seldom provide access methods to the data they render, making that data invisible to the LabVIEW environment and impossible to retrieve. Hence resulting the need for additional programs that can extract data form these components and make it accessible to LabVIEW. This method, although seemingly very intricate, could provide excellent results in time. This paper provides our experimental results regarding the components that can be used and how the data can be recovered.

2. ActiveX Technology

ActiveX is the general name for a set of Microsoft Technologies that allows code reuse and the connecting of individual programs together. Based on COM (Component Object Model) technologies, ActiveX is an extension of a previous technology called OLE (Object Linking and Embedding). Each program does not regenerate components, but rather, reuses components allowing for the combination of applications together.

The most common usage of ActiveX is via ActiveX controls, which are embeddable components that exist inside ActiveX containers. From these containers, the ActiveX controls have their own methods and properties that can be used in order to manipulate the control. LabVIEW constitutes such an ActiveX container and can therefore house ActiveX controls.

In general, LabVIEW can embed any ActiveX control. Using the control's properties and methods, LabVIEW can programmatically interact with the control.

2.1. SopCast ActiveX

Sopcast is a Streaming Direct Broadcasting System based on P2P, SoP being the abbreviation for Streaming over P2P. Its core communication protocol which is named sop://, or SoP technology.

Due to the fact that it uses P2P technology, channels are easily available and more stable, having a minimal delay in the P2P streaming market and fast buffering (10 – 30 seconds). It is capable of real-time streaming, supporting two main streaming transport protocols (mms, http), as well as streaming media files such as .asf, .wmv, .rm, .rmvb, mp3, .spl. It can also redirect streams towards other players such as Windows Media Player, RealPlayer, VLC, etc.

SopCast has 2 main components a SopCast Client, also named SopPlayer, and a SopCast Server, also named SopServer or SO (Stream Originator). The Sop Player has an ActiveX component that allows it to be embedded into a webpage or any software applications. This feature allowed the author to embed it into a LabVIEW application (Figure 1).

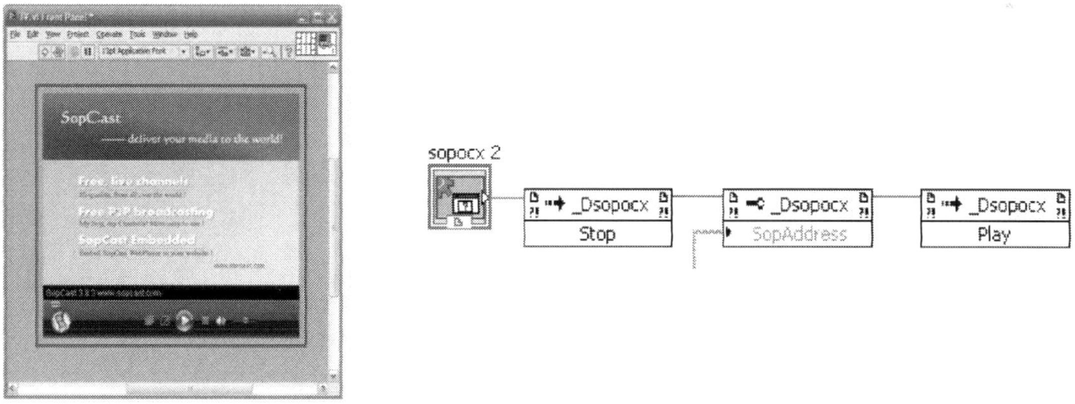

a) b)

Figure 1. SopCast embedded in LabVIEW

a) Front Panel featuring the SopCast ActiveX Control

b) Use of methods and properties of the SopCast Control (code for streaming video content from the address given as a String)

Its most important feature, which constitutes the main reason why it was included in this study, is the fact that it allows the users to build their own channels and broadcast it over the Internet, being

IOP Conf. Series: Materials Science and Engineering **85** (2015) 012023 doi:10.1088/1757-899X/85/1/012023

able to broadcast both video and audio content. It, therefore, constitutes a server and a source of high quality data. It contains security features for the broadcasters and the viewers, the broadcaster having full control on his channels. It supports multiple channels broadcast on the same server (5-10 channels). It also has an extremely low memory footprint and CPU load [9].

2.2. Windows Media Player ActiveX

The reason why Windows Media Player in included this study is the fact that it is capable of receiving streams through 3 major technologies: the User Datagram Protocol (UDP), the Transmission Control Protocol (TCP), and the Hypertext Transfer Protocol (HTTP). It can also receive a multicast stream. UDP and TCP are basic networking protocols that manage more fundamental tasks such as network connectivity and packet error correction. These are used in combination with the Microsoft Media Server (MMS) and Real Time Streaming Protocol (RTSP) protocols, which manage the high-level exchange of data. Hence, the protocols which are used in managing the high-level data exchange when streaming Windows Media-based content are MMS, RTSP and HTTP [3]. Windows Media-based content consists of a large number of formats: .asf, .asx, .avi, .wav, .wma, .wax, .wm, .wmv, .wvx (Windows Media formats), .ra, .ram, .rm, .rmm (RealAudio and RealVideo), m3u, mp2v, .mpg, .mpeg, .m1v, .mp2, .mp3, .m3u, .mpa, .mpe, mpv2 (MPEG), .mid, midi, .rmi (MIDI), .mov (Apple QuickTime), .aif, .aifc, .aiff (Macintosh AIFF Resource), .ivf (Intel Video File), .avi (Audio Visual Interleave), .swf (Macromedia Flash).

Windows Media Player also has an ActiveX component, making it possible to embed it into LabVIEW (Figure 2).

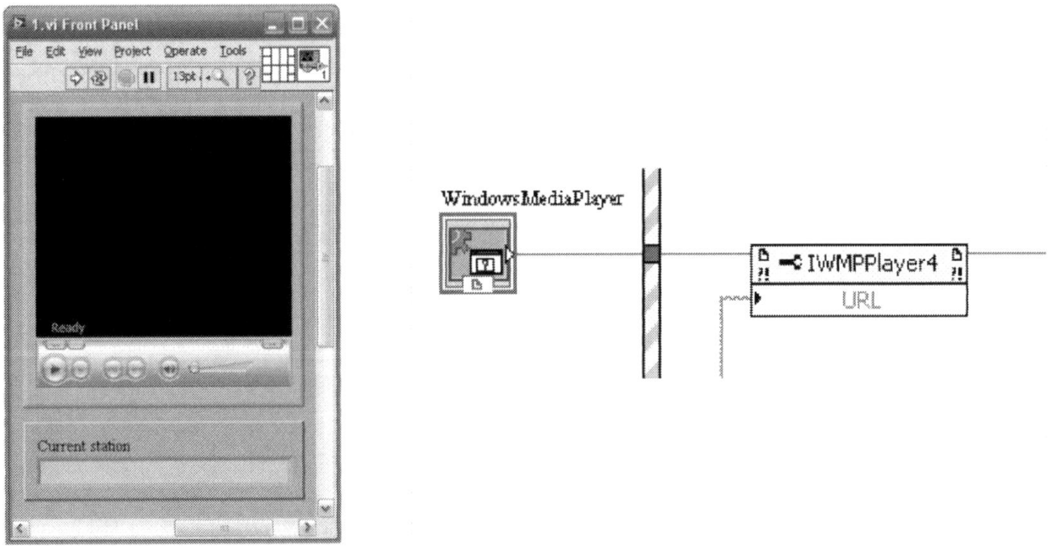

a) b)

Figure 2. Windows Media Player embedded in LabVIEW

a) Front Panel featuring the Windows Media Player ActiveX Control

b) Use of methods and properties of the Windows Media Player Control (code for streaming video content from the URL address given as a String)

2.3. *VLC ActiveX*

VLC media player (commonly known as VLC) is a portable, free and open-source, cross-platform media player and streaming media server written by the VideoLAN project. VLC media player supports many audio and video compression methods and file formats, including streaming protocols.

It is able to stream media over computer networks, supporting the following streaming protocols: RTP/UDP, RTSP, RTP/DCCP, Raw UDP, (RTP or raw) Multicast, File, HTTP, and MMSH. It also plays m2t MPEG transport streams, (.TS) files while they are still being digitized from an HDV camera via a FireWire cable, making it possible to monitor the video as it is being played.

Because VLC is a packet-based media player, it plays almost any type of video content, even if it is damaged, incomplete, or unfinished; such as files that are still downloading via a peer-to-peer (P2P) network.

It is also capable of transcoding multimedia files. VLC can transcode or stream audio and video into several formats including: ASF, AVI, FLV, MP4, Ogg, WAV, MPEG-2 (ES, PS, TS, PVA, and MP3), MPJPEG, FLAC, QuickTime File Format, Matroska, WebM, H.263, H.264/MPEG-4 AVC, MJPEG, MPEG-1, MPEG-2, MPEG-4 Part 2, VP5, VP6, VP8, and VP9 [4], [10].

VLC is able to receive streamed data form such sources as SopCast and it can transcode it into an AVI format which can be read with NI Vision components in LabVIEW. This method is not ideal for real-time image processing because of the delay in writing the .avi file and reading it. There are also limits to how large the .avi file can be; making it mandatory to occasionally interrupt the process and create new storage files. Another reason why this method is inconvenient is because it takes up much disk space.

Also, VLC's ActiveX components are not functional and cannot be embedded into LabVIEW. There is a way of incorporating it into LabVIEW using dynamic libraries, but the method becomes extremely complicated. [11]

2.4. *Other ActiveX Components*

There are a number of other ActiveX Components that, although are not expressly used for video streaming, can be used to render streamed data. We have explored this option because of the possibility that there might not be a stand-alone ActiveX Component that can render a certain data stream. Our purpose is to represent the options available while adapting to the source, but we considered the notion that there might not always be an elegant way to do this. Hence, we present a few universal means that serve the purpose. One of these is the Microsoft Web Browser ActiveX Component (Figure 3). It is a simple browser that can be very useful when it comes to live streaming. Another option would be the ShockwaveFlash ActiveX (Figure 4), which is basically the Shockwave Player. This is a component that is usually required in computers that work with the Internet. Literature indicates that it is capable of streaming data but there is not much information on how this is achieved [12]. This component was mentioned because it represents the best solution for rendering flash videos, which are a convenient option for transferring video data over the internet, being, therefore, extremely widespread.

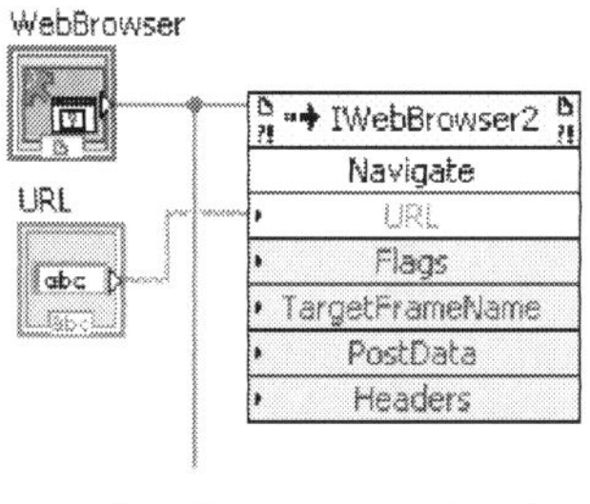

Figure 3. WebBrowser ActiveX Control - code for streaming video content from the URL address given as a String.

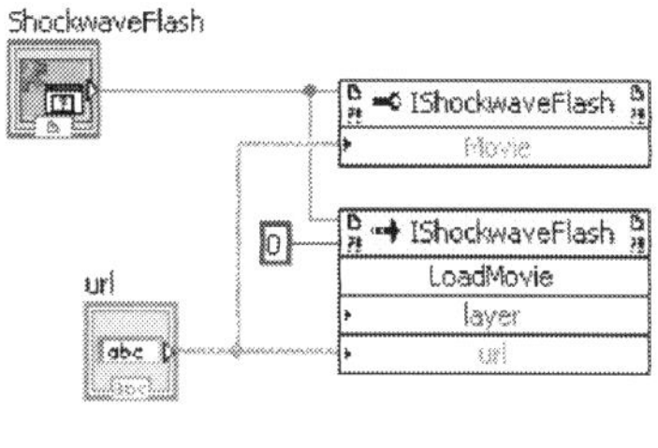

Figure 4. ShockwaveFlash ActiveX Control - code for streaming video content from the URL address given as a String.

IOP Conf. Series: Materials Science and Engineering **85** (2015) 012023 doi:10.1088/1757-899X/85/1/012023

3. Transcoding

Although ActiveX components can be embedded into LabVIEW and programmatically controlled by it, the access to them is not complete: they do not have access methods to the data they display and they do not usually have transcode methods that output in a format LabVIEW can recognize, namely AVI. The lack of these things becomes a serious impediment when the data rendered must be extracted for further processing. Therefore, the question of how to make the data available to the environment in which we want to process it comes to mind. We will analyze the options for each of the aforementioned technologies.

The advantage of using SopCast is that it makes broadcasting data simple and efficient, but it does not accept many formats and AVI is not among the ones it accepts. SopCast does not have a transcode method, but it is capable of redirecting steams towards VLC and Windows Media Player.

VLC can transcode it to AVI, but Window Media Player can only play the stream. VLC, as a streaming program, has the advantage of being able to recognize a large number of steaming technologies, as well as being able to stream itself, being, in this sense, both a server and a client. Yet it has the great disadvantage of not having functional ActiveX components, making it a challenge to embed it into LabVIEW. There are solutions, but they are considerably intricate [11]. Hence, the control of VLC has to be done either manually or through programmatic command line within LabVIEW (Figure 5).

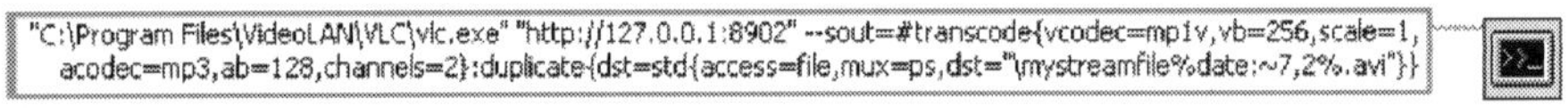

Figure 5. LabVIEW code for a programmatic command line used to work with VLC

Windows Media Player has an ActiveX component that can receive a number of stream technologies, but it cannot transcode these into AVI files. Neither can many of the other ActiveX components mentioned in this paper. So, in order to process the data they received in LabVIEW, we resort to a more indirect way of getting the data. This consists of programmatic PrintScreen functions, as can be seen in Figure 6 a) and b). The first one uses an Activex component named ScreenCapture.ocx, the other presses the PrintScreen key programmatically [13]. Both these methods can be used for image acquisition. These methods usually take a picture of the entire screen, but the region of interest can be easily extracted from the obtained image and further processed. This method has the advantage of acquiring the image directly in LabVIEW and being able to process it strait away. Moments of interest can also be saved as .bmp, .jpeg, .png, .tiff images or as .avi videos.

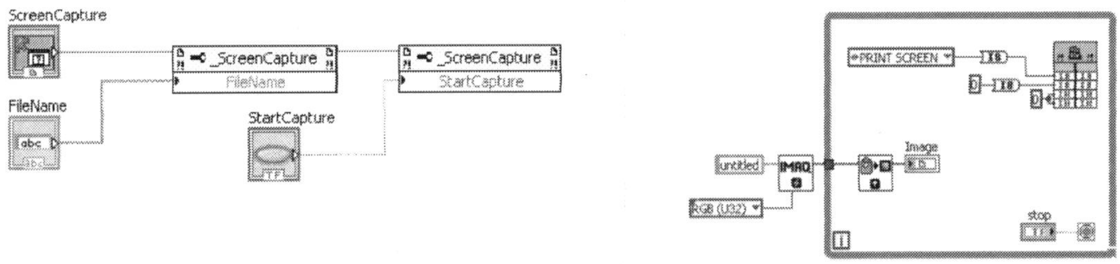

a) b)

Figure 6. Programmatic PrintScreen functions

a) PrintScreen function that uses ScreenCapture.ocx

b) PrintScreen function that presses the PrintScreen key programmatically

4. Conclusions

The goal of this paper is to present means of interaction and data exchange between two programming technologies: LabVIEW and ActiveX.

The LabVIEW environment itself provides little if any possibilities for video streaming, and the methods it does offer are usually not high performance, but it does offer high performance toolkits and modules specialized in image processing. Hence, we chose to use existing software, specialized in video streaming along with LabVIEW and to capture the data provided by them, for further use, within LabVIEW. The software we studied (the ActiveX controls of a series of media players that utilize streaming technology) provide high quality data and a very small transmission delay, ensuring the reliability of the results of the image processing.

References

[1] Ewald H and Page G F 2000 Performing Experiments by Remote Control Using the Internet, *Global Journal of Engineering Education* **4**(3) 287-292

[2] Krehbiel D, Zerger R and Piper J K 2003 A Remote-Access LabVIEW-based Laboratory for Environmental and Ecological Science, International Journal of Engineering Education **19**(3) 495-502

[3] Chen Z J, Lin C and Wei X G 2009 Enabling on-demand Internet Video Streaming Services to Multi-terminal Users in Large Scale, *IEEE Transactions on Consumer Electronics* **55**(4) 1988-1996

[4] Bermudez I, Mellia M and Meo M 2011 Investigating Overlay Topologies and Dynamics of P2P-TV Systems: The Case of SopCast, *IEEE Journal on Selected Areas in Communications* **29**(9) 1863-1871

[5] Kiong T K, Rong W, Htet K K, Narayanan A S and Chung C K 2013 *Adaptive Streaming Framework for Control and Monitoring Applications*, Proceedings of 2013 IEEE International Conference of Mechatronics and Automation, Takamatsu, Japan, August 4-7, pp 207 – 212

[6] Tanwer A and Singh P R 2010 *ReelEffects Of Threshold Of Hard Cut Based Technique For Advertisement Detection In TV Video Streams*, Proceedings of the 2010 IEEE Students' Technology Symposium, IIT Kharagpur, India, April 3-4, pp 211-216

[7] http://www.ni.com/white-paper/2983/en/

[8] http://www.ni.com/vision/

[9] http://www.sopcast.com/

[10] http://en.wikipedia.org/wiki/VLC_media_player

[11] https://decibel.ni.com/content/docs/DOC-24114

[12] http://help.adobe.com/en_US/Director/11.0/help.html?content=25_using_shockwave_01.html

[13] https://decibel.ni.com/content/docs/DOC-23213

IOP Conf. Series: Materials Science and Engineering **85** (2015) 012024 doi:10.1088/1757-899X/85/1/012024

Applications of energy harvesting for ultralow power technology

A Pop-Vadean[1], P P Pop[1], C Barz[2] and O Chiver[2]

[1] Technical University of Cluj Napoca, Department of Mechatronics and Machine Dynamics, Muncii str., no. 103-105, 400641 Cluj Napoca, Romania
[2] Technical University of Cluj Napoca, Center University of North from Baia-Mare, Department of Electrical Engineering Electronics and Computers, dr. Victor Babes str., no.62A, 430083 Baia Mare, Romania

E-mail: adinaluciapopvadean@yahoo.com

Abstract. Ultra-low-power (ULP) technology is enabling a wide range of new applications that harvest ambient energy in very small amounts and need little or no maintenance-self-sustaining devices that are capable of perpetual or nearly perpetual operation. These new systems, which are now appearing in industrial and consumer electronics, also promise great changes in medicine and health. Until recently, the idea of micro-scale energy harvesting, and collecting miniscule amounts of ambient energy to power electronic systems, was still limited to research proposals and laboratory experiments.Today an increasing number of systems are appearing that take advantage of light, vibrations and other forms of previously wasted environmental energy for applications where providing line power or maintaining batteries is inconvenient. In the industrial world, where sensors gather information from remote equipment and hazardous processes; in consumer electronics, where mobility and convenience are served; and in medical systems, with unique requirements for prosthetics and non-invasive monitoring, energy harvesting is rapidly expanding into new applications.This paper serves as a survey for applications of energy harvesting for ultra low power technology based on various technical papers available in the public domain.

1. Introduction

We've all heard the familiar saying "there's no free lunch." It seems to apply to so many situations in life as well as in circuit design. But still, we all like the idea of getting something for nothing. The concept of "Energy Harvesting" has been getting some attention lately because of newly available technologies that allow us to take advantage of free energy sources like sunlight, ambient heat, or the vibration from traffic on a roadway (Figure 1). So do we finally get our free lunch?
You may also be familiar with the adage "If it sounds too good to be true, it probably is."

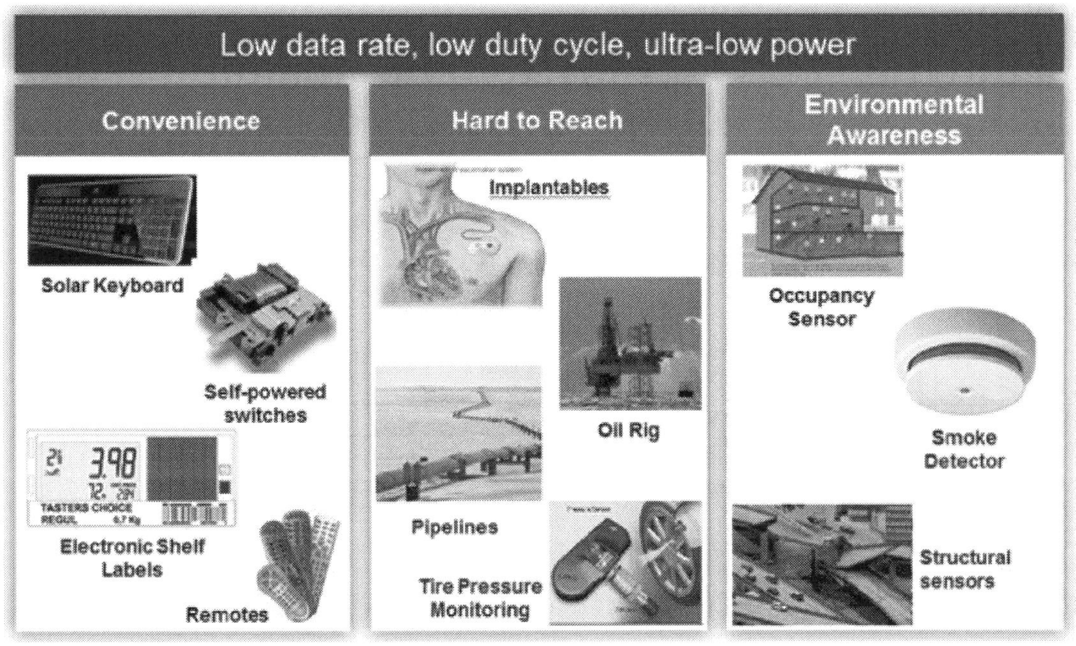

Figure 1. Typical Energy Harvesting Applications.

2. Harvesting energy system

Radio Frequency (RF) communication has of course been used for over a hundred years, and microprocessor technology has been common for many decades. In the last few years, low power RF technology, combined with ultra-low power microcontrollers, have enabled many new end-applications like wireless sensor and control systems [4], [12]. But the limiting factor has often been the power source. If a remote sensor needs a wired electrical power supply, then it can't really be very remote, after all. And if we use a battery, that battery will eventually need to be recharged or replaced, a hindrance, at best, for something that's truly "remote" [19]. How can we build an intelligent remote device that can operate indefinitely once installed in a hard-to-reach location?

Energy harvesting can help provide part of the solution to this problem. But while the long-term source of fuel, sunlight, heat or vibration, may indeed be free, that source is unreliable and highly variable. Unlike the AC mains or a gas powered generator, the availability of the power source is not guaranteed. There are two main steps to determining the rest of the solution [17].

First, we need power conversion circuitry that can convert the low voltage signal from a transducer (such as solar panel or thermoelectric generator) into a stable, regulated output to operate the rest of the system (microcontroller, radio, indicators, etc.) [1].

Second, we need energy storage elements, batteries or large capacitors, that provide a stable source of energy to the system when the ambient environment is not cooperating. These storage elements must be chosen to provide enough capacity to keep the system operating through the time periods when the "free" energy source may not be providing any output. For critical applications, multiple storage elements might be needed for reliability. But we hope that the "dark times" don't last forever, otherwise we will need a really large storage element.

The low-input-voltage power conversion aspect can be handled by a number of integrated boost converter devices [2]. However since the ultimate source of the energy may have very weak / limited output power capability, the converter also needs to operate with extremely low quiescent current and high conversion efficiency. Even one wasted milliamp can render an energy harvesting source useless. Furthermore, to minimize design complexity from the system designer's point of view, we would like the boost converter to keep the storage element charged up automatically, as well as provide a regulated output voltage at the appropriate level to operate the rest of the system. For example the bq25570 device satisfies these requirements; while the bq25505 allows for a second (backup) storage element if needed. The TPS627xx devices are also suitable for external ultra-low power DC-DC regulation if additional output rails are required (Figure 2).

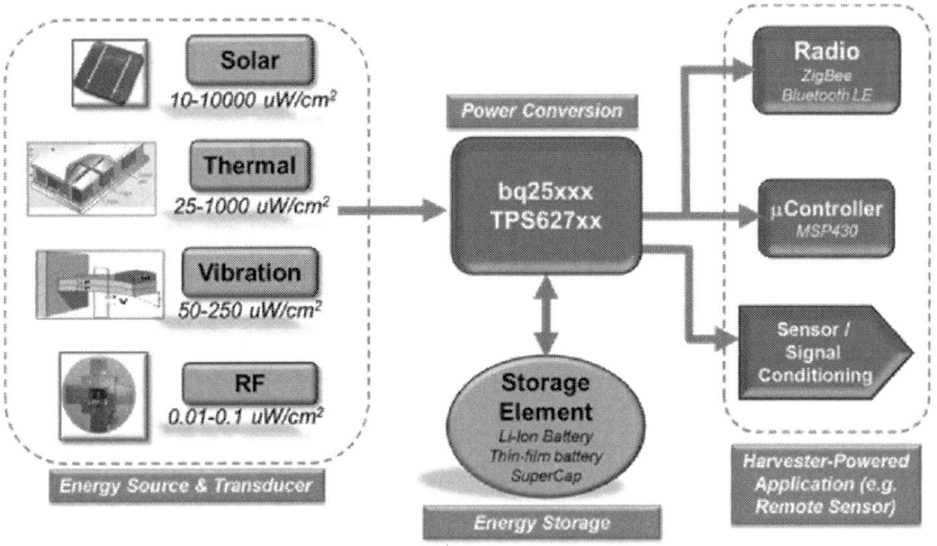

Figure 2. Components of an intelligent remote sensor system.

Getting back to the free lunch concept, we can see that some initial cost is required for the added circuitry and storage elements needed to take advantage of the free energy that might be available from the environment. But once this initial investment is made, assuming the transducer and storage elements are sized properly, it is actually possible to implement a remote sensor / controller device that can operate for long periods of time (theoretically forever) with no maintenance (removal / replacement of batteries or refueling).

As mobile phones morphed from wireless analog telephones to handheld computers, users kept demanding more and more energy-consuming features, such as web browsing, videos, gaming, and email, while still requiring extended battery life. Since battery makers were not much help, semiconductor manufacturers devised numerous energy-saving techniques to make it all possible. They have been wildly successful [18].

Low power has been the most important electronic design criterion for at least the last ten years. Thanks to Moore's Law and a lot of smart engineers, semiconductor power levels have dropped dramatically, often consuming milliwatts in run mode and nanowatts in standby mode. As a direct result, ultra-low-power wireless sensorless networks finally became possible and their adoption has been dramatic. Now, sensors stand alone in remote or hard-to-reach areas to warn of building and bridge stresses, air pollution, forest fires, pending landslides, worn bearings, and wing vibration. Low-power wireless sensor networks are at the heart of numerous industrial, medical, and commercial applications [10], [14].

However, off-grid, as well as portable sensor nodes, rely on batteries for power and face the same problem as cell phones. In such cases, it is advisable to prolong battery life by harvesting environmental energy sources – most often available as light, heat, vibration, motion, or ambient RF [6], [9]. If a device's energy requirement is low enough and battery replacement would be difficult or expensive, it may be possible to dispense with the battery altogether and rely exclusively on harvesting ambient energy sources for power [12]. The combination of ultra-low-power MCUs and energy harvesting have given rise to a wealth of applications that previously were not possible [15].

The energy harvesting market is large and growing rapidly. According to analysts at IDTechEx, energy harvesting was a $0.7 billion market in 2012 and is expected to exceed $5 billion by 2022; by then 250 million sensors will be powered by energy harvesting sources. The market for thermoelectric energy harvesting alone will reach $865 million by 2023.

3. Current technologies and applications

There is several energy harvesting technologies in common use, with some innovative techniques just over the horizon. The most common energy sources are light, heat, vibration, and RF. Short of rooftop solar panels none of them generate a great deal of energy (Figure 3), but one or more of them may be more than adequate to power low-power devices in a particular environment [3], [13].

Source	Source Power	Harvested Power
Light		
Indoor	0.1 mW/cm²	10 μW/cm²
Outdoor	100 mW/cm²	10 mW/cm²
Vibration/Motion		
Human	0.5m at 1 Hz	
	1m/s² at 50 Hz	4 μW/cm²
Machine	1m at 5 Hz	
	10m/s² at 1 kHz	100 μW/cm²
Thermal		
Human	20 mW/cm²	30 μW/cm²
Machine	100 mW/cm²	1-10 mW/cm²
RF		
GSM BSS	0.3 μW/cm²	0.1 μW/cm²

Figure 3. Power available from energy harvesting sources.

3.1. Solar harvesters

There is hardly a home or office that does not have at least one solar-powered calculator – actually, a calculator with a coin-cell battery and a small front panel photovoltaic (PV) cell to top it up. These polycrystalline silicon or thin-film cells convert photons to electrons with a typical efficiency of about 15 to 20% for polycrystalline and 6 to 12% for thin film cells. Since the power available from indoor lighting is typically only about 10 μW/cm², their usefulness depends on the size of the module plus the spectral composition of the light [7].

Small solar cells are frequently used in consumer and industrial applications, including toys, watches, calculators, street lighting controls, portable power supplies, and satellites. Since light sources tend to be intermittent, solar cells are used to charge batteries and/or supercapacitors to provide a stable energy source [16].

IOP Conf. Series: Materials Science and Engineering **85** (2015) 012024 doi:10.1088/1757-899X/85/1/012024

3.2. Thermoelectric harvesters

Thermoelectric harvesters exploit the Seebeck effect, where a voltage is created when a temperature differential exists at the junction of two dissimilar metals. Thermoelectric generators (TEGs) consist of an array of these thermocouples connected together in series to a common heat source such as an engine, water heater, or even the back of a solar panel [21]. Output depends on the size of the TEG and the temperature differential that can be maintained. TEGs are typically used to power wireless sensor nodes in high-temperature environments such as industrial heating systems. A TEG mounted between a power transistor and its heatsink can recycle some of the energy that would otherwise be lost as heat.

Micropelt's TE-CORE7 Thermal Energy Harvesting Modules convert locally available waste heat to provide long-life operation for low-power devices. The TE-CORE TEG converts heat to an electrical charge which is then boosted, stored in a 100μF capacitor, and regulated to supply up to 5.5V. Running at 50°C the TE-CORE7 can deliver 6.424mAh annually, the equivalent of three to four AA batteries – at that rate the batteries would need to be changed every few months [22].

Forcing a current to flow through the junction of dissimilar metals will cause heat to transfer from the hot to the cool junction – the Peltier effect, essentially the opposite of the Seebeck effect. The Peltier effect is the basis for thermoelectric heat pumps [20].

3.3. Piezoelectric harvesters

Piezoelectric transducers generate electricity when stressed, which make them good candidates for vibration sensors when they are used in energy harvesting modules that detect motor bearing noise and the vibration of aircraft wings. The MidéVolture™ V-20W Vibration Energy Harvester employs a cantilever that attaches to a piezoelectric crystal. When vibrations set the cantilever in motion it generates an AC output voltage that is rectified, regulated, and stored in a supercapacitor or thin-film battery.

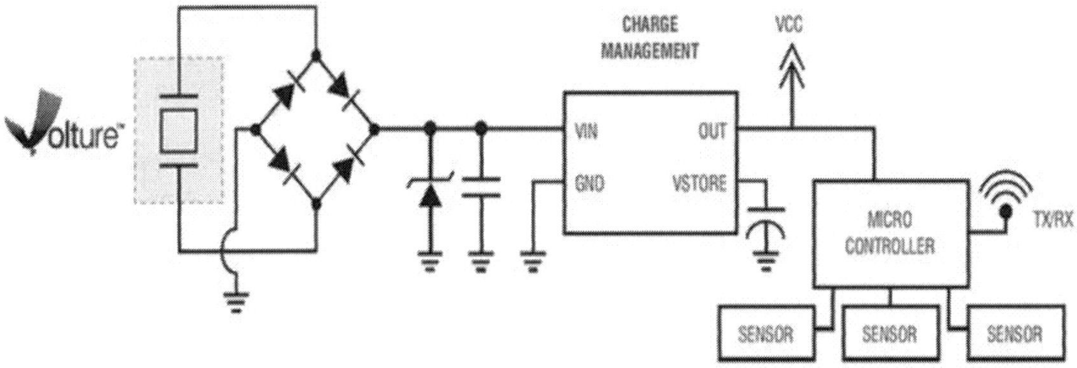

Figure 4: MidéVolture piezoelectric energy harvester (Courtesy of Midé).

3.4. Radio Frequency – RF harvesters

RFID works by rectifying a strong local signal (not ambient RF) aimed directly at the sensor [8]. Similarly Powercast's P2110 RF Powerharvester receiver converts low-frequency RF signals to 5.25V, providing up to 50mA output current. In conjunction with a low-power MCU, sensors, and a radio module, the P2110 can provide a complete, battery-free wireless sensor node that can operate with as little as -11.5dBm RF input. Applications for the device include battery-free wireless sensors for industrial monitoring, building automation, smart grid, agriculture, and defense applications. Mouser carries Powercast development kits for battery charging andwireless sensors.

Side-by-side figures. Where possible, try to place figures side-by-side to reduce the amount of space used. Use a table to do this. For example, to put two figures side-by-side create a table with three columns and two rows. Make the middle column narrow to provide some space between the graphics, as shown below. Note that the table borders are shown as broken lines for guidance only; they should not, of course, be shown in your actual paper.

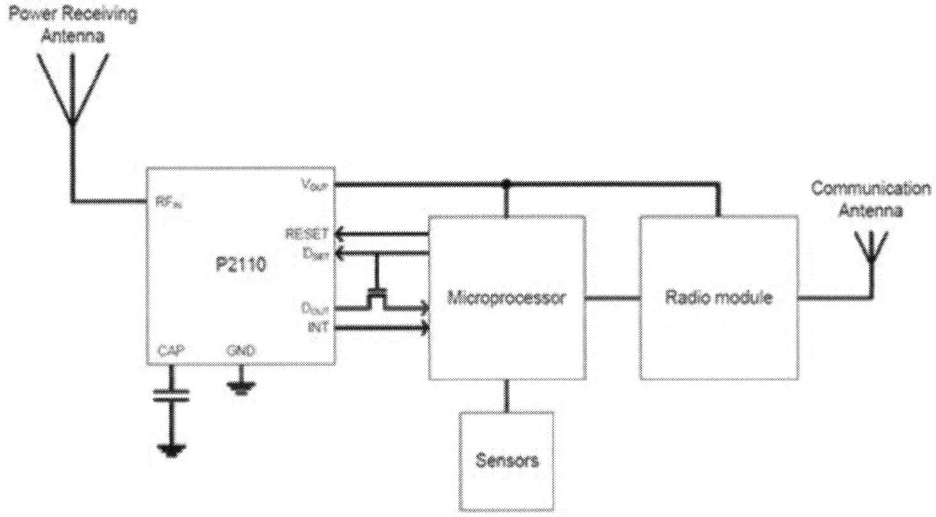

Figure 5. Powercast P2110 in a batteryless wireless sensor (Courtesy of Powercast).

3.5. MEMS pyroelectric generator

Oak Ridge National Laboratories has developed a unique pyroelectric generator that can cool electronic devices, photocells, computers, and even large waste-heat producing systems while generating electricity. The device is based on a MEMS pyroelectric capacitor at the end of a bimetal cantilever that oscillates between hot and cold surfaces. The tip of the hot cantilever comes into contact with a cold surface, the heat sink, where it rapidly loses its heat and causes the cantilever to move back and make contact with the hot surface. The oscillation continues as long as long as there is a sufficient temperature differential – anywhere from a few degrees to several hundred degrees – between the two surfaces.The cantilever structures are only about 1 mm^2 and generate 1 to 10mW per device; however 1,000 of them can be attached to a one square inch substrate, creating a relatively high output power source. Due to the fast cycle time of the cantilever, the developer projects 10 to 30% efficiency – far better than current thermoelectric and piezoelectric energy harvesting devices.

3.6. Nantennas

Photovoltaic cells are the most widely used energy harvesting source, but they are not very efficient. The best monocrystalline PV cells – with a theoretical maximum efficiency of 30% – do well to top 20% efficiency. Now scientists at the University of Missouri and the Idaho National Laboratory have developed a flexible solar film that can theoretically achieve 90% efficiency.

In contrast to conventional photovoltaic cells, the film is essentially an array of nanoantennas (or "nantennas"), each tuned to a specific frequency of light. Rather than generating single electron-hole pairs, as in the case of PVs, the incoming electromagnetic field from the sun induces a current in the antenna that is then collected at the feed point, rectified, and stored. Nanoelectronic electromagnetic collectors (NECs) can be configured as frequency selective surfaces to efficiently absorb the entire solar spectrum. Or NECs can be configured as a reflective bandpass filter centered at a wavelength of 6.5μm; this would enable them to absorb infrared rays, thus recycling waste heat from engines, furnaces, and other high-temperature power sources.

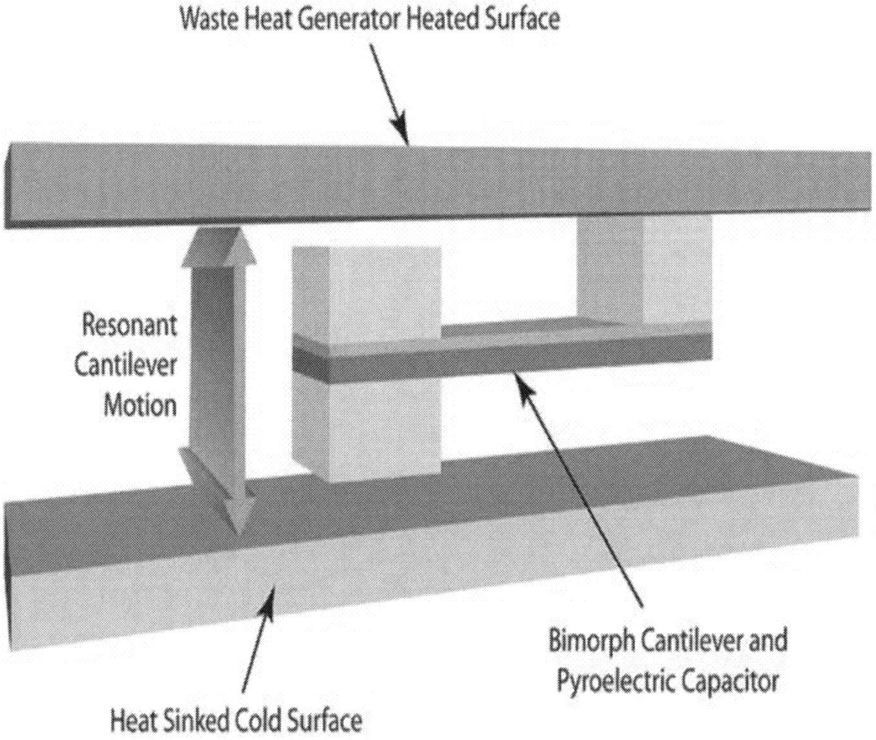

Figure 6. MEMS pyroelectric generator (Courtesy of Oak Ridge National Laboratories).

NEC devices have been successfully prototyped on both silicon and polyethylene substrates, however developing economical mass production processes will require further funding and time. The researchers foresee a product that complements conventional PV solar panels by capturing currently unused infrared energy. As a film, it could be incorporated into building materials and infrastructure. NECs can be integrated into polymer materials so they might also be incorporated into the skin of consumer electronic devices to continuously charge batteries [16].

3.7. Medical and Fitness Devices

Some novel uses for piezoelectric energy harvesting are starting to emerge. Researchers at the University of Michigan have developed a device that harvests energy from the reverberation of heartbeats through the chest and converts it to electricity to run a pacemaker or an implanted defibrillator, hopefully obviating the need for periodic battery replacement. Research is also under way looking for ways to scavenge body heat, movement, and vibration to power other implantable devices.

RF is already being used experimentally to recharge the batteries in pacemakers and implanted transcutaneous electrical nerve stimulation (TENS) devices. The patient sits in a chair that contains a low-frequency RF source whose output is received, rectified, and stored by the device.

Researchers at MIT and Harvard have developed a chip that can be implanted into the inner ear, with power provided by harvesting the energy in sound waves. The chip is designed to monitor biological activity in the ears of people with hearing or balance impairments.

Engineers at MIT are also developing glucose fuel cells to power neural implants. The fuel cell operates by stripping electrons from glucose molecules to create a small electric current. Fuel cells integrated with ultra-low power circuitry, onto a silicon chip, enables entirely self-powered devices such as brain implants.Such implants are being developed to help people with spinal cord injuries or who have suffered strokes. Advances in neuromodulation have resulted in implants that influence the nervous system to control pain, and help eliminate the tremor in patients suffering from Parkinson's disease [11].

Figure 7. Imaging. Ingestible electronic devices.

The ingestible Pillcam (Figure 7) developed by Given Imaging.Ingestible electronic devices, using energy harvesting and/or tiny solid-state batteries, can perform a number of tasks. The 'Pillcam', the size of a large vitamin capsule, is used like an endoscope to visualize the digestive system, detecting abnormalities as it passes through, avoiding the alternative lengthy and uncomfortable procedures.

Targeted drug delivery for certain types of cancerous tumors is another important application for ingestible implantable devices. The ability to direct an active device to a precise location, minimizing the drug dosage administered and avoiding damaging adjacent cells, is proving particularly efficacious. Exploiting body heat to power electronic devices is an obvious candidate for energy harvesting. Chip-based thermoelectric generators (TEGs) are now envisioned that can be inserted under the skin, or in the skull, exploiting the small temperature differences between the brain and skin tissue.

Fitness buffs will be happy to learn that they can recapture some of the energy they expend at the gym. Three British universities have teamed up to develop a piezoelectric energy harvesting device that attaches to the knee, generating power as they walk or run on the treadmill [5].

Riga Technical University offers a mechanical energy harvester that requires magnets to be sewn into the sleeves and coils into the pockets of a jacket; swinging the arms past the pockets while walking generates a current that can be stored in a battery. Anything to keep that iPhone charged!

4. Conclusion

The development of ultra-low-power has created a huge and rapidly expanding energy harvesting market on which they are becoming increasingly dependent. The first wave of energy harvesting has given rise to low-power wireless sensors, which are turning up seemingly everywhere. But the ripple effect will continue throughout consumer, industrial, and medical markets, creating new applications that we can only begin to imagine. Whether planning portable battery-powered devices or the desire to improve the energy efficiency of larger ones, all design engineers should consider incorporating energy harvesting techniques into their products.

This study provides an overview of ultra-low power energy harvesting application, especially recent technology developments in medical domain and existing barriers.

References

[1] Hsu J, Zahedi S, Kansal A, Srivastava M and Raghunathan V 2006 *Adaptive duty cycling for energy harvesting systems*, Proceedings of the 2006 International Symposium on Low Power Electronics and Design (ISLPED'06), Tegernsee, Germany, October 4-6, pp 180-185

[2] Raju M 2008 Energy harvesting. ULP meets energy harvesting: agame-changing combination for design engineers, http://focus.ti.com/lit/wp/slyy018/slyy018.pdf
[3] Randall J 2005 *Designing Indoor Solar Products*, John Wiley & Sons
[4] Roudy S and Frechette L 2005 *Energy scavenging and nontraditional power sources for wireless sensor networks*, in Handbook of Sensor Networks: Algorithms and Architectures, I. Stojmenovic, John Wiley & Sons
[5] Mitcheson P D, Yeatman E M, Kondala Rao G, Holmes A S and Green T C 2008 Energy Harvesting From Human and Machine Motion for Wireless Electronic Devices, *Proceedings of IEEE* **96**(9) 1457-1486
[6] Roundy S, Rabaey J M and Wright P K 2004 *Energy Scavenging for Wireless Sensor Networks, with Special Focus on Vibrations*, Kluiwer Academic Publishers, Norwell, Massachusetts
[7] Randall J F 2005 *Designing Indoor Solar Products. Photovoltaic Technologies for AES*, John Wiley & Sons, West Sussex
[8] Want R 2006 An introduction to RFID technology, *IEEE Pervasive Computing* **5**(1) 25–33
[9] Paradiso A 2008 *Energy Scavenging for Mobile and Wireless Electronics,* Massachusetts Institute of Technology Media Laboratory Thad Starner, Georgia Institute of Technology, GVU Center
[10] ***Energy-Harvesting Active Networked Tags (EnHANTs) Project, Columbia University, http://enhants.ee.columbia.edu
[11] ***http://www.bccresearch.com/market-research/healthcare/microelectronic-medical-implants-produts-hlc016d.html
[12] ***http://www.ecnmag.com/articles/2011/12/energy-harvesting-enables-ultra-low-power-applications
[13] ***http://www.idtechex.com/events/presentations/harnessing-free-energy-ultra-low-power-energy-harvesting-circuits-for-smart-buildings-wearable-devices-and-factory-automation-005283.asp
[14] ***Texas instruments, http://www.ti.com/
[15] ***http:// www.ti.com/energyharvesting.
[16] ***http://news-service.stanford.edu/news/2008/january9/nanowire-010908.html
[17] ***IEEE 802.15.4, IEEE Standard for Information Technology, 2006, http://standards.ieee.org/getieee802/download/802.15.4-2006.pdf
[18] ***IEEE 802.15.4a, IEEE Standard for Information Technology, 2007, http://standards.ieee.org/getieee802/download/802.15.4a-2007.pdf
[19] ***http://www.wibree.com
[20] ***http://www.poweredbythermolife.com
[21] ***http://www.nextreme.com/pages/products/teg.html
[22] ***http://www.micropelt.com/down/datasheet_mpg_d602_d751.pdf

Considerations regarding the concept of the level of reality in the process of learning the transdisciplinary approach

P P Pop[1], A Pop-Vadean[1], C Barz[2] and O Chiver[2]

[1] Technical University of Cluj Napoca, Department of Mechatronics and Machine Dynamics, Muncii str., no. 103-105, 400641 Cluj Napoca, Romania
[2] Technical University of Cluj Napoca, Center University of North from Baia-Mare, Department of Electrical Engineering Electronics and Computers, dr. Victor Babes str., no.62A, 430083 Baia Mare, Romania

E-mail: poppaulpetrica@yahoo.com

Abstract. According to the transdisciplinary approach [15], the leap from one level of reality to another requires the generation of a third party [9], on a certain level of reality, which leads to the reaching of a superior level, through knowledge and complexity.In technical terms, what exactly is the third party included? How does it manifest? The way through which, on a reality level of the technical system one can aquire extra knowledge is called integration [10], [12].

1. The integration and reality

The integration is visible in reality through two abilities: integration through the generation of extra knowledge in the system and integration beside this extra knowledge and of one third party included on that level. The third party included, in a phylosophical sense, means the existence of an A and a Non A, on the same level [9]. In technical terms, this third party cannot be represented but by one component, an equipment or integrated subsystem.

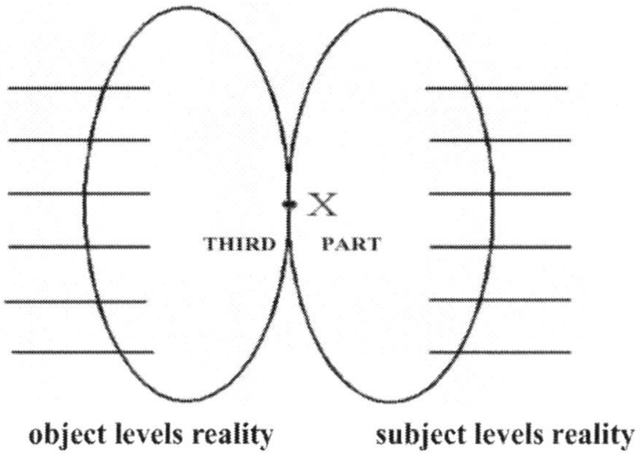

Figure 1. Level reality.

We have already know above what the third party has meant in a hystorical meaning, on each and every level. We have specified on the superior level the third party, from the inferior level, which has lead to the occuring of another type of reality. A level in itself is not superior because it is above others, or because it has more components or subsystems which are integrated, but because it generated, on the basis of the included third party, a bit of extra knowledge that corresponds to a new reality, that is that of the coexistence between A and Non A, in the sense of matter=substance +energy+ information +entropy [1], [2], [5], [6]. The integration can generate extra knowledge without the support of a third party. But this third party does not divide on each level, but it changes on an informational manner [22], [23], [24]. To demonstrate this, we should analyse now the mecatronic system, fom a superior level, down to the basis. Supposing that from the interface (situated on the information level, which according to my opinion is not the last level,as I am about to show), we edit a command in the system. On the level of the interface this is, from a physical point of view, a digital sign (joint if we use the bivalent logic of the usual systems of calculations or with more means if we use a system with Fuzzy logic or a quantifying calculator). (Here is the level of the artificial intelligence that exceeds the informational level. That is to say, another level). On the sensitive level, the sensor-the third party included-gets the signal with the logical possibility of true or false and, according to the stimulus it allows me or not the passing of the command, that is of the signal with which it operates.

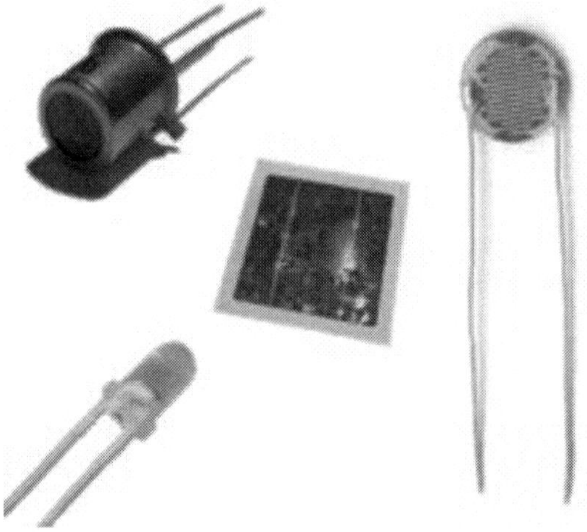

Figure 3. Sensors and transistors.

2. The integration and reality

This signal is recognized by the electronic level , which on its way through the transistor- the third included party- becomes electrical impulse of power, lack of power. The rotating actuator - a third party on the electro mechanical level, gets the electrical sign and changes it ina a movement moment, which generates movement for the executional element on the physical level. Now if we look from downside to upside we notice that the signal which starts from a level aquires, through the third pary , a plus of quality, more quantity and extra energetic, informational power, specfic to the realities of every level. This signal aquires an extra knowledge that is superior to the inferior reality. From an informational point of view, the digital signal is a superior reality, compared with the electrical or the electronical systems. That is, we could always change any kind of text into a digital one and then in power impulses, or no powered impulses, but you can't change, on the other hand, any kind of power into a digital one and then into a text. So, the integration that brings up in the mecatronic science, a third party, besides extra knowledge, generates, in a technical sense, another level of reality.

On the basis of what is mentioned above, we can sustain the fact that, starting from a rudimentary system, on a physical level, we can manage, through integration and a third party, to obtain a system which is technically changing and which can suffer transformations of a reality that rules a system, in a superior reality that governs another level: physical in electromachanical, electromechanical in electronic, electronical in sensitive, sensitive in informational, in the artificial intelligence and so on. The third party may be built up from subsystems, as well. The rotating actuator is an example of fixed integrated elements, which, through coercion, generates movement in the system, with the help of an exterior signal [7], [13].

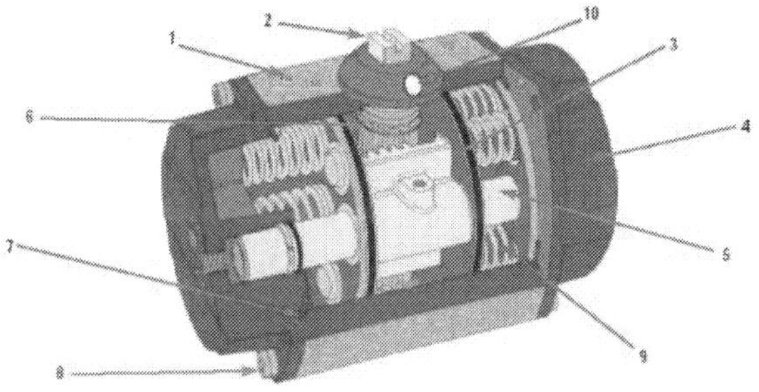

Figure 2. Rotating actuator. Note 1 to 10 - integrated components through constraint.

If this signal comes from a superior level of reality, compared to a mecatronic system, as i have shown above, then the actuator is the third party here. But if you look at it in isolation, it isn't. But, can we consider nowadays the actuator, a separate part of the system? Of course we can't. All the actuators bearing mechanical characteristics function on the basis of the laws that rule the electromechanical level [8]. Even those that are part of the real, those that we haven't invented yet. According to the transdisciplinary approach, the hidden third party, permits that the reality levels won't self distruct, but to transmit through the generation of the third party, from each level, something superior to the next superior level. Thus, in tecnical terms, this means balance in the mecatronical system [19]. But who refines this balance? The answer is, with no doubt, the human being, who, through, integration, as we have mentioned above, ensures the reality levels in mecatronic [20], [21], [24]. Man is always present, both integrated in the system and outside it. This is a concept taken from quantum mechanics of the particles that, for a short time we can notice, but which transcend another reality [16], [17], [18]. After a man set a system into function, he is already integrated with it, even if he went out for a beer in the meantime. How can this be possible? The engineer has obtained, on the basis of the information from his superior level of reality, a reality for each level of the mecatronic system. From an informational point of view, he established a balance between the levels of the system. Its balance is governed by the informational area, created by him for the system, that is, he created a hidden third party for the system. The side that embodies human knowledge, that the system needs, is thus, incorporated. This is just like an informational fingerprint on the part of the man, for the system. If I am allowed to make a comparison, it's like the omnipresence of god related to our world. As much as we should or must know. This doesn't mean that we will send God to have a beer. Can a mecatronic system function without a superior level? Of course it can.

The actuator generates movement even without an interface, sensors or transistors. It only needs external energy, that should be from the same level. Can a superior level exist without an inferior level? No.

3. Conclusion

We don't have an interface without information, we don't have information without sensors, we don't have sensors without transistors, we don't have transistors if we don't have an actuator, that is to whom we can order signal and actuator without fixed components. This finally means superiority on the part of upside, compared to the downside, which means an informational autonomy on the part of the superior towards the inferior.

References

[1] Amerongen J and Jongkano W 1996, Mechatronics in the Netherlands, *IEEE/ASME Transactions on Mechatronics* **1**(2) 106-110

[2] Auslander D M 1996 *Mechatronics: Mechanical System Interfacing*, Prentice Hall Inc., New Jersey, USA

[3] Auslander D M 1996, What is Mechatronics? *IEEE/ASME Transactions On Mechatronics* **1**(1) 5-9

[4] Basarab N and Michel C 2004 *Rădăcinile libertății*, „Polirom" Publishing House, Iasi, Romania

[5] Balan R 2001 *Microcontrolere. Structura si aplicatii,* „Todesco" Publishing House, Cluj-Napoca, Romania

[6] Balan R 2006 *Delphi. Aplicații în mecatronică*, „Todesco" Publishing House, Cluj-Napoca, Romania

[7] Bergh F C, Kita A and Ume C I 1999, *Development of Mechatronics Course in the School of Mecanical Engineering at Georgia Tech,* Proceedings of the 1999 IEEE/ASME International Conference on Advanced Intelligent Mechatronics, Atlanta, USA, Septembre 19-23, pp 459-465

[8] Draganescu M 1980 *A doua revoluție industrială. Microelectronica, automatica, informatica – factori determinanți*, „Tehnica" Publishing House, București, Romania

[9] Lupasco S 1982 *Logica dinamică a contradictoriului*, „Politica" Publishing House, București, Romania

[10] Maties V 1998 *Mecatronică*, „Dacia" Publishing House, Cluj-Napoca, Romania

[11] Maties V, Mandru D, Balan R 2000 *Tehnologie și educație în societatea informațională*, Lucrările Sesiunii Jubiliare de Comunicări Științifice "Creștere economică, dezvoltare, progres", Universitatea Babeș-Bolyai Cluj-Napoca, Romania

[12] Maties V, Mandru D, Balan R,Tatar O and Rusu C 2001 *Tehnologie și educație mecatronică*, „Todesco" Publishing House, Cluj-Napoca, Romania

[13] Maties V, Mandru D, Tatar O, Maties M and Csibi V 2000 *Actuatori în mecatronică*, „Mediamira" Publishing House, Cluj-Napoca, Romania

[14] Maties V, Mirescu C, Mandru D, Balan R, Tatar O and Rusu C 2002 *Tehnologie și educație mecatronică. Auxiliar curricular*, „Economica" Publishing House, Bucuresti, Romania

[15] Nicolescu B 1999 *Transdisciplinaritatea. Manifest*, „Polirom" Publishing House, Iași, Romania

[16] Nicolescu B 2002 *Noi, particula și lumea*, „Polirom" Publishing House, Iași, Romania

[17] Nicolescu B 2009 *Ce este realitatea?*, „Polirom" Publishing House, Iași, Romania

[18] Nicolescu B 2009 *In oglinda destinului*, „Ideea Europeană" Publishing House, Bucuresti, Romania

[19] Restian A 1989 *Integronică*, „Științifică și Enciclopedică" Publishing House, București, Romania

[20] Rzevski G 1991 *Mechatronics at The Open University*, Proceedings of the ISATA'91 Conference, Florence, Italy, May 20-24, pp 1-9

[21] Salminen V 1996 Ten Years of Mechatronic Reserch and Industrial Applications in Finland, *IEEE/ASME Transactions on Mechatronics* **1**(2) 103-105

[22] Van Brussel H 2003 Mechatronics Network of Excellence, Propunere de proiect în PC 6, Bruxelles

[23] Visa I 2001 *Structural Analysis of Linkages as Multibody Systems*, Simpozion SYROM 01,

București, Romania
[24] Visa I et all. 2004 *Proiectarea funcțională a mecanismelor. Metode clasice și moderne*, „Lux Libris" Publishing House, Brașov, Romania

IOP Conf. Series: Materials Science and Engineering **85** (2015) 012026 doi:10.1088/1757-899X/85/1/012026

The influence of the cast iron structure upon the hardness of brake shoes meant for the rolling sock

A Socalici[1], L Pascu[1], E Popa[1] and T Heput[1]

[1] Politehnica University of Timisoara, Department of Engineering and Management, 5 Revolutiei Street, 331128 Hunedoara, Romania

E-mail: virginia.socalici@fih.upt.ro

Abstract. An important characteristic with a high impact upon the exploitation durability of the brake shoes is hardness. The paper introduces the influence of the phosphorous cast iron structure upon the hardness of the brake shoes meant for the tractive and trailing rolling stock. The results presented show the variation of hardness on the surface and the cross section of the braking shoe

1. Introduction

The On manufacturing brake shoes meant for the rolling stock, phosphorous pig irons are largely used. They are pearlitic pig irons, with a high content of phosphor. A phosphorous ternary eutectic also appears in the structure, namely Steatite, which is made of pearlite, cementite and iron phosphide (Fe_3P). The microstructure of a phosphorous pig iron [1] is shown in Figure1.

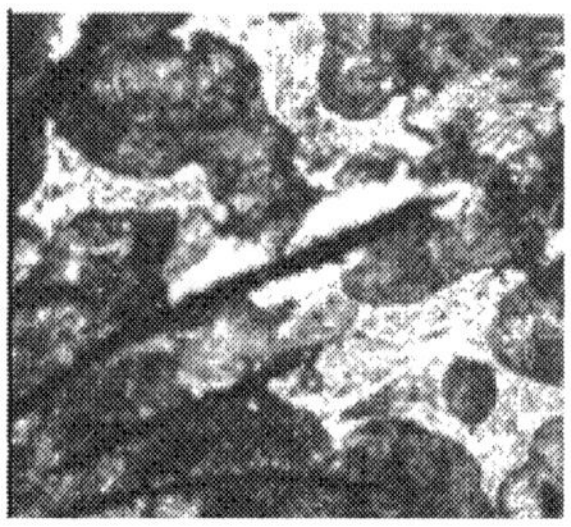

Figure 1. Microstructure of iron phosphorous, attack NITAL 2%, 400x [1].

The microstructure of the material used to cast brake shoes has to be homogenous, pearlitic with lamellar graphite and phosphorous eutectic, evenly distributed.

The phosphorous content of the cast iron used in manufacturing brake shoes is important, but the influence of other components upon the metallographic structure should not be ignored.

Also, an important role in the quality of the cast iron meant for brake shoes is played by the dimensions of the eutectic grains, by the distribution of the phosphide lattice within the structure, the shape, dimension and distribution of the graphitic lattice, as well as the quantity of graphite released during the graphitic crystallizing.

The P10 type brake shoes size S1, S2 and S3 are cast out of phosphorous cast iron and are used for the traction and trailing stock. Brake shoes are manufactured along a technological flux (Figure2) consisting in the elaboration aggregate – an induction electric furnace, an automated air-flow-pressure molding machine and a casting line.

Figure 2. Flow technology for producing brake blocks.

The iron casting process is carried out as required clogs provider must comply with the specifications nr.1/SFMR/SDT/2000 [2].

The body of the brake shoe for the traction and trailing stock is made of cast iron allied with P10 type phosphor, having the chemical structure according to the standards in force [2], [3], [4]:

- The content of phosphor must range within 0,8% and 1,1%;
- The content of manganese must be calculated so as: 1,72%S + 0,30% < %Mn < 1%, (where S% represents the content of sulphur in the cast iron);
- The total content of carbon 2,90 % < % C_t < 3,30 %;
- The content of silicon 1,20 % < %Si < 2,00 %

The braking shoe backing for the traction and trailing stock is made of laminated steel [5-7]. The content of carbon, sulphur and phosphor of this steel must be: % C < 0,130 %; % S < 0,0062 %; % P < 0,0062 % [6], [9]. The braking shoe backing must have a cross section of at least 100 mm^2 and a length of at least 80 % of the shoe length.

2. Experimental regarding

An important characteristic with a significant influence upon the useful life of brake shoes if hardness. Hardness is tested according to SR EN ISO 6506-1:2006. Hardness is measured at the extreme points of the brake shoe (*1* and *2* – Figure 3.a), on its front and at three points located diagonally (*s*, *m* and *j* – Figure 3.b). The brake shoe must have on its lateral surface, as well as on the cross section, a Brinell hardness ranging within 197-225HB [2].

The industrial research has been done by analyzing 25 charges of phosphorous cast iron produced by a company specialized in brake shoe manufacturing [8]. The analyses focused on the content of the metallic charge, the chemical structure of the elaborated cast iron, the physical and chemical characteristics of the shoes, respectively the metallographic examination of the shoe cast iron, according to SR EN ISO 945-1:2009 and file UIC 832, with no chemical treatment and with a G magnifying of 100x.

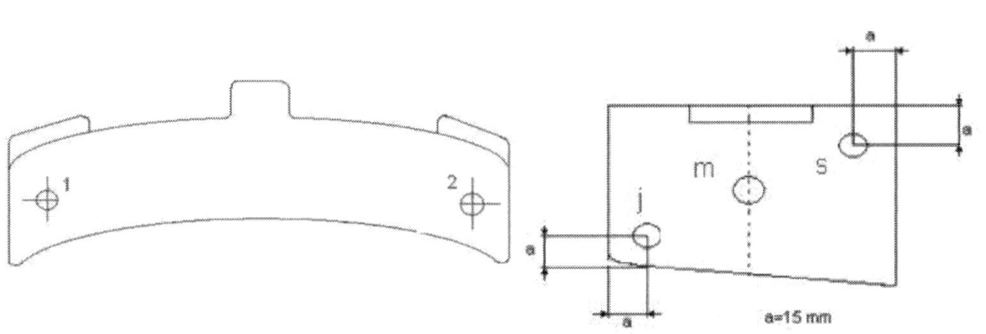

a. shoe b. sample

Figure 3. Sampling areas hardness shoe [2].

The data pertaining to the standard chemical structure of the cast iron elaborated and the experimental data are shown in Table1. Table2 shows the data related to the brake shoe hardness variations, determined according to SR EN ISO 6506-1:2006, both on the surface of the shoe (HB_1, HB_2) and in its cross section (HB_s, HB_m, HB_j) at the points of hardness sampling (Figure 3). The mean values of the brake shoe hardness on the surface $(HB_1+HB_2)/2$ and in cross section $(HB_s+HB_m+HB_j)/3$ as well as their difference are given in Table3. The chemical structure of the charges under analysis is given in Figure 4.

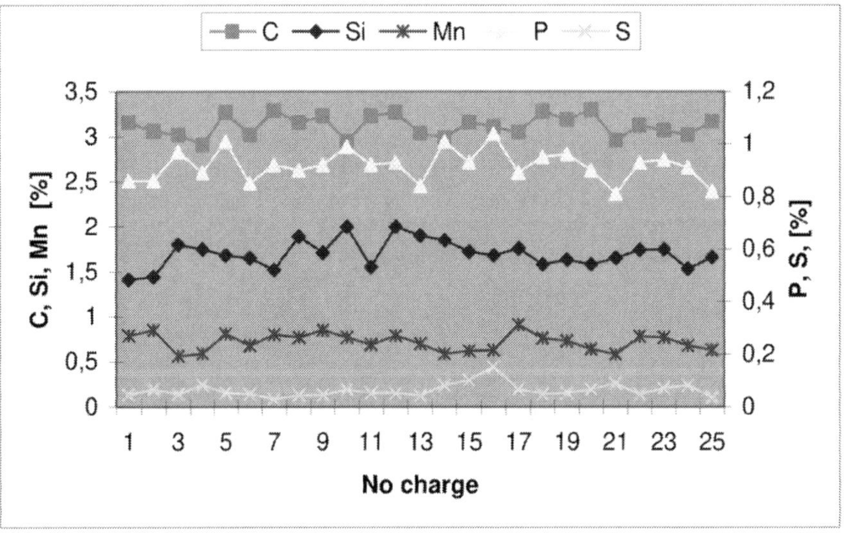

Figure 4. The chemical structure of the charges.

Table 1. Cast iron chemical composition

Terms	Values	Chemical composition, %					
		C	Si	Mn	P	S	Cr+Mo+Ti+Nb+V+W
	Maximum	3.30	2.00	1.00	1.10	0.155	-
Standard	Minimum	1.20	1.20	0.33	0.80	0.03	-
	Average	2.25	1.60	0.665	0.95	0.065	-
	Maximum	3.30	2.00	0.91	1.04	0.153	0.344
Exp.	Minimum	2.91	1.41	0.56	0.81	0.03	0.109
	Average	3.12	1.70	0.72	0.919	0.065	0.213

Table 2. Hardness tests

Values	According to standard	Hardness determined as shown in fig.3 [HB]				
	HB	HB_1	HB_2	HB_s	HB_m	HB_j
Maximum	255	249	246	245	232	241.66
Minimum	197	208	199	202	197	202.67
Average	226	222	220.62	220.44	208.60	217.02

Table 3. Average values of surface hardness and shoe section

Values	Hardness values and the average lateral cross-sectional area		The side surface hardness differences in cross-section	
	$(HB_1+Hb_2)/2$	$(HB_m+HB_m+HB_j)/3$	$\lvert HB_1-HB_2 \rvert$	$\lvert HB_s-HB_j \rvert$
Maximum	247.5	241.66	9	11
Minimum	203.5	202.66	0	0
Average	221.61	217.03	3.77	4.07

Figure 5 presents metallographic examination for iron phosphorous developed. The analysis of the diagrams given in Figures 6 and 7 shows that there are slight variations of the hardness measured on the lateral surface and in the cross section of the brake shoe, respectively of their average. This is due to the narrow-limited variation of the shoe chemical structure. As to the difference between the hardness values measured on the lateral surface and in the cross section, whose graphic representation is given in Figures 8 – 13, the resulting values are relatively small, which points to a good chemical homogeneity of the brake shoes.

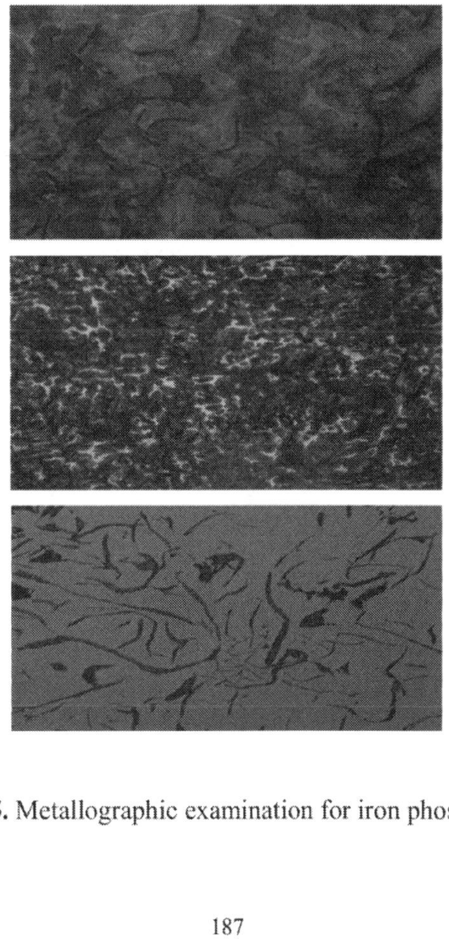

Figure 5. Metallographic examination for iron phosphorous.

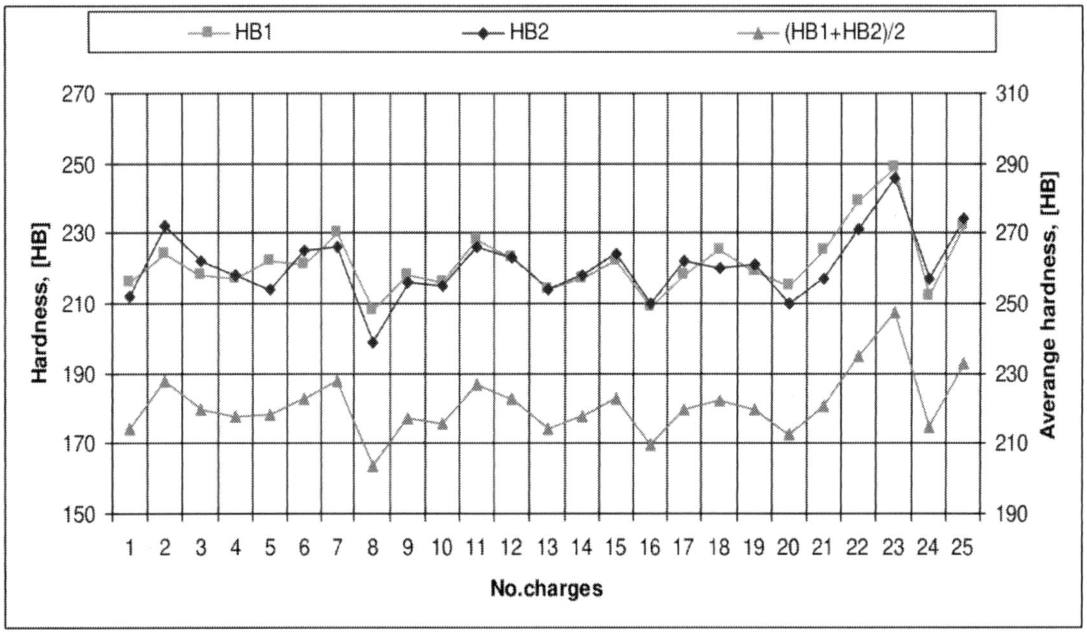

Figure 6. The variation of hardness and its average on the lateral surface of brake shoes.

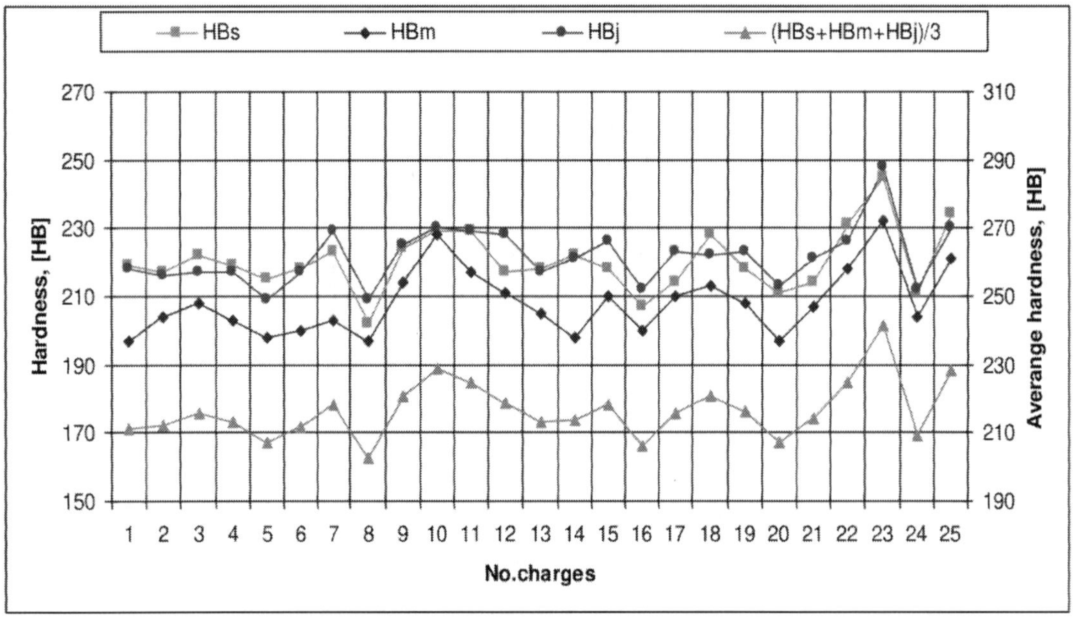

Figure 7. The variation of the hardness and its average on the cross-sectional surface of brake shoes.

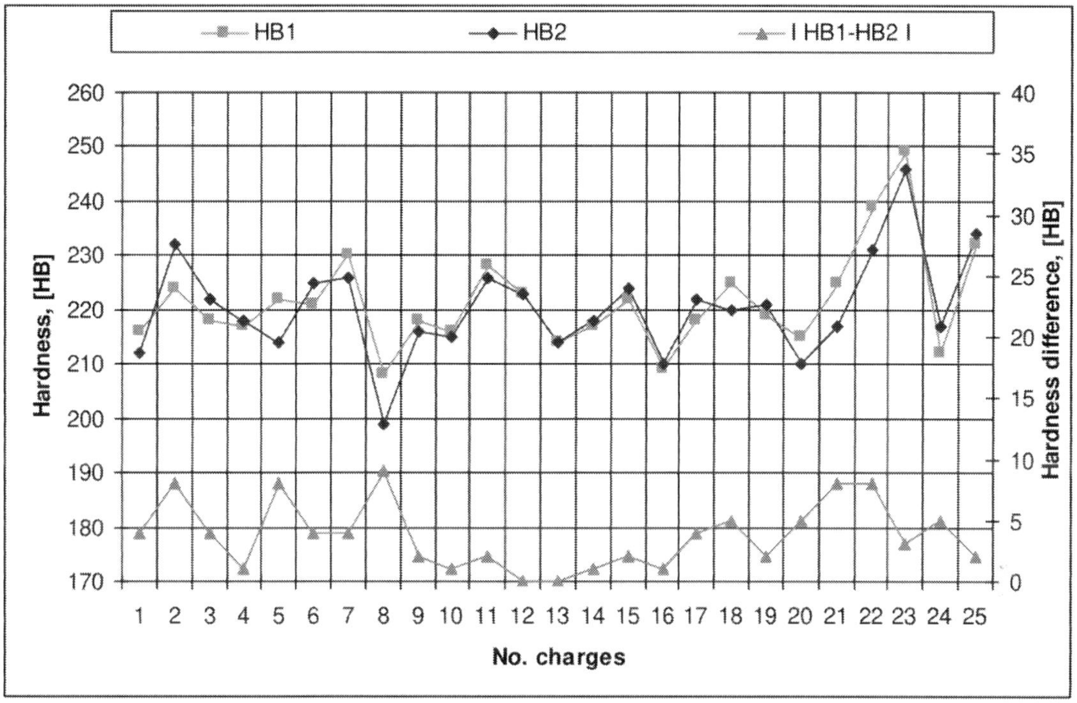

Figure 8. The variation of the hardness and hardness difference on the lateral surface of brake shoes.

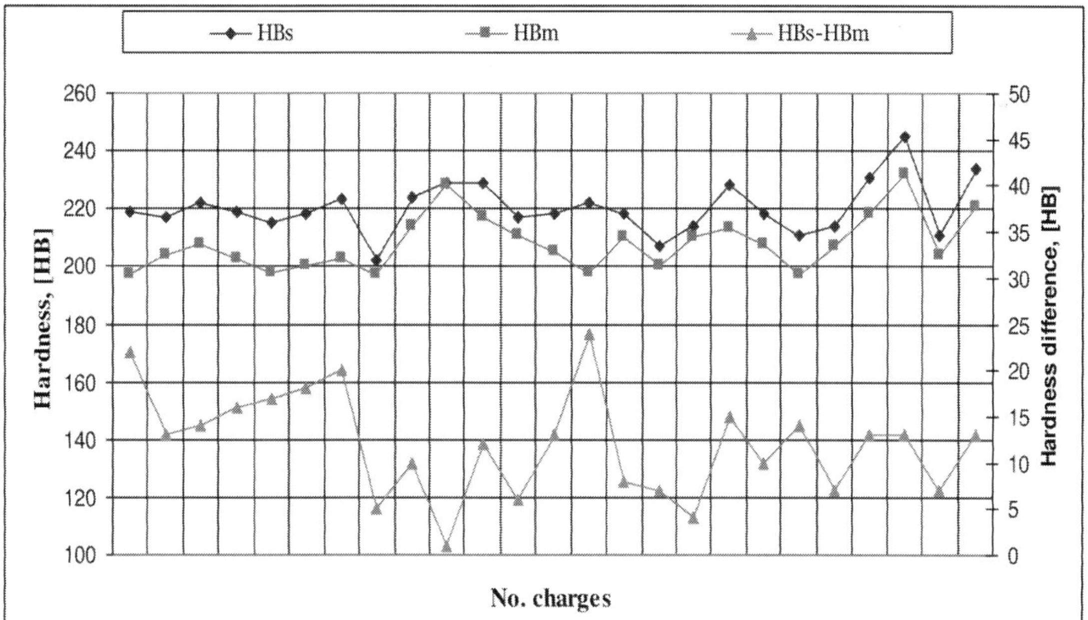

Figure 9. The variation of the hardness and hardness difference on the cross-sectional surface of brake shoes (the upper and central area).

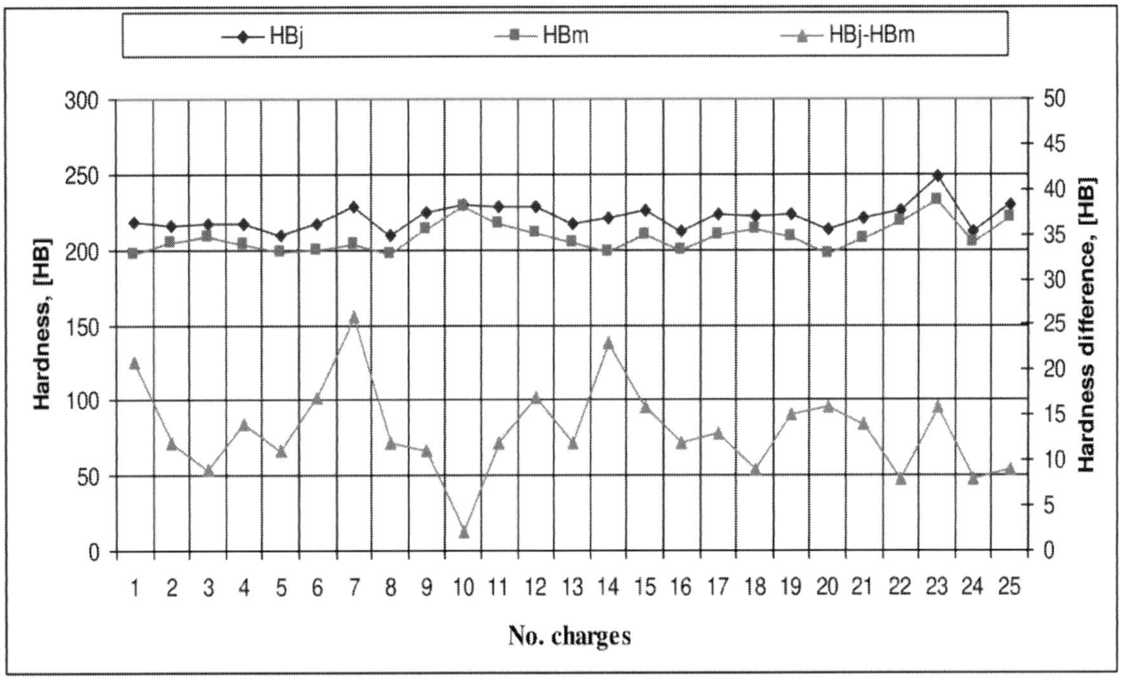

Figure 10. The variation of the hardness and hardness difference on the cross-sectional surface of brake shoes (the lower and central area).

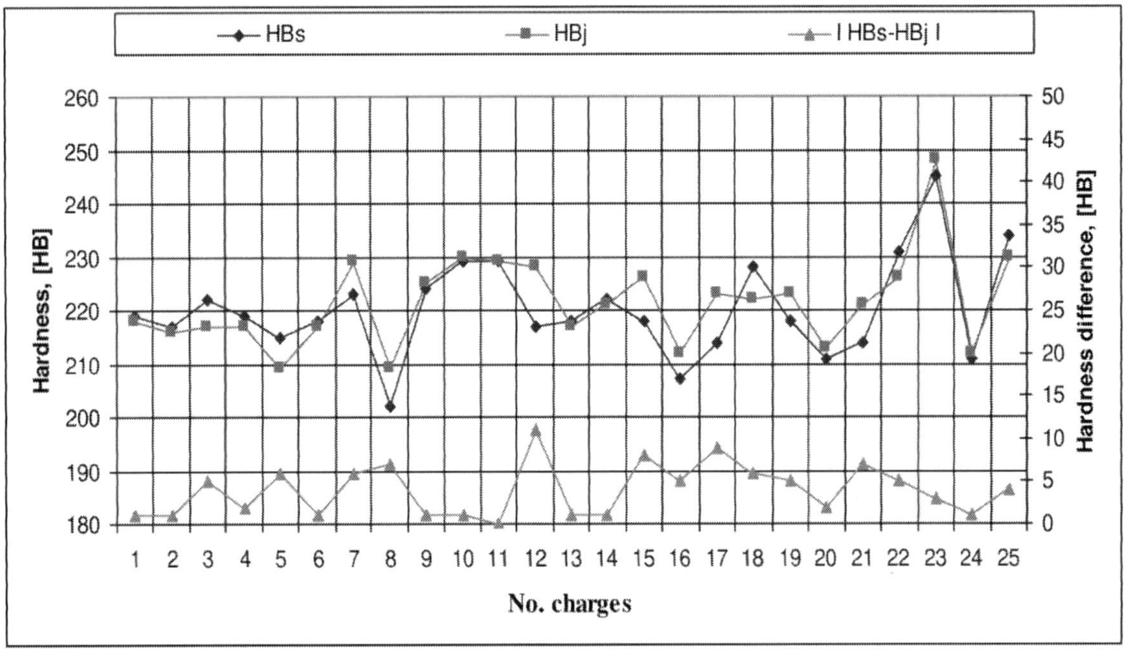

Figure 11. The variation of the hardness and hardness difference on the cross-sectional surface of brake shoes (the lower and upper area).

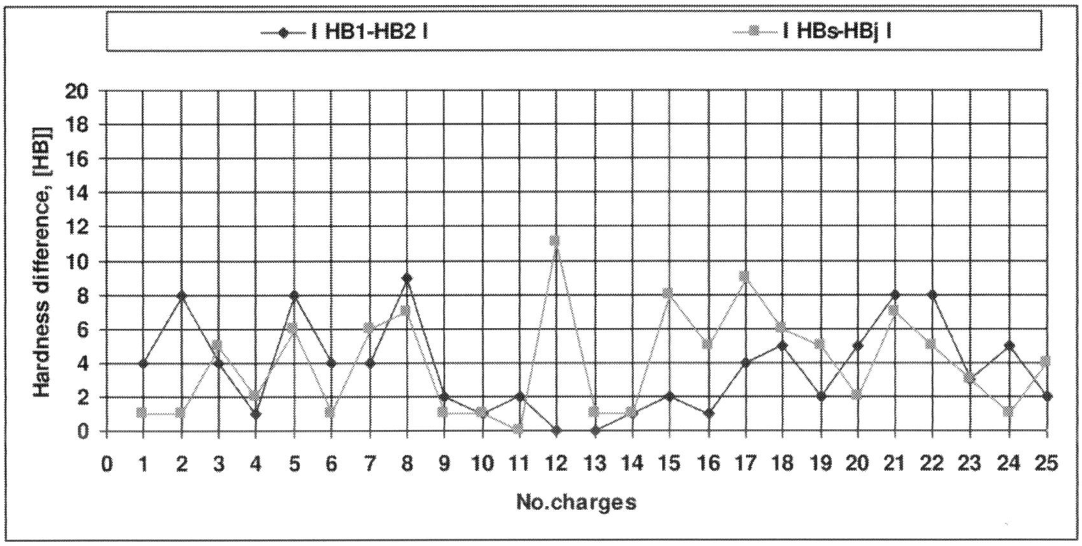

Figure 12. The variation of the hardness difference on the lateral surface and in the cross sectional area of the brake shoes.

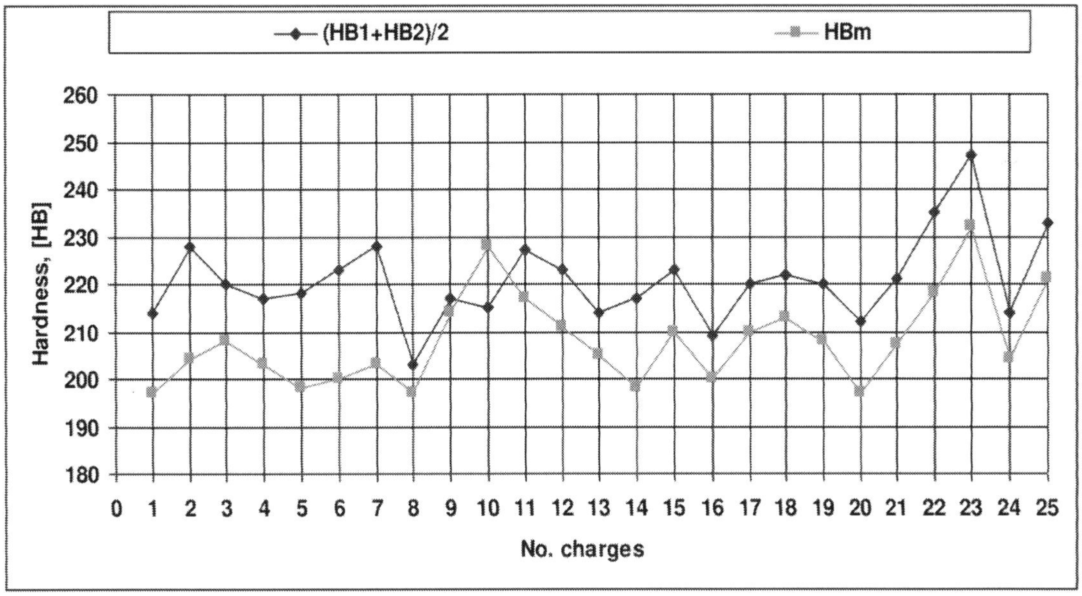

Figure 13. The variation of the average hardness on the lateral area and the hardness in the central area of the cross section of the brake shoes.

3. Conclusions

The values of hardness obtained during the industrial experiments range within the interval 197-240HB, which is in accordance with the international standards. The chemical and structural homogeneity of the shoe brakes leads to slight variations of the hardness values, both on the lateral surface and in the cross-sectional one, which will make for a good exploitation durability of the brake shoes.

Acknowledgments
This work was partially supported by the strategic grant POSDRU/159/1.5/S/137070 (2014) of the Ministry of National Education, Romania, co-financed by the European Social Fund – Investing in People, within the Sectoral Operational Programme Human Resources Development 2007-2013.

References

[1] ***Specification, nr.1/SFMR/SDT/2000, *Brake shoes for the rolling sock,* approved by Romanian Railway Authority.
[2] ***www.tef.com
[3] Cojocaru V and Filipiu C 2006 *Preparing load for the development of iron in induction furnaces*, Samia, Iaşi, Romania
[4] ***Commission Decision 2006/66/EC of 23 December 2005 concerning the technical specification introperabilitate on the subsystem "rolling stock - noise" of the conventional rail system, OJ L 37, 8.2.2006
[5] ***www.uic.asso.fr (Noise Reduction in Rail Freight - a 2007 report on the state of the art)
[6] ***Staff Working Document European Commission, Brussels, 8.7. 2008, SEC (2008) 2204
[7] Socalici A, Popa E, Heput T and Putan V 2014 Contributions to steel semi-finished parts quality improuvements, *IOP Conference Series: Material Science and Engineering* **57** 012003
[8] Pascu L V, Putan A, Socalici A and Heput T 2013 Researches on the quality of pig iron to be used in making rail braking shoes, *Metalurgia International* **18**(8) 193-199
[9] Socalici A, Popa E, Heput T and Vilceanu L 2014 Contributions related to the control of steel ingot solidification, *IOP Conference Series – Material Science and Engineering* **57** 012004

Spotting the earth connection and short circuits between the electric conductors, using D.C. bridges for resistance measurements

I Popa[1], G N Popa[1], S I Deaconu[1] and A Iagăr[1]

[1] Politehnica University of Timisoara, Department of Electrical Engineering and Industrial Informatics, 5 Revolution Street, 331128 Hunedoara, Romania

E-mail: gabriel.popa@fih.upt.ro

Abstract. The paper establishes the necessary connections meant to spot the earth connections and short circuits between the conductors of a power line, using the DC bridges meant for measuring resistances between conductors at the ends of the power line. Since it is a relative method, it imposes an exact knowledge of the faulty power line setting. For values of the resistances measured between the conductors of the power line having over 1 Ω at one end, the measurement will be carried out with a Wheatstone bridge, and for values below 1 Ω with a Thomson bridge. In order to accurately determine the place of the fault, it measured the distances from the end of the line up to the fault and then we performed a correction calculation for this distance.

1. Introduction

During the exploitation of power lines, there may arise simple faults (permanent short circuits or conductor breaks) or complex ones (short circuits or earth connections with a high value of the fault resistance, conductor breaks with a low fault resistance, short circuits accompanied by earth connections, etc.) [1-4]. Before spotting the fault, it has to be brought to a simple form, using the procedure of "burning" the faulty line. After this operation, we used one of the specific methods of spotting the fault [2, 5-8].

The paper introduces an original method of spotting the short circuits and earth connections of the cable or aerial power lines, using the d.c. bridges for measuring resistances [9]. In this respect, it determined relations for the calculation of the distance between the end of the line and the place of the fault, the fault resistance not implied.

2. Spotting the earth connection of a conductor in an electric line, using D.C. bridges

2.1. Spotting the earth connection without an auxiliary conductor

The method given henceforth can only be used if the power line has at least one good conductor. It is supposed that the power line of length l in Figure 1, was affected by an earth connection at distances l_x and l_y with respect to the ends of the line [2].

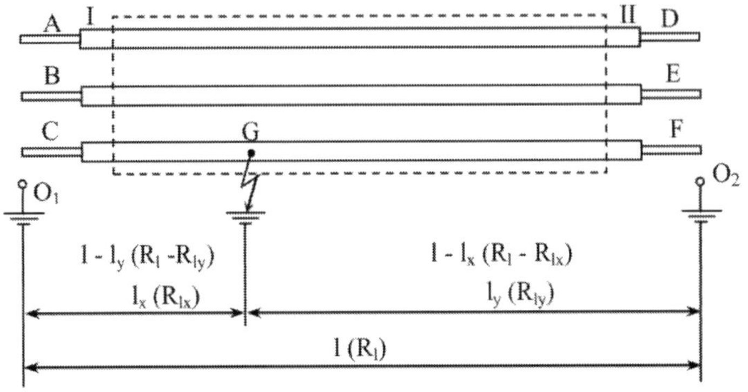

Figure 1. Three-phase power line with one earth connected conductor.

In order to spot the earth connection of conductor *CF* we connect the *E* and *F* ends of conductors *B* - *E* and *C* - *F* and then, with a d.c. bridge we measure at the end of cable *I* resistances R_{BD} and R_{COI}. If we ignore the contact resistances and the resistances of the connecting conductors between the *I* end of the cable, plot O_I and the bridge, the result is:

$$R_{BOI} = R_{BE} + R_{FG} + R_{GOI} \quad [\Omega] \tag{1}$$

$$R_{COI} = R_{CG} + R_{GOI} \quad [\Omega] \tag{2}$$

In these relations, R_{GOI} is the equivalent resistance of the circuit made of the resistance of the earth at the end of line *l* of the cable, used in spotting the fault, the resistance of the circuit through the earth and the fault resistance of the cable. Resistances R_{BE}, R_{FG} and R_{CG} are determined by means of relations:

$$R_{BE} = \rho_c \cdot \frac{k_p \cdot l}{s_c}; R_l = R_{BE} \quad [\Omega] \tag{3}$$

$$R_{FG} = \rho_c \cdot \frac{k_p \cdot (l - l_x)}{s_c} \quad [\Omega] \tag{4}$$

$$R_{CG} = \rho_c \cdot \frac{k_p \cdot l_x}{s_c} \quad [\Omega] \tag{5}$$

where k_p is a correction coefficient for the length of the conductors of the cable power lines.

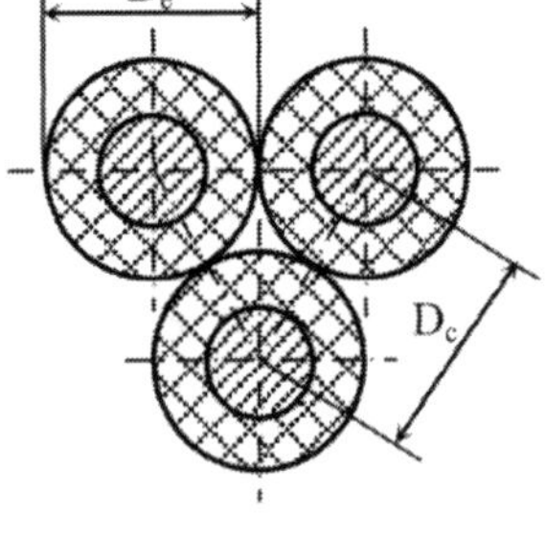

Figure 2. The placing of conductors in a three-phase cable.

Coefficient k_p represents the ration between the real length l_{pr} [m] of the conductor along a distance equal to the *p* [m] twist step:

$$k_p = \frac{l_{pr}}{p} \tag{6}$$

IOP Conf. Series: Materials Science and Engineering **85** (2015) 012027 doi:10.1088/1757-899X/85/1/012027

Coefficient k_p can be determined according to distance D_c [m] between the conductors (Figure 2), if we know step p, by means of relation [7]:

$$k_p = \sqrt{1 + \frac{4}{3} \cdot \left(\frac{\pi \cdot D_c}{p} \right)^2} \tag{7}$$

For the circular cross-section conductors, without space between the insulation layers, distance D_c [m] between the conductors is equal to the exterior diameter D_e [m] of the insulation. By means of formulae (1), …, (5) we obtain distance l_x:

$$l_x = l - \frac{S_c}{2 \cdot \rho_c \cdot k_p} \cdot \left(R_{BO1} - R_{CO1} \right) \quad \text{[m]} \tag{8}$$

If we short circuit the B and C ends of conductors EB and FC, and we measure resistances R_{EO2} and R_{FO2} in the same way, we obtain:

$$l_y = l - \frac{S_c}{2 \cdot \rho_c \cdot k_p} \cdot \left(R_{EO2} - R_{FO2} \right) \quad \text{[m]} \tag{9}$$

By means of formulae (4.7) of [7], (8) and (9) we have:

$$l_x = \frac{S_c}{2 \cdot \rho_c \cdot k_p} \cdot \left(R_{EO2} - R_{FO2} \right) \quad \text{[m]} \tag{10}$$

$$l_y = \frac{S_c}{2 \cdot \rho_c \cdot k_p} \cdot \left(R_{BO1} - R_{CO1} \right) \quad \text{[m]} \tag{11}$$

Expressions (8), …, (11) can also be written as [7]:

$$l_x = \left[1 - \frac{1}{2 \cdot k_p \cdot R_l} \cdot \left(R_{EO2} - R_{FO2} \right) \right] \cdot l \text{ [m]} \tag{12}$$

$$l_y = \left[1 - \frac{1}{2 \cdot k_p \cdot R_l} \cdot \left(R_{BO1} - R_{CO1} \right) \right] \cdot l \text{ [m]} \tag{13}$$

$$l_x = \frac{1}{2 \cdot k_p \cdot R_l} \cdot \left(R_{EO2} - R_{FO2} \right) \cdot l \quad \text{[m]} \tag{14}$$

$$l_y = \frac{1}{2 \cdot k_p \cdot R_l} \cdot \left(R_{BO1} - R_{CO1} \right) \cdot l \quad \text{[m]} \tag{15}$$

For the aerial power lines $k_p = 1$. In this case, the spotting of the earth connection for a conductor of the line can be carried out using the pairs of relations:

$$l_x = l - \frac{S_c}{2 \cdot \rho_c} \cdot \left(R_{BO1} - R_{CO1} \right) \quad \text{[m]} \tag{16}$$

$$l_y = l - \frac{S_c}{2 \cdot \rho_c} \cdot \left(R_{EO2} - R_{FO2} \right) \quad \text{[m]} \tag{17}$$

$$l_x = \frac{S_c}{2 \cdot \rho_c} \cdot \left(R_{EO2} - R_{FO2} \right) \quad \text{[m]} \tag{18}$$

$$l_y = \frac{S_c}{2 \cdot \rho_c} \cdot \left(R_{BO1} - R_{CO1} \right) \quad \text{[m]} \tag{19}$$

$$l_x = \left[1 - \frac{1}{2 \cdot R_l} \cdot (R_{BO1} - R_{CO1})\right] \cdot l \quad \text{[m]} \tag{20}$$

$$l_y = \left[1 - \frac{1}{2 \cdot R_l} \cdot (R_{EO2} - R_{FO2})\right] \cdot l \quad \text{[m]} \tag{21}$$

$$l_x = \frac{1}{2 \cdot R_l} \cdot (R_{EO2} - R_{FO2}) \cdot l \quad \text{[m]} \tag{22}$$

$$l_y = \frac{1}{2 \cdot R_l} \cdot (R_{BO1} - R_{CO1}) \cdot l \quad \text{[m]} \tag{23}$$

Distances l_x and l_y will be corrected using the procedure given in [7].

2.2. Spotting the earth connection using an auxiliary conductor

If all the conductors of the power line are earth connected this method can be used if we replace conductor $B - E$ (Figure 1) by an auxiliary conductor having the parameters R_{la}, ρ_{ca} and the dimensions l_a and s_{ca}. So: $R_{BE} = R_t$. This equality, as well as relations (1), …, (5) and (4.2) of [7], if one of the power lines in the cable has failed, lead to the conclusion that:

$$l_x = \frac{l}{2} \cdot \left[l + \frac{s_c}{k_p \cdot \rho_c} \cdot (R_{la} + R_{COl} - R_{BOl})\right] \text{[m]} \tag{24}$$

With the results of the measurements made at end *II* of the cable, we can obtain in the same way:

$$l_y = \frac{l}{2} \cdot \left[l + \frac{s_c}{k_p \cdot \rho_c} \cdot (R_{la} + R_{FO2} - R_{EO2})\right] \text{[m]} \tag{25}$$

For the aerial power lines $k_p = 1$ and relations (24) and (25) become:

$$l_x = \frac{l}{2} \cdot \left[l + \frac{s_c}{\rho_c} \cdot (R_{la} + R_{COl} - R_{BOl})\right] \text{[m]} \tag{26}$$

$$l_y = \frac{l}{2} \cdot \left[l + \frac{s_c}{\rho_c} \cdot (R_{la} + R_{FO2} - R_{EO2})\right] \text{[m]} \tag{27}$$

Irrespective of the type of power line, we obtain, in the same way the relations:

$$l_x = \frac{l}{2} \cdot \left[l + \frac{l}{R_l} \cdot (R_{la} + R_{COl} - R_{BOl})\right] \cdot l \text{ [m]} \tag{28}$$

$$l_y = \frac{l}{2} \cdot \left[l + \frac{l}{R_l} \cdot (R_{la} + R_{FO2} - R_{EO2})\right] \cdot l \text{ [m]} \tag{29}$$

The next step is to correct the distances obtained above.

3. Spotting the short circuits between power lines, using D.C. bridges

3.1. Spotting short circuits without an auxiliary conductor

Unlike the procedures used at present in spotting short circuits, the method given henceforth can be applied using the apparatus the power line servicing teams are endowed with. Spotting the short circuit between two conductors of a power line can be done by measuring the electric resistances at the two ends of the line.

We suppose there arose a permanent short circuit between the two conductors of a cable power line (Figure 3) so, $R_s = 0\ \Omega$. In this case:

$$R_{BC} = 2 \cdot \rho_c \cdot k_p \cdot \frac{l_x}{s_c}; R_{BC} = 2 \cdot \rho_c \cdot k_p \cdot \frac{l - l_y}{s_c}\ [\Omega] \tag{30}$$

$$R_{EF} = 2 \cdot \rho_c \cdot k_p \cdot \frac{l - l_x}{s_c}; R_{EF} = 2 \cdot \rho_c \cdot k_p \cdot \frac{l_y}{s_c}\ [\Omega] \tag{31}$$

The groups of relations (30) and (31) result in:

$$l_x = \frac{R_{BC}}{R_{BC} + R_{EF}} \cdot l\ [\text{m}] \tag{32}$$

$$l_y = \frac{R_{EF}}{R_{BC} + R_{EF}} \cdot l\ [\text{m}] \tag{33}$$

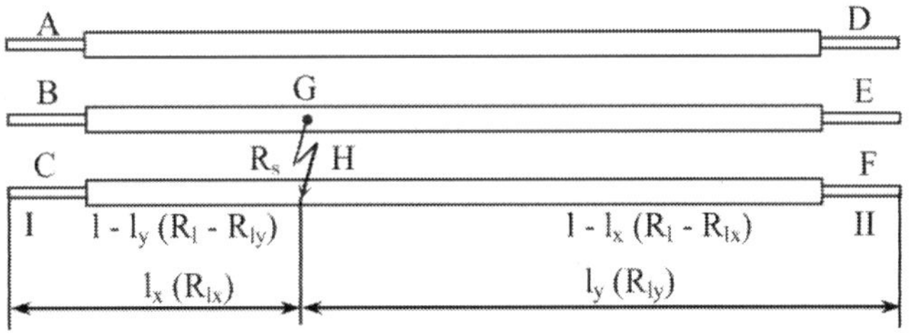

Figure 3. Three – phase power line with a short circuit between two conductors.

Generally $R_s \neq 0$, therefore the values of resistances R_{BC} and R_{EF} are influenced by the resistances at the place of the short circuit, according to the relations:

$$R_{BC} = 2 \cdot \rho_c \cdot k_p \cdot \frac{l_x}{s_c} + R_s; R_{BC} = 2 \cdot \rho_c \cdot k_p \cdot \frac{l - l_y}{s_c}\ [\Omega] \tag{34}$$

$$R_{EF} = 2 \cdot \rho_c \cdot k_p \cdot \frac{l - l_x}{s_c} + R_s; R_{EF} = 2 \cdot \rho_c \cdot k_p \cdot \frac{l_y}{s_c}\ [\Omega] \tag{35}$$

With formulae (4.103) of [7], (34) and (35), by eliminating the fail resistance R_s, we obtain the distances from the two ends of the power line to the place of the fail:

$$l_x = \frac{1}{2} \cdot \left(1 + \frac{R_{BC} - R_{EF}}{2 \cdot R_l} \right) \cdot l\ [\text{m}] \tag{36}$$

$$l_y = \frac{1}{2} \cdot \left(1 + \frac{R_{EF} - R_{BC}}{2 \cdot R_l} \right) \cdot l\ [\text{m}] \tag{37}$$

Spotting the short circuit can also be done by two measurements at one end of the cable. If the measurements are carried out at end *I*, we short circuit conductors *A* – *D* and *B* – *E* at end *II*. We measure the resistances R_{BC} and R_{AC} with a d.c. bridge. They can be determined with expressions (34) and (38):

$$R_{AC} = 2 \cdot \rho_c \cdot k_p \cdot \frac{l}{s_c} + R_s\quad [\Omega] \tag{38}$$

With formulae (4.103) of [7], (34) and (38) we obtain:

$$l_x = \frac{2 \cdot R_l + R_{BC} - R_{AC}}{2 \cdot R_l} \cdot l \quad [\mathrm{m}] \tag{39}$$

$$l_y = \frac{R_{AC} - R_{BC}}{2 \cdot R_l} \cdot l \ [\mathrm{m}] \tag{40}$$

Similarly, we can determine the relations for spotting short circuit (39) and (40), using the values of resistances R_{DF} and R_{EF} at end *II* of the cable, when points *B* and *C* are connected:

$$l_y = \frac{2 \cdot R_l + R_{EF} - R_{DF}}{2 \cdot R_l} \cdot l \ [\mathrm{m}] \tag{41}$$

$$l_x = \frac{R_{DF} - R_{EF}}{2 \cdot R_l} \cdot l \qquad [\mathrm{m}] \tag{42}$$

3.2. Spotting short circuits using an auxiliary conductor

When all the conductors of the power line are in short circuit, the method of fail spotting by two measurements at the same end of the cable can only be used with the help of an auxiliary conductor, having resistance R_{la}. In order to use this method, instead of conductor $A - D$ we use the auxiliary conductor. Measuring resistances R_{AC} and R_{BC} at end *I*, with a bridge between points *D* and *E*, having the values:

$$R_{AC} = R_{la} + R_l + R_s \qquad [\Omega] \tag{43}$$

$$R_{BC} = 2 \cdot R_{lx} + R_s \qquad [\Omega] \tag{44}$$

by eliminating the fail resistance we obtain:

$$l_x = \frac{R_{la} + R_l + R_{BC} - R_{AC}}{2 \cdot R_l} \cdot l \ [\mathrm{m}] \tag{45}$$

We then short circuit conductors $A - D$ and $B - E$ at end *I* and measure resistances R_{DF} and R_{EF}, used to determine l_y:

$$l_y = \frac{R_{la} + R_l + R_{EF} - R_{DF}}{2 \cdot R_l} \cdot l \ [\mathrm{m}] \tag{46}$$

4. Experimentation of the method

The experimental verification of the method of fail spotting in power lines, which this paper deals with, has been carried out on a laboratory model, having the parameters given in table 1, and the diagram in figure 3. The simulation of the effect resistance R_s has been done with a laboratory variable resistance connected between points *G* and *H*.

The experimental data and the results of the calculations done with formulae (36) and (37) in order to spot the short circuits in power lines are given in table 1. The relative short circuit spotting errors have been calculated by means of formulae:

$$\varepsilon_{rx} = \frac{l_x - l_{xn}}{l_{xn}} \cdot 100[\%] \tag{47}$$

$$\varepsilon_{ry} = \frac{l_y - l_{yn}}{l_{yn}} \cdot 100[\%] \tag{48}$$

where l_x and l_y are the distances from ends *I* and *II* up to the place of the fail, obtained by calculation, using the experimental data, and l_{xn} and l_{yn} are the same distances very precisely measured from the two ends to the short circuit artificially created.

Table 1. The determination of distances l_x and l_y from the ends of the power line to the place of the short circuit l=2m; l_{xn}=0.6m; l_{yn}=1.4m; R_l=123.528Ω.

R_s (Ω)	R_{BC} (Ω)	R_{EF} (Ω)	l_x (m)	l_y (m)	ε_{rx} (%)	ε_{ry} (%)
0	28.04	65.38	0.6002	1.3998	+0.0333	+0.0333
20	47.98	85.35	0.5998	1.4001	-0.0333	-0.0333
40	67.90	105.20	0.6006	1.3994	+0.1000	+0.1000
60	87.80	125.50	0.5962	1.4036	-0.6333	-0.6333
80	107.60	145.50	0.5942	1.4058	-0.9667	-0.9667
100	128	165.45	0.5990	1.4010	-0.1667	-0.1667

5. Conclusions

The analysis of the calculation results referring to the short circuit spotting leads to the conclusion that for fail resistances whose values are below or equal to the resistance of the conductor, the spotting errors are very small and therefore relations (36) and (37), as well as the method we suggested can be used in practice.

References

[1] Bottril G, Cheyne D and Vijayaraghavan G 2005 *Practical Electrical Equipment and Installations in Hazardous Areas*, Elsevier, Netherlands

[2] Cruceru C 1976 *Procedures for Fault Location in Power Cables Networks*, Technical Publishing House, Bucharest, Romania

[3] Kock J and Strauss C 2004 *Practical Power Distribution for Industry*, Elsevier, Netherlands

[4] Vijayaraghavan G, Brown M and Barnes M 2004 *Practical Grounding, Bonding, Shielding and Surge Protection*, Elsevier, Netherlands

[5] Lucache D D 2009 *Low Voltage Electrical Installations*, PIM Publishing House, Iași, Romania

[6] Pop E and Chivu M 1976 *Electric and Magnetic Measurements*, vol.II, "Traian Vuia" Politechnical Institute, Timişoara, Romania

[7] Popa I and Popa G N 2000 *Electrical Installations*, vol.I, Mirton Publishing House, Timişoara, Romania

[8] Saimac A and Cruceru C 1981 *Electrotechnics*, Didactic and Pedagogical Publishing House, Bucharest, Romania

[9] Regtien P P L, Heijden F, Korsten M J and Olthuis W 2004 *Measurement Science for Engineers*, Elsevier, Netherlands

IOP Conf. Series: Materials Science and Engineering **85** (2015) 012028 doi:10.1088/1757-899X/85/1/012028

Functional performance testing of the universal super absorbing air filters FSU 70 „Air by Corneliu"

S Rațiu[1], C Birtok–Băneasă[2], V Alexa[1] and I Kiss[1]

[1] University Politehnica Timisoara, Faculty of Engineering Hunedoara, Department of Engineering and Management, 5 Revolution Street, 331128 Hunedoara, Romania
[2] CorneliuGroup Association, Romania, www.corneliugroup.ro

E–mail: sorin.ratiu@fih.upt.ro

Abstract. This paper presents the experimental methodology to carry out functional performance tests for an air filter with a particular design of its housing, generically named *Universal super absorbing FSU 70 „Air by Corneliu"*. The tests were carried out in the Internal Combustion Engines Laboratory, within the specialization "Road automotives" belonging to the Faculty of Engineering Hunedoara, component of "Politehnica" University of Timişoara. We present some comparative values of various operating parameters of the engine fitted, in the first measuring session, with the original filter, and then with the studied filter.

1. Introduction

The proper filtration of the air entering the cylinders of an internal combustion engine is essential for extending its service life [1], [2], [3]. Preventing various impurities from entering the engine along with the atmospheric air reduces significantly the wear over time of the engine moving parts.

Unfortunately, in addition to the purpose of filtering the air drawn from the atmosphere, the air filter – as a distinct part of the engine – represents a significant gas–dynamic resistance, interposed on the suction route. If it is not regularly cleaned, and the vehicle travels frequently within a dusty environment, the suction pressure (p_a) is consistently decreasing and the air–filling coefficient (η_V) suffers penalties.

Therefore, we can say that the innovative design of the air filtration system is essential for a high performance filtration and air flow improvement, with effects on increasing of the engine performance and service life [1], [6], [7].

The design proposed by the authors aims at reducing the shortcomings described above by designing filter housing with particular geometry, enabling the improvement of the gas–dynamic performance at the engine air inlet [1], [8] . We note that this study does not deal with the nature of the filtering element of the filter, which is a standard one [4], [5].

2. Presentation of the experimental equipment

For performing the experimental measurements, we had an engine test stand, produced by Christiani, consisting of a multi–point fuel injection engine, brand VW, 1.4 MPI, used by the *Golf VI 5K1* models, cylinder capacity: 1390 cm^3, power: 59 KW/80 CP, 4 cylinders in line, manufacturing years: 2009–2012.

Figure 1. Overview of the test stand.

The stand offers the possibility to monitor the engine during its operation. It enables the connection to the OBD port of all the testers usually found in garages, and the measurement of signals and electrical quantities, with the possibility of assessing the obtained parameters.

The related software contains a professional interface that allows the PC to be transformed, via an USB port, into a device to be used for diagnostics and visualization of the functional parameters. Also, it enables the visualization and simultaneous storing of three measuring blocks, and can graphically display the essential features of the engine.

The tested super absorbing filter is cylindrical and contains an internal universal spreader. This has the additional function of accelerating the air speed at the output of the filter. Due to the constructive geometry, it provides a significant increase in the air–filling coefficient of the engine cylinders.

The filter (Figure 2 and Figure 3) is provided with a perforated housing (1) that has an internal spreader (2) fitted by removable mounting, at one end of the housing. The spreader has a cylindrical connector (3) attached in the opposite side of the perforated housing (1), that provides the connection with the intake manifold of the internal combustion engine, which is not illustrated in the drawing. In the opposite side of the internal spreader (2), the perforated housing (1) has a cover (4) attached through removable mounting, fitted with an internal cone (5). This one is provided on the outside with four or more targeting and swirling wings "C". Inside the perforated housing (1), it is placed the filtering element (6), cylindrically shaped.

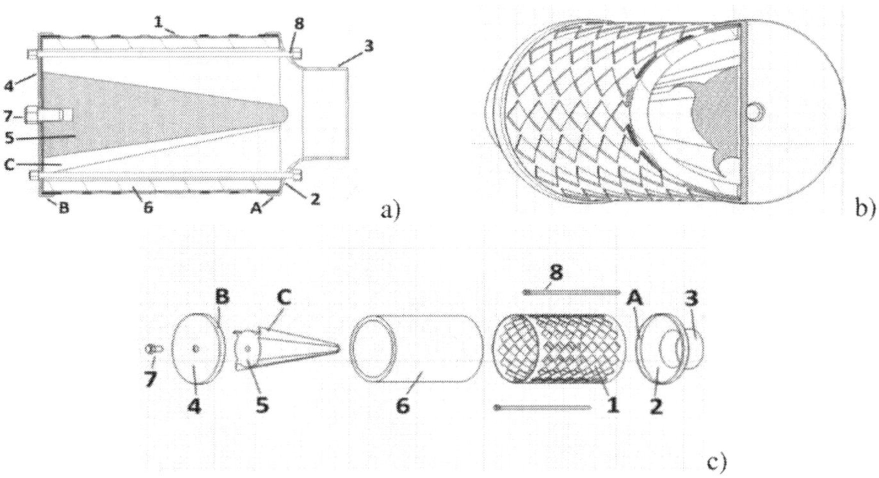

Figure 2. Super absorbing filter FSU 70 „Air by Corneliu", the exploded model [1]
a, b – drawings of the installed filter; c – drawing of the exploded model;

The perforated housing (1) is cylindrical, being concentrically framed by the centring edge "A" of the internal spreader (2), and on the opposite side by the centring edge "B" of the cover (4).

The internal cone (5) is attached to the cover (4) by means of a screw (7).

To maintain in contact the internal spreader (2), the perforated housing (1) and the cover (4), two threaded connecting rods (8) have been provided.

The targeting and swirling wings "C" have a concave or convex curvilinear profile, and a straight or spiral orientation over the lateral area of the internal cone (5).

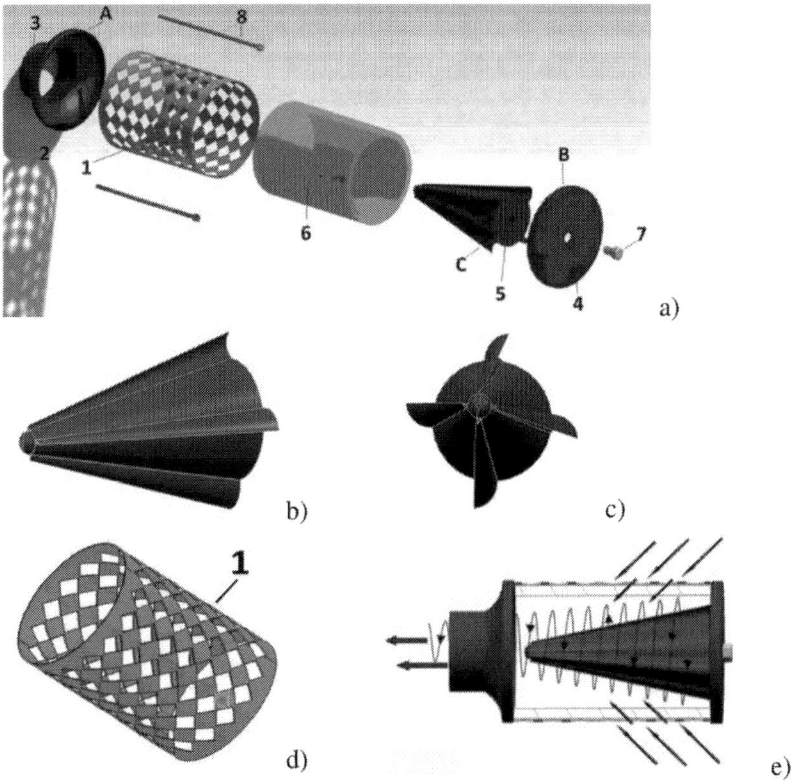

Figure 3. Super absorbing filter FSU 70 „Air by Corneliu", virtual model [1]
a – exploded virtual model, realised in AutoDesk Inventor; b, c – virtual model of the internal cone; d – perforated housing; e – illustration of the operating principle.

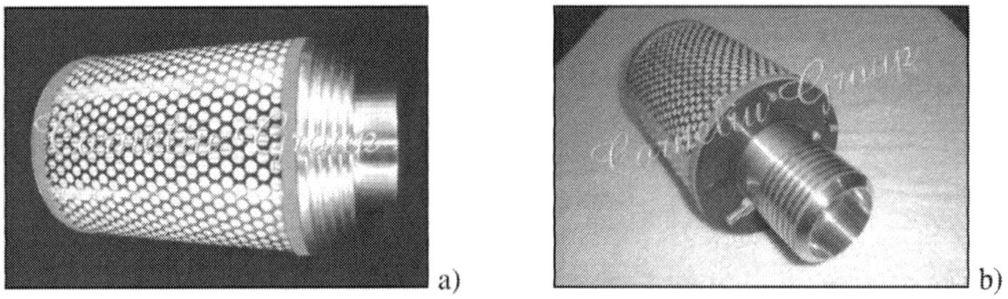

Figure 4. Super absorbing filter FSU 70 „Air by Corneliu" [1]
a, b – physical models

The components of this type of filter are removable, being able to be installed and removed as needed. The internal spreader has variable size, depending on the engine cylinder capacity. A higher cylinder capacity requires a larger spreader capacity, and vice versa.

3. Testing session and the experimental methodology

The tests consisted in determining the key operational parameters of the engine, at various speeds at idle, the engine being equipped, at a time, with the original filter or with FSU 70. The measuring sessions were repeated to confirm the results, and the test conditions for the two variants (engine equipped with original filter and with FSU 70) were maintained strictly constant. The filtering element is the same for both filters.

For the first measuring session, the experimental engine was equipped with the original filter (Figure 5). The data acquisition was made at various engine speeds at idle, beginning with the idling speed and continuing with 1500, 2000, 2500, 3000, 3500 and 4000 rpm. We measured:

- engine speed;
- throttle position;
- manifold air pressure;
- coolant temperature;
- manifold air temperature;
- hourly fuel consumption;
- ambient temperature;
- atmospheric pressure.

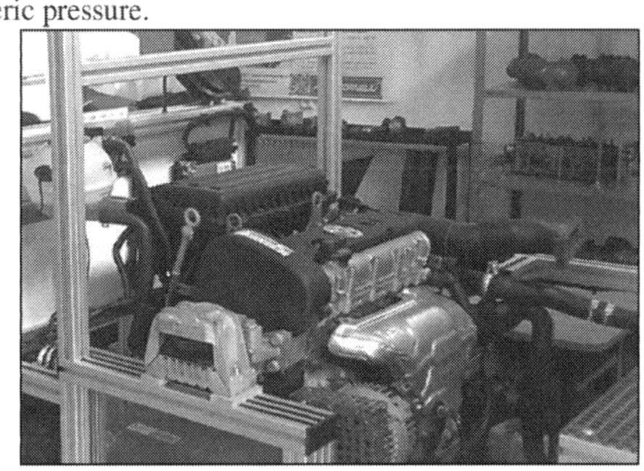

Figure 5. The experimental engine equipped with original filter.

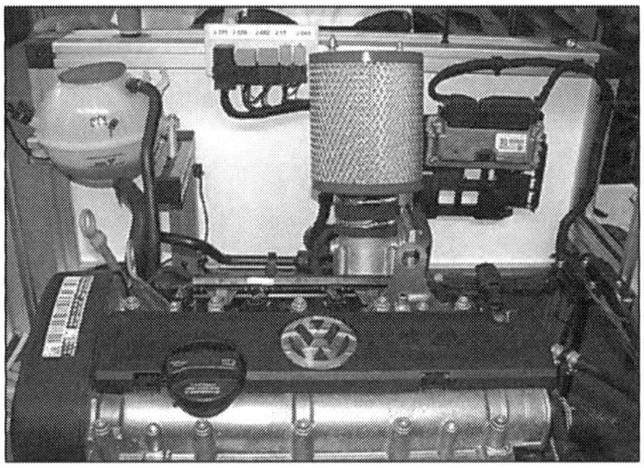

Figure 6. The experimental engine equipped with FSU 70.

The next session of measurements was carried out with the engine equipped with FSU 70 (Figure 6). The same above–specified parameters have been determined, under the same conditions.

We mention that the measuring sessions were conducted inside the laboratory, where the temperature and pressure remained strictly constant.

4. Results and conclusions

We present below the comparative values of the above–specified parameters on the performance of the engine equipped with original filter and FSU 70, respectively.

The chart presented in Figure 7 shows that, at engine speeds below 2000 rpm, when using the FSU 70, the throttle valve opens less, which means less gas–dynamic resistance due to this filter installation, situation beneficial for the process of filling the engine cylinder with fresh load. At higher speed values, it can be seen that the throttle valve position is identical in both cases.

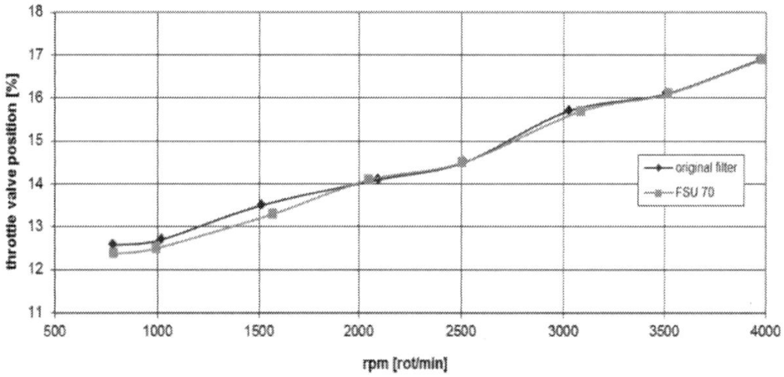

Figure 7. Position of the throttle valve versus engine speed at idle.

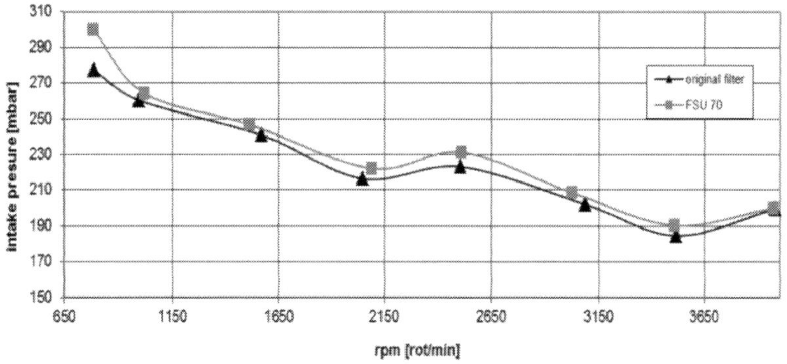

Figure 8. Pressure in the intake manifold versus engine idle speed.

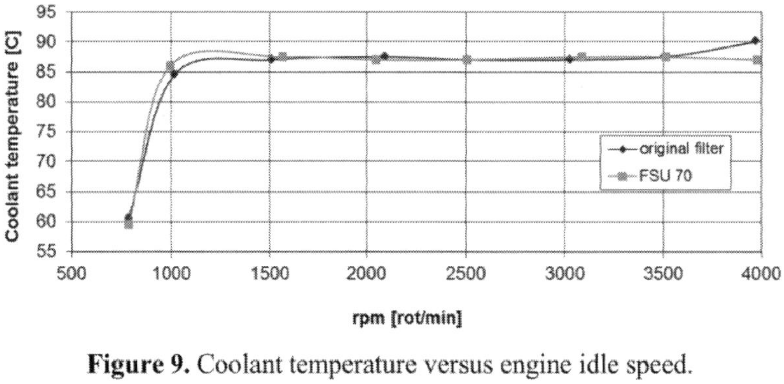

Figure 9. Coolant temperature versus engine idle speed.

From Figure 8, we can deduct that, throughout the range of speeds at which the tests have been performed, the pressure in the intake manifold is higher when using FSU 70 than when using the original filter, situation beneficial for the process of filling the engine cylinder with fresh load.

From Figure 9, we can deduct that there are no major differences in the coolant temperatures when using the two filters.

From Figure 10, it can be concluded that, up to a speed of 3300 rpm, the air temperature in the intake manifold, when using the FSU 70, is lower than that obtained when using the original filter, situation beneficial for the cylinder filling process, because a lower air temperature when entering the engine means a higher density, and therefore a larger amount introduced into the cylinders, with beneficial consequences for the engine power.

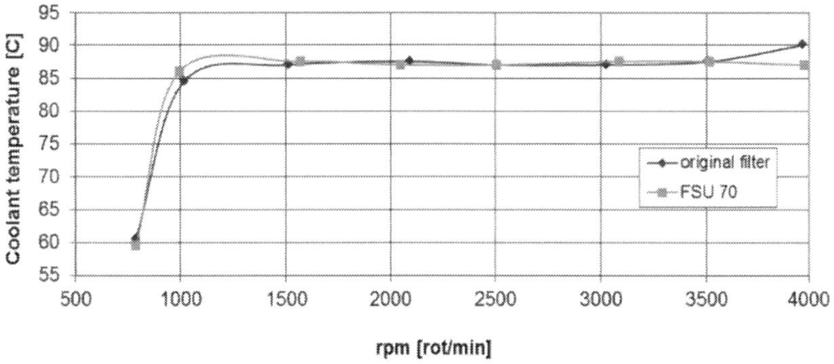

Figure 10. Air temperature in the intake manifold versus engine idle speed.

The most important parameter monitored during the tests was the hourly fuel consumption, in litres/hour. In Figure 11, we can see the variation of the hourly fuel consumption with the engine speed, for the cases when the engine is running equipped with the original filter and with FSU 70, respectively. It appears that, from speed values around 1000 rpm, the fuel consumption begins to decrease considerably when using the FSU 70.

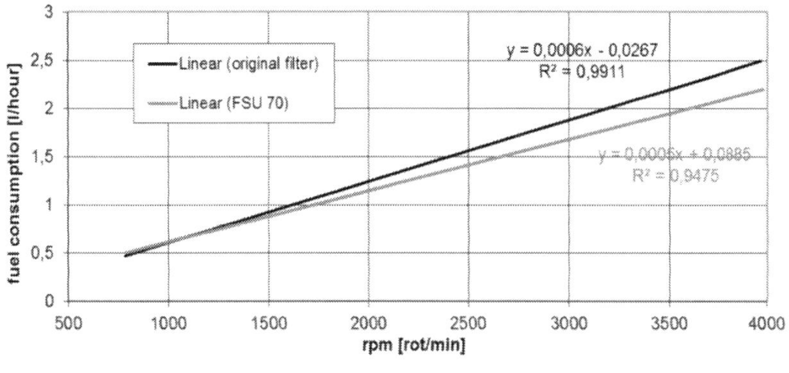

Figure 11. Hourly fuel consumption versus engine idle speed.

The percentage difference between the hourly consumption with original filter and the hourly consumption with FSU 70 is calculated using the relation:

$$d = 100 - \frac{Ch_{FSU70}}{Ch_{original\,filter}} \cdot 100 \ [\%] \quad (1)$$

In Figure 12, we can see that the percentage difference is positive at all the engine speeds, with the exception of 2000 rpm (more precisely: 2049 rpm). The positive values show lower fuel consumption than in case of the original filter.

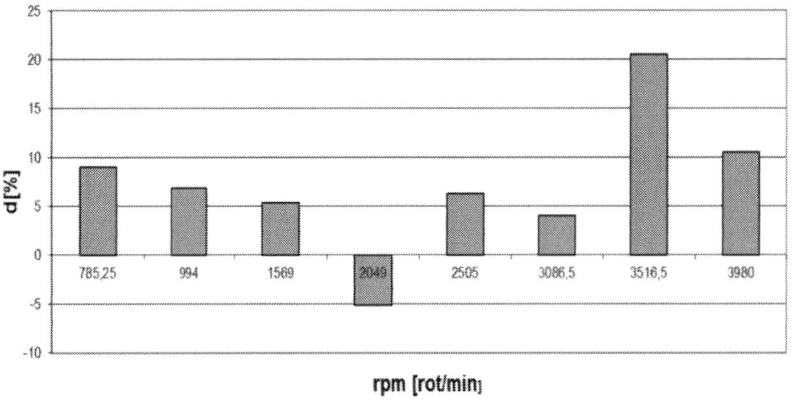

Figure 12. The percentage consumption difference.

Throughout the speed range used for conducting the tests, the reduction of fuel consumption by using the universal super absorbing FSU 70 is 7.15%.

Following the conducted researches and monitoring of the behaviour when installing the filter on various types of engines, we concluded that this one has a number of advantages, such as:

- being in contact with the air, the filter element ensures a minimum gas–dynamic resistance to the absorbed air, increasing thereby the absorption and capture rates;
- the self–cleaning possibility of the filter element;
- visualisation of the filter element, without the prior removal of the filter, for checking its pollutant loading level;
- the capacity of air filter housing to significantly increase the speed of absorbed air, both at the input and output of the filter;
- the ability to "pre–cool" the air drawn inside it;
- the ability to create a slightly boost, which increases proportionally to the vehicle movement speed.

The tests were carried out in the Internal Combustion Engines Laboratory, within the specialization "Road automotives" belonging to the Faculty of Engineering Hunedoara, component of University Politehnica Timişoara.

References

[1] Gieras J F, Wang R J and Kamper M J 2008 *Axial Flux Permanent Brushless Machines*, Second Edition, Springer Science, Boston, MS USABirtok–Băneasă C and Raţiu S 2011 *Admisia aerului în motoare cu ardere internă – Filtre supraaspirante – Sisteme dinamice de transfer* [*Air intake in internal combustion engines – Super absorbing filters – Dynamical Air Transfer Systems*], Publisher: Politehnica, Timişoara, Romania

[2] Raţiu S and Mihon L 2008 *Motoare cu ardere internă pentru autovehicule rutiere – Procese şi caracteristici* [*Internal combustion engines for road vehicles – Processes and Features*], Publisher: Mirton, Timişoara, Romania

[3] Raţiu S 2009 *Motoare cu ardere internă pentru autovehicule rutiere – Procese şi caracteristici, Experimente de laborator* [*Internal combustion engines for road vehicles – Processes and Features – Laboratory experiments*], Publisher: Mirton, Timişoara, Romania

[4] *** International Standard ISO 5011, *Inlet air cleaning equipment for internal combustion engines and compressors – Performance testing*, Second edition 2000–12–01, corrected and reprinted on 2001–07–15

[5] *** *Bosch Automotive Hand–book*, 8th Edition, May 2011
[6] Stahl U and Heinz R 2006 *New Non-woven Media for Engine In–take Air Filtration with Improved Performances*, SAE technical paper series 2006–01–0272, 2006 Word Congress, Detroit, Michigan, April 3–6, SP–2014
[7] Bugli N J 2001 *Automotive Engine Air Cleaners – Performance Trends*, SAE technical paper series 2001–01–1356, SAE 2001 World Congress, Detroit, Michigan, March 5–8
[8] Raţiu S et. al. 2009 *New concepts in modelling air filters for internal combustion engines*, The 20th International DAAAM SYMPOSIUM "Intelligent Manufacturing & Automation: Theory, Practice & Education", 25–28 November, Vienna, Austria

IOP Conf. Series: Materials Science and Engineering **85** (2015) 012029 doi:10.1088/1757-899X/85/1/012029

Study regarding the power quality improvement in functioning of nonlinear loads

R Rob[1], C Panoiu[1] and M Panoiu[1]

[1] Politehnica University of Timisoara, Department of Electrical Engineering and Industrial Informatics, 5 Revolution Street, 331128 Hunedoara, Romania

E-mail: raluca.rob@fih.upt.ro

Abstract. This paper presents a study regarding the variation of the electrical parameters acquired during the functioning of a nonlinear load – an electrothermal installation with electromagnetic induction. Two different situations are presented: the installation works without harmonic compensation devices and when a three phase passive filters are connected at the point of common coupling.

1. Introduction

Electrothermal equipment usually represents important electric energy consumers. Many electrothermal equipment generate significant distortion regime into the power system that affects the power quality for other consumers connected to the same power network.

In developing studies about the variation of the current and voltages in some specified power distribution areas, must be analysed if the power quality indicators are maintaining between standard limits.

Usually, the aspects about the harmonic pollution of the power network are considered from the beginning of designing the electrical installations. By adopting these measures established within reason in technical and economic terms, at the point of common coupling with nonlinear load must be connected filtering devices to bring the system in normal limits.

2. Description of electrical power analysis system

The electrothermal installation with electromagnetic induction that is studied in this paper is composed by an electronic converter CTC100K15 and a hardening inductor. The electrical characteristics are the followings: supplying voltage 3x400V, frequency of the supplying voltage 50Hz, rated current 27A, control voltage 24Vdc, consumed power at high frequency 15kW, voltage at medium frequency 500Vac.

In order to study the distorting regime effect induced into the power network (0.4kV) by the functioning of the electrothermal installation, 10 measuring sets were accomplished by increasing with 10% the consumed power of the installation. In this scope an acquisition application designed in LabVIEW together with a data acquisition board were used. Communication with data acquisition board was accomplished taking into consideration the maximum sampling rate of 250kS/s. Using 10 analogical channels, the maximum sampling rate for a single channel is 25kS/s.

Instantaneous values of the three phase currents and voltages at the point of common coupling were acquired in order to study the variation of active, reactive, distorted and apparent powers and the total harmonic distortion for the phase currents and voltages. In Table 1 the values of these electrical parameters accomplished during the functioning of the electrothermal installation without compensation devices are synthesized.

Studying the variation of the electrical parameters presented in Table 1, the increasing of the effective values of the phase current with the increasing of the consumed power of the installation can be observed. As well, the effective values of the phase voltages are balanced. The total harmonic distortion of the phase voltages stands between the standard limits, but the total harmonic distortion of the phase currents riches very high values, 140%. Due to the current harmonic distortion, the distorted power has high values, as well.

In order to compensate the harmonic distortion, the electrothermal installation must be equipped with harmonic compensation devices for bringing the parameters into standard limits and a power quality improvement system was designed. This system contains an adapting interface. The main function of the interface is to ensure the compatibility between voltage levels of the acquired signals and also the data acquisition board requirements.

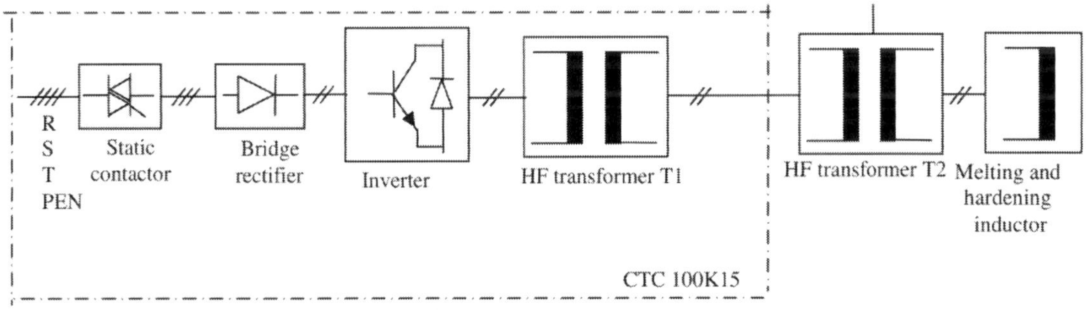

Figure 1. Block scheme of electrothermal installation.

Table 1. Variation of electrical parameters (without compensation devices).

Installation power [kW]	Phase voltage [V]	Phase current [A]	Active power [W]	Reactive power [VAr]	Distorted power [VAd]	Apparent power [VA]	THD voltages [%]	THD currents [%]
1.5	226.5	3.5	1300	180	1750	2187	2.95	140
3	227	6	2700	300	3100	4122	2.8	120
4.5	226.5	10	4700	500	4800	6737	2.65	105
6	226.6	10.5	4800	500	4900	6878	2.7	105
7.5	226.5	11	5500	750	5300	7675	2.65	100
9	226.5	13	6200	800	6000	8665	2.5	98
10.5	226.3	15.5	7500	1000	7000	10308	2.4	95
12	227	19	9500	1500	8500	12835	2.35	92
13.5	227	24	11500	1700	10000	15334	2.1	90
15	226.5	27	14000	2000	12000	18547	2	85

Another condition that must be accomplished is the galvanic isolation between power supply system and data acquisition board. The improvement system also uses NI-6221 data acquisition board for acquiring samples from phase voltages and phase currents.

In order to decrease the total harmonic distortion of the phase currents, the system contains a three phase passive filtering system accorded on 5th, 7th, 11th and 13th harmonic currents.

In Figure 2 the passive filter connection to the point of common coupling is presented.

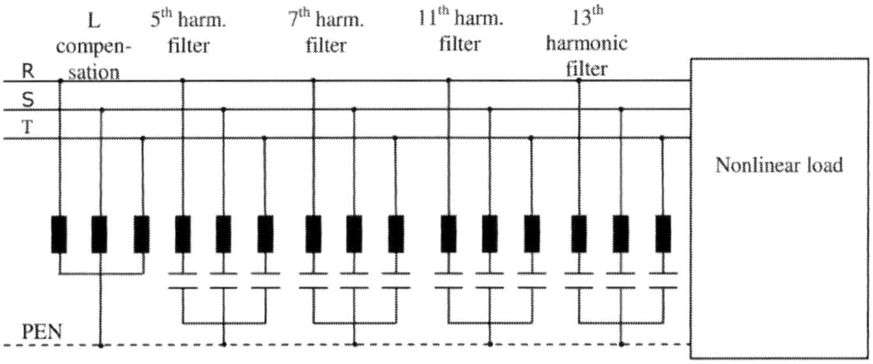

Figure 2. Passive filter system connection.

Description of the passive filtering system

The passive filtering system are composed by serial circuits with inductivities and capacities which are connected in parallel with the nonlinear load and also with the power distribution. These circuits accomplish a reduced impedance at one or more frequency values. The achievement of these filters is to shortcircuit the harmonic current as close to the distorting source is possible. In this way the harmonic penetration into the local network can be avoided and also the voltage and current distortion. In presented situation, the algorithm for designing the passive filters are based on the nominal values of the 5th, 7th, 11th and 13th harmonic currents:

- nominal current for 5th harmonic: $I_n^5 = 20A$
- nominal current for 7th harmonic: $I_n^7 = 15A$
- nominal current for 11th harmonic: $I_n^{11} = 5A$
- nominal current for 13th harmonic: $I_n^{13} = 5A$

Present algorithm is following four methods.

The first method permits the dimensioning based on the conditions concerning the voltage and thermal requirement of the capacitors. The first condition imposes that the value of filter voltage (due to the fundamental component and also to the *k* harmonic rank) to be lower that the admitted voltage of the chosen capacitor:

$$U_{C50} + U_{Ck} \leq U_{adm} \tag{1}$$

The second condition imposes that the filter power to be lower that the admitting power:

$$\omega \cdot C \cdot U_{C50}^2 \cdot tg\delta + \frac{I_n^k}{k\omega C} \cdot tg\delta \leq \omega \cdot C \cdot U_{adm}^2 \cdot tg\delta \tag{2}$$

Filter inductivity based on *k* harmonic rank can be calculated as the following:

$$L = \frac{1}{k^2 \omega^2 C} \tag{3}$$

IOP Conf. Series: Materials Science and Engineering **85** (2015) 012029 doi:10.1088/1757-899X/85/1/012029

In presented relation, U_{C50} represents the voltage loss due to the fundamental component, U_{Ck} - the voltage loss due to k harmonic rank, U_{adm} - admitted voltage of the chosen capacitor, $tg\delta$ - loss angle tangent.

The second dimension method consists in determining of the capacitor number in parallel connection with respect to the conditions:

$$a)\ U_{C50} + U_{Ck} \leq U_0$$

$$b)\ \sqrt{\left(I_{50}^2 + \left(I_n^k\right)^2\right)} \leq 1{,}3 \cdot I_{adm} \tag{4}$$

U_0 - admitted voltage on one connection side.

The third method is based on the admitted voltage and current conditions. In presented situation must be accomplished the condition presented in (3) and also:

$$I_{50}^2 + \left(I_n^k\right)^2 \leq (1{,}5I_0)^2 \tag{5}$$

The forth method is based on the admitted voltage and power conditions. In presented situation the condition presented in (3) must be accomplished and also:

$$\omega C \cdot U_{C50}^2 \cdot tg\delta + k\omega C \cdot U_{Ck}^2 \cdot tg\delta \leq \omega C \cdot U_0^2 \cdot tg\delta \tag{6}$$

Taking into consideration the conditions presented in (1)-(6), the values of inductivities and capacities of the passive filtering system are the followings:

- passive filter accorded on 5th harmonic current: L_5 = 16,21mH, C_5 = 25μF
- passive filter accorded on 7th harmonic current: L_7 = 2,5mH, C_7 = 80μF
- passive filter accorded on 11th harmonic current: L_{11} = 2,1mH, C_{11} = 40μF
- passive filter accorded on 13th harmonic current: L_{13} = 1,5mH, C_{13} = 40μF

After connecting the passive filtering system, another 10 measurement set were accomplished, by increasing with 10% the consumed power of the installation. These measurements permit to study the electrical parameters during the functioning of the electrothermal installation using passive filters for harmonic compensation [2], [3], [4]. The active filtering system is described in the following paragraph.

Using passive compensation, the acquisition system acquires samples from phase voltage and phase current at the point of common coupling and saves them into .txt documents. The computing system designed in LabVIEW is able to calculate the electrical parameters and to display their variation on the front panel of the application.

Presented application also contains an active filtering system. The input data of the active filter application represent the .txt files that contain the samples acquired from phase voltages and phase currents. The purpose of this application is to improve the electrical parameters, the phase currents are expected to be more sinusoidal, so the THD for phase currents will be decreased.

The algorithm of the present subroutine computes the electrical parameters using the acquired samples during the functioning of the electrothermal installation equipped with the passive filtering system. The voltage and current samples are assumed from the loading vector in the first sequence [7], [8], [9]. Using appropriate functions from LabVIEW library, the algorithm separates the current from the voltage samples, computing the effective values of the phase currents and voltages on each period of the power system.

The effective values of the voltage on each phase are computed with the following relation:

$$U_{ef}(i) = \sqrt{\frac{1}{N} \cdot \sum_{n=0}^{N-1} U(i,n)^2} \tag{7}$$

The effective values of the current on each phase are computed with the following relation:

$$I_{ef}(i)=\sqrt{\frac{1}{N}\cdot\sum_{n=0}^{N-1} I(i,n)^2} \tag{8}$$

where N is the number of samples/period, N = 500, i is the phase, $i = 1..3$.

So, U(i, n) represents the amplitude of the voltage sample on i phase, on n position in the vector having the length equal with N.

The total values of the voltages and currents can be computed as the followings:

$$U_{ef}=\sqrt{\frac{U_{ef}^2(1)+U_{ef}^2(2)+U_{ef}^2(3)}{3}} \tag{9}$$

$$I_{ef}=\sqrt{\frac{I_{ef}^2(1)+I_{ef}^2(2)+I_{ef}^2(3)}{3}} \tag{10}$$

The filtering analyser 1 contains other subroutines that calculate the active and reactive powers, and the harmonic currents and voltages as well, using appropriate functions from LabVIEW library.

The active power of i phase is computed with the relation:

$$P(i)=\sum_{k=1}^{50} U_{efk}(i)\cdot I_{efk}(i)\cdot \cos\Phi(i) \tag{11}$$

The reactive power of i phase is computed with the relation:

$$Q(i)=\sum_{k=1}^{50} U_{efk}(i)\cdot I_{efk}(i)\cdot \sin\Phi(i) \tag{12}$$

In relations (11) and (12), k represents the harmonic rank.

The harmonic voltage on i phase is computed as in the relation:

$$U_{thd}(i)=\frac{\sqrt{\sum_{n=2}^{50} U_{harm}(i,n)^2}}{U_{harm}(i,1)} \tag{13}$$

The harmonic current on i phase is computed as in the relation:

$$I_{thd}(i)=\frac{\sqrt{\sum_{n=2}^{50} I_{harm}(i,n)^2}}{I_{harm}(i,1)} \tag{14}$$

where N is the number of samples / period and i is the phase.

The total values of the active and reactive powers are computed as in the following relations:

$$P=\sum_{i=1}^{3} P_i\ ,\ Q=\sum_{i=1}^{3} Q_i \tag{15}$$

The total apparent power is computed as:

$$S=3\cdot U_{ef}\cdot I_{ef} \tag{16}$$

The distorted power is computed as in the relation:

$$D=\sqrt{S^2-P^2-Q^2} \tag{17}$$

The total values of the harmonic voltages and currents are computed as in the followings:

$$U_{thd} = \sqrt{\frac{U^2_{thd(1)} + U^2_{thd(2)} + U^2_{thd(3)}}{3}}$$

$$I_{thd} = \sqrt{\frac{I^2_{thd(1)} + I^2_{thd(2)} + I^2_{thd(3)}}{3}} \quad (18)$$

Presented computing application is depicted in Figure 3.

Table 2 presents the variation of the electrical parameters during the functioning of the nonlinear load when the passive filtering system is connected to the point of common coupling.

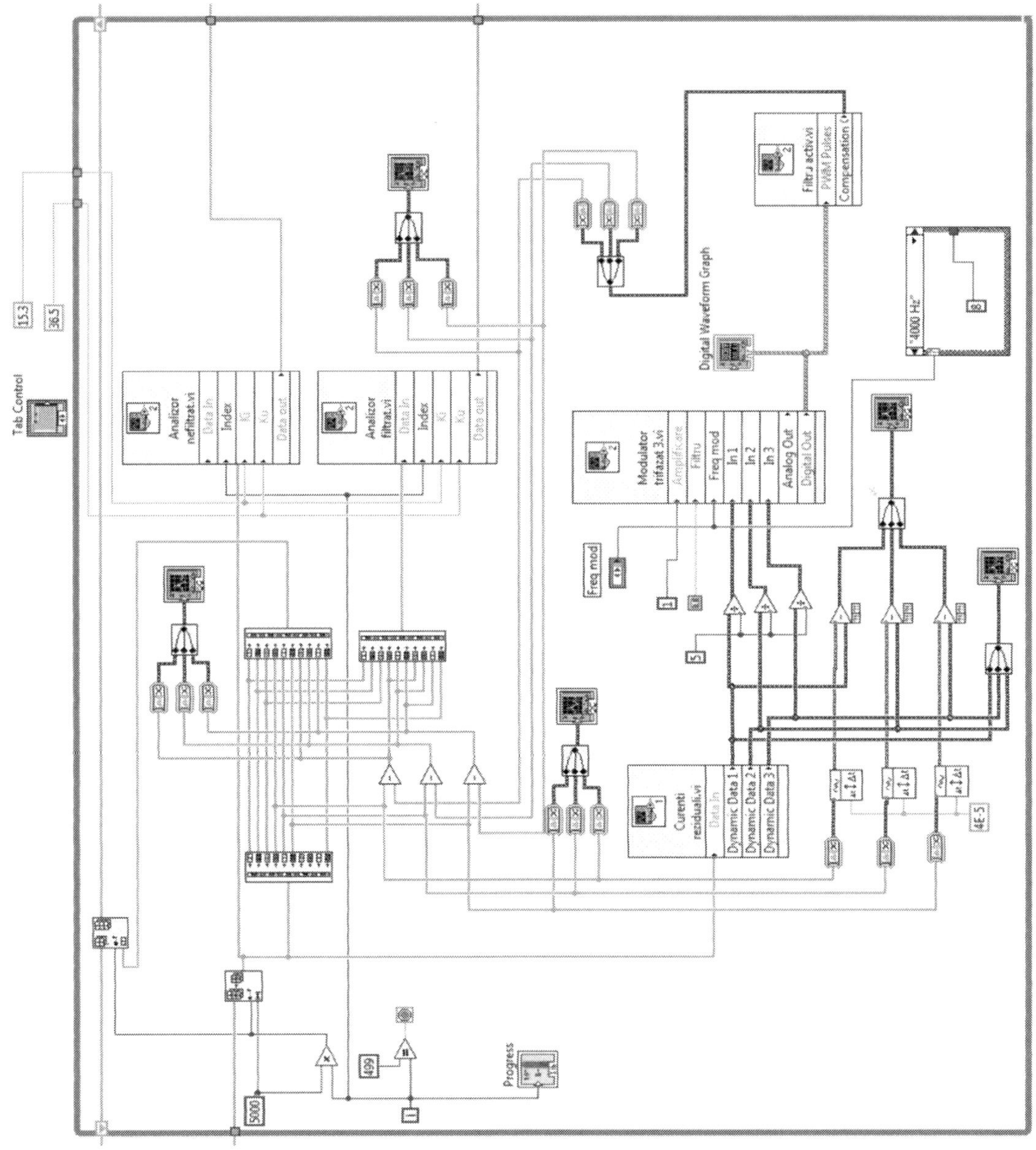

Figure 3. Computing application LabVIEW code.

Table 2. Variation of electrical parameters (when the passive filtering system is connected).

Installation power [kW]	Phase voltage [V]	Phase current [A]	Active power [W]	Reactive power [VAr]	Distorted power [VAd]	Apparent power [VA]	THD voltages [%]	THD currents [%]
1.5	227.6	17	1300	-11000	3000	11476	3.15	28.6
3	228	17.4	2700	-11000	3000	11717	3	33
4.5	228	18	4000	-10800	4800	12477	2.9	39
6	228	18.1	4000	-10800	4800	12477	2.9	37
7.5	228	18.7	5000	-10800	5000	12909	2.85	42.5
9	228.4	19.3	5800	-10700	5000	13158	2.85	45
10.5	228.8	19.5	6000	-10700	6000	13656	2.8	45
12	228.2	22.2	8800	-10000	7000	15048	2.65	55
13.5	228	25	11000	-10000	8000	16882	2.5	60
15	227.5	28	13000	-9500	10000	18954	2.3	65

3. Conclusions

This paper represents a comparative study regarding the variation of electrical parameters during the functioning of an electrothermal installation with electromagnetic induction. The phase voltage, active, reactive, distorted and apparent powers were studied when the nonlinear load works without harmonic compensation devices.

A very important conclusion represents the voltage balancing. The phase voltages are approximately constant during the increasing with 10% of the consumed power of the installation and they are situated between 226-227V in Table 1 and between 227-228V in Table 2 when the passive filtering system is connected. The RMS values of phase currents are increased from 27A to 28A when the filtering system is connected. The absolute values of reactive powers are decreased as well.

Another important conclusion is that the total harmonic distortion of the phase currents is decreased from 140% to 65% when the filtering system is connected.

Following the results, the distortion of the current variation falls below the standard limits and the current harmonics must be compensated using appropriate devices [1], [5], [6].

References

[1] Akagi A, Watanabe E and Aredes M 2004 The P-Q theory for active filter control: some problems and solutions, *Controle & Automacao* **15**(1) 78-84

[2] Panoiu C, Rob R, Panoiu M and Popa G 2011 Active Power Influence on the Electrical Parameters of an Electrothermal Installation with Electromagnetic Induction, *WSEAS Transactions on Power Systems* **6** 63-72

[3] Rob R, Sora I, Panoiu C and Panoiu M 2010 *Using Active Filters in Reducing of the Total Harmonic Distortion on Electro Thermal Installation with Electromagnetic Induction*, International Symposium of Power Electronics. Electrical Drives. Automation and Motion. SPEEDAM 2010, Pisa, Italy, June 4-6, pp. 233-238

[4] Panoiu C, Rob R, Panoiu M and Popa G 2010 *Measurements of the Electrical Parameters of an Electro Thermal Installation with Electromagnetic Induction*, Proceedings of the 14th

WSEAS International Conference on System, Corfu, Greece, July 10-12, pp. 216-221
[5] Fujita H and Akagi H The Unified Power Quality Conditioner: the Integration of Series and Shunt Active filters, *IEEE Tran. Power Electronics* **13**(2) 315-322
[6] Jacobina C, de Rossiter Correa M, Ferreira Pinheiro. R, de Silva E and Nogueira Lima A 2004 Modeling and control of unbalanced three phase systems containing PWM converters, *IEEE Transactions on Industry Applications* **6**(37) 1807-1816
[7] Tirian G.O, Pinca-Bretotean C, Cristea D and Topor M 2010 Applications of fuzzy logic in continuous casting, *Transactions on Systems and Control* **5**(3) 133-142
[8] Iordan A 2012 Oriented object development of an interactive software designed for the study of A* method, *AWER Procedia Information Technology & Computer Science* **2** 127-131
[9] Zamora I, Mazon A J, Eguia P, Albizu I, Sagastabeitia K J and Fernández E 2003 *Simulation by MATLAB/Simulink of active filters for reducing THD created by industrial systems*, IEEE Bologna Power Tech. Conference, Bologna, Italy, June 23-26 DOI 10.1109/PTC.2003.1304433